BASIC MATHEMATICS

2nd Edition

Lawerence A. Trivieri, Ph.D.
Georgia Perimeter College, GA

Contributing Editor
Heidi Reich, M.A.

Collins
An Imprint of HarperCollinsPublishers

Basic Mathematics. Copyright © 1992, 2006 by HarperCollins Publishers. All rights reserved. Printed in the United States. No part of this book may be used or reproduced in any manner whatsoever without written permission except in the case of brief quotations embodied in critical articles and reviews. For information address HarperCollins Publishers, 10 East 53rd Street, New York, N. Y. 10022.

An American BookWorks Corporation Production

HarperCollins books may be purchased for educational, business, or sales promotional use. For information please write: Special Markets Department, HarperCollins Publishers, 10 East 53rd Street, New York, N.Y. 10022.

Library of Congress Cataloging-in-Publication Data has been applied for.

ISBN: 13: 978-0-06-088146-7
ISBN: 10: 0-06-088146-1

10 BRR 10 9 8 7 6 5 4 3

Contents

	Preface	v
	A Note on the Use of Calculators	vii
1.	The Whole Numbers	1
2.	Primes, Factors, and Multiples	45
3.	Fractions	67
4.	Decimals	117
5.	Ratio and Proportion	151
6.	Percent	171
7.	Measurement	195
8.	Geometric Measure	217
9.	Signed Numbers	253
10.	Exponents and Scientific Notation	299
11.	An Introduction to Algebra	317
12.	Statistics	363
13.	Consumer Mathematics	391
	Index	423

Preface

This text can be used in one of two ways:
- It can be used to supplement the text or materials that you are currently using in your course, if you are enrolled in one.
- Or, it can be used as a stand-alone text.

If you are using it as a supplement, you can start at the beginning and continue sequentially throughout or you can use the table of contents to determine which topics you wish to study and in which order you wish to study them. In each section of the text, there are many solved exercises. Study as many of them as you think necessary to understand the topic well. Note that if you have difficulty with a particular topic, study the brief instructional material given in the section. Examine carefully the first couple of solved exercises until you understand what is being discussed, and for the remainder of the exercises, write down the statement of the exercise on your own paper. Solve the exercise on your own, and then compare your solution with that given in the text. Do not leave a topic until you understand it completely. And keep in mind that you can always receive additional help from your instructor or from the staff in the mathematics lab (if there is one) at your college.

If you have been away from the study of basic mathematics for some time and are thinking about returning to or starting college, this text will help you to review some basic mathematics prior to enrolling in a pre-algebra or elementary algebra course.

What is new with this edition of ***Basic Mathematics?*** First, several exercises and applications have been updated to reflect current situations. In addition, all chapters now have the same number of supplemental exercises at the end. Also, instead of just having answers for the supplemental exercises, complete solutions are now included.

A major change in this edition is the inclusion of exercises for the handheld calculator for most of the chapters (following the supplemental exercises), together with their complete solutions.

Several people are owed my sincere thanks for the completion of this project. Foremost, I wish to thank Fred N. Grayson of American BookWorks Corporation for encouraging me to undertake this revision. Sincere thanks also go to the individuals who worked on this project and helped to bring it to fruition. Finally I wish to thank my wife, Joyce B. Trivieri, for her loving support, encouragement, patience, and technological support throughout this project.

This text is dedicated to the loving memory of my daughter, Andrea Trivieri Herman.

A NOTE ON THE USE OF CALCULATORS

INTRODUCTION

Calculators made their way into mathematics classrooms in the early 1970s. At first, teachers were reluctant to use the new technology due to their unfamiliarity with them. Since then, calculators have become as common in the classrooms as are textbooks. Today, they are found even at the elementary grade levels.

The early calculators were relatively expensive. Those that could do little more than perform the four basic arithmetic operations sold for more than $100. And, there weren't too many different calculators from which to choose. Today, there are many different calculators, coming in all variations of size, type, and price. For most purposes, a calculator that can perform the four basic arithmetic operations is adequate. Of course, for students who are studying mathematics, physics, or engineering, more sophisticated calculators may be required. A basic calculator sells for about $10. For more advanced courses, sophisticated calculators sell for about $100. Because calculators differ among the various manufacturers, always refer to the owner's manual when using your calculator.

We now examine performing some operations with calculators. If decimals are involved in your calculations, you don't have to be concerned about the decimal point in your final answer. Most calculators use what is known as a **floating decimal,** which locates the decimal point in your final answer exactly where it should be.

ADDITION:
To add 23 and 54, enter:
[2][3] [+] [5] [4] [=]. Read the result: 77

SUBTRACTION:
To subtract 1.27 from 56.8, enter:
[5] [6] [.] [8] [–] [1] [.] [2] [7] [=]. Read the result: 55.53

MULTIPLICATION:
To multiply 97 and 89, enter:
[9] [7] [x] [8] [9] [=]. Read the result: 8,633

DIVISION:
To divide 47.88 by 38, enter:
[4] [7] [.] [8] [8] [÷] [3] [8] [=]. Read the result: 1.26

SQUARING:

To square 54:

a. On the TI-30Xa, enter:

[5] [4] [×] [5] [4] [=]. Read the result: 2,916

or, use the y^x and enter:

[5] [4] [y^x] [2] [=]. Read the result: 2,916

b. On the TI-83, use the x^2 key and enter:

[5] [4] [x^2] [=]. Read the result: 2,916

POWERS:

To compute 82^4:

a. On some calculators, you use the y^x key and enter:

[8] [2] [y^x] [4] [=]. Read the result: 45,212,176

b. On other calculators, use the ^ key and enter:

[8] [2] [^] [4] [=]. Read the result 45,212,176

PARENTHESES:

When using a calculator, parentheses play a very significant role. For example, to compute $\frac{25 \times 54}{9 \times 5}$, use two sets of parentheses as follows: (25 x 54) ÷ (9 x 5) and enter:

[( ] [2] [5] [×] [5] [4] [ )] [÷] [( ] [9] [×] [5] [ )] [=]. Read the result: 30, which is correct. Without parentheses, you would have:

$$25 \times 54 \div 9 \times 5$$

which produces a result of 750, which is incorrect.

SCIENTIFIC NOTATION:

When using your calculator to perform various operations, very large or very small answers may be obtained. Your calculator will express the answer in a form known as **scientific notation.** For example, multiplying 564,897 by 726,384, your result may be displayed as

$$4.103321424 \quad 11$$

or, 4.103321424 E 11

The 11 or E 11 means that your result must be multiplied by 10^{11}. The actual result would be approximated by 410,332,142,400.

Multiplying 0.0089945 by 0.000258, the result may be displayed as

$$2.320581 \quad -6$$

or, 2.320581 E –6

Here, the result must be multiplied by 10^{-6}. The actual result would be approximated by 0.000002320581

CHAPTER 1

The Whole Numbers

In this chapter, we consider the whole numbers, together with the arithmetic operations on them. Various properties associated with these operations and ordering of the whole numbers are also examined. Applications involving whole numbers are included throughout the chapter.

1.1 NUMBERS, NUMERALS, COUNTING, AND PLACE VALUE

In this section, we discuss the value of each digit in a numeral, write the word name for a number, and write the numeral for the word name of a number.

NUMERALS

Counting involves the use of only ten symbols or numerals called **digits.** These are 0, 1, 2, 3, 4, 5, 6, 7, 8, and 9. The counting numbers from 1 through 9 involve a single digit. The numbers from 10 through 99 have two digits. The numbers from 100 through 999 have three digits, and so on.

As the numbers get larger, we need longer numerals to represent them. The digits in longer numerals are separated by commas into groups of three, such as 759,189,304.

PLACE VALUE

Each digit has **place value.** The location of a digit in a numeral indicates the value of the digit. Starting from the *right* and moving to the left, the digits represent *ones, tens, hundreds, thousands*, and so on.

Exercise 1.1

What numbers are represented by the following numerals?
a) 48,902
b) 367,518

Solution 1.1

a) In 48,902:
- 4 is in the ten thousands place; it represents $4 \times 10{,}000 = 40{,}000$.
- 8 is in the thousands place; it represents $8 \times 1{,}000 = 8{,}000$.
- 9 is in the hundreds place; it represents $9 \times 100 = 900$.
- 0 is in the tens place; it represents $0 \times 10 = 0$.
- 2 is in the ones place; it represents $2 \times 1 = 2$. Thus, 48,902 represents $40{,}000 + 8{,}000 + 900 + 2$ and is read, "forty-eight thousand, nine hundred two."

b) In 367,518:
- 3 is in the hundred thousands place; it represents $3 \times 100{,}000 = 300{,}000$.
- 6 is in the ten thousands place; it represents $6 \times 10{,}000 = 60{,}000$.
- 7 is in the thousands place; it represents $7 \times 1{,}000 = 7{,}000$.
- 5 is in the hundreds place; it represents $5 \times 100 = 500$.
- 1 is in the tens place; it represents $1 \times 10 = 10$.
- 8 is in the ones place; it represents $8 \times 1 = 8$.

Thus, 367,518 represents $300{,}000 + 60{,}000 + 7{,}000 + 500 + 10 + 8$ and is read, "three hundred sixty-seven thousand, five hundred eighteen."

Exercise 1.2

Write the word names for the following numerals:
a) 333,222,111
b) 43,598,567,104
c) 329,862,107
d) 4,444,444,444
e) 4,832,367,756,523

Solution 1.2

a) The word name for the numeral 333,222,111 is three hundred thirty-three million, two hundred twenty-two thousand, one hundred eleven.
b) The word name for the numeral 43,598,567,104 is forty-three billion, five hundred ninety-eight million, five hundred sixty-seven thousand, one hundred four.
c) 329,862,107 is read, "three hundred twenty-nine million, eight hundred sixty-two thousand, one hundred seven."
d) 4,444,444,444 is read, "four billion, four hundred forty-four million, four hundred forty-four thousand, four hundred forty-four."
e) 4,832,367,756,523 is read, "four trillion, eight hundred thirty-two billion, three hundred sixty-seven million, seven hundred fifty-six thousand, five hundred twenty-three."

Exercise 1.3

Write the numeral for each of the following:

a) Five thousand, six hundred seventy-eight.

b) Four million, eight hundred seventy-three thousand, six hundred.

c) Two hundred thirty-seven trillion, eight hundred fifty-six billion, seven hundred nineteen million, five hundred eleven thousand, one hundred twelve.

Solution 1.3

a) 5,678

b) 4,873,600

c) 237,856,719,511,112

1.2 ORDERING AND ADDITION OF WHOLE NUMBERS

In this section, we discuss ordering of whole numbers and addition of whole numbers. We also use the operation of addition of whole numbers to determine the solutions to certain word problems.

NATURAL AND WHOLE NUMBERS

The counting numbers are also called the **natural numbers.** The least of the counting numbers is 1. If you include 0 with all of the counting numbers, you have the **whole numbers.**

To count numbers, some must come before others. For example, 5 comes before 9, 9 comes before 17, and 17 comes before 29. The process of arranging the counting numbers from the smallest to the largest is called **ordering.**

GREATER THAN AND LESS THAN

The symbol $>$ is used to mean, "is greater than." In counting, 9 comes after 5. Therefore, 9 is greater than 5, and you can write $9 > 5$. Because 17 comes after 9, it is greater than 9, and we write $17 > 9$.

The symbol $<$ is used to mean "is less than." In counting, 5 comes before 9. Therefore, 5 is less than 9, and you can write $5 < 9$. Because 17 comes before 25, it is less than 25, and we write $17 < 25$.

Exercise 1.4

Read each of the following:

a) $14 < 18$

b) $44 > 23$

c) $0 < 39$

d) $304 > 289$

e) $199 < 205$

Solution 1.4

a) 14 < 18 is read, "14 is less than 18."
b) 44 > 23 is read, "44 is greater than 23."
c) 0 < 39 is read, "0 is less than 39."
d) 304 > 289 is read, "304 is greater than 289."
e) 199 < 205 is read, "199 is less than 205."

ADDITION

To add whole numbers, add together the digits in the numbers that represent the same place value. That is, add the ones together, add the tens together, add the hundreds together, and so on. The result in an addition problem is called the **sum.**

PROPERTIES OF ADDITION

Note the following properties of addition of whole numbers.

Commutative property: The order in which you add two whole numbers is not important. That is, if you let a and b represent any two whole numbers, then
$$a + b = b + a$$

Associative property: To add three whole numbers, you must group them first, because only two numbers can be added at one time. However, the order in which you group the numbers is not important. That is, if you let a, b, and c represent any three whole numbers, then
$$(a + b) + c = a + (b + c)$$

If 0 is added to a whole number, or if a whole number added to 0, the sum is equal to the given whole number. That is, if a is any whole number, then
$$a + 0 = a \quad \text{and} \quad 0 + a = a$$

The whole number 0 is called the **additive identity.**

Exercise 1.5

Add:

a) 213 + 42 + 634
b) 35 + 49

Solution 1.5

a) 213 + 42 + 634

Step 1: Line up the numbers from the *right*:

```
 213
  42
+634
```

Step 2: Add the **ones:**

```
 213
  42
+634
   9
```

Step 3: Add the **tens:**

```
 213
  42
+634
  89
```

Step 4: Add the **hundreds:**

```
 213
  42
+634
 889
```

b) 35 + 49

Step 1: Line up the numbers from the **right:**

```
 35
+49
```

Step 2: Add the **ones:**

```
 35
+49
```

Note that if you add 5 ones and 9 ones, you get 14 ones, but 14 is a two-digit numeral and cannot be written in the ones column. However, you can rewrite 14 as 10 + 4. Because 10 ones is equal to 1 ten, you can **carry** the 10 ones over to the tens column as 1 ten:

```
 1
 35
+49
  4
```

Step 3: Add the **tens:**

```
 1
 35
+49
 84
```

In Exercise 1.5b, observe that the 4 ones of the 14 were placed in the ones column, and the 1 ten of the 14 was carried over to the tens column and added to the tens already there. Use this procedure when adding whole numbers. However, remember that when carrying, you always start with units on the right and work to the left.

Exercise 1.6
Add 326 + 37 + 1,049

Solution 1.6
Step 1: Line up the numbers from the *right:*

$$\begin{array}{r} 326 \\ 37 \\ +1,049 \\ \hline \end{array}$$

Step 2: Add the **ones:**

$$\begin{array}{r} {}^{2} \\ 326 \\ 37 \\ +1,049 \\ \hline 2 \end{array}$$ (6 + 7 + 9 = 22 ones or 2 tens plus 2 ones)

Step 3: Add the **tens:**

$$\begin{array}{r} {}^{1\,2} \\ 326 \\ 37 \\ +1,049 \\ \hline 12 \end{array}$$

Step 4: Add the **hundreds:**

$$\begin{array}{r} {}^{1\,2} \\ 326 \\ 37 \\ +1,049 \\ \hline 412 \end{array}$$

Step 5: Add the **thousands:**

$$\begin{array}{r} {}^{1\,2} \\ 326 \\ 37 \\ +1,049 \\ \hline 1,412 \end{array}$$

The "carry" procedure shown in the preceding exercises is also used when adding like units such as hours, minutes, pounds, ounces, inches, meters, quarts, liters, and so forth.

Exercise 1.7

Michelle worked three days to make a bookshelf. The first day, she worked 4 hours and 20 minutes. The second day, she worked 2 hours and 45 minutes. The third day, she worked 6 hours and 17 minutes. What was the total time spent making the bookshelf?

Solution 1.7

The total time is the sum of the time spent the first day, the second day, and the third day. We can arrange this as follows:

First day:	4 hr + 20 min
Second day:	2 hr + 45 min
Third day:	+6 hr + 17 min
Total time:	12 hr + 82 min
	= 12 hr + 60 min + 22 min (rewriting)
	= 12 hr + 1 hr + 22 min (renaming)
	= 13 hr + 22 min (adding hours)

Hence, the total time spent was 13 hours and 22 minutes.

Exercise 1.8

Liz ordered four items from a catalog store. The weights of these items were 3 lb 8 oz, 1 lb 10 oz, 5 lb 9 oz, and 11 oz. What was the total weight of these four items? (Hint: 16 oz = 1 lb)

Solution 1.8

The total weight of these four items can be represented as follows:

First item:	3 lb + 8 oz
Second item:	1 lb + 10 oz
Third item:	5 lb + 9 oz
Fourth item:	+ 11 oz
Total	= 9 lb + 38 oz
	= 9 lb + 16 oz + 16 oz + 6 oz (rewriting)
	= 9 lb + 1 lb + 1 lb + 6 oz (renaming)
	= 11 lb + 6 oz (adding the pounds)

Therefore, the total weight of the four items was 11 lb 6 oz.

8 Basic Math

Exercise 1.9
Four members of a baseball team weigh 165 lb 7 oz, 156 lb 10 oz, 167 lb 8 oz, and 153 lb 5 oz. Determine their combined weight.

Solution 1.9
The combined weight of the four players can be represented as follows:

First player:	165 lb + 7 oz
Second player:	156 lb + 10 oz
Third player:	167 lb + 8 oz
Fourth player:	+153 lb + 5 oz
Total weight:	641 lb + 30 oz

$$= 641 \text{ lb} + 16 \text{ oz} + 14 \text{ oz (rewriting)}$$
$$= 641 \text{ lb} + 1 \text{ lb} + 14 \text{ oz (renaming)}$$
$$= 642 \text{ lb} + 14 \text{ oz (adding the pounds)}$$

Therefore, the combined weight of the four players is 642 lb 14 oz.

Exercise 1.10
A maintenance worker has four containers of different sizes. One container has a capacity of 4 gal 2 qt 1 pt; one has a capacity of 2 gal 2 qt; one has a capacity of 3 qt 1 pt; and one has a capacity of 2 qt 1 pt. What is the total capacity of the four containers? (Hint: 1 gal = 4 qt; 1 qt = 2 pt)

Solution 1.10
The total capacity of the four containers can be represented as follows:

First container:	4 gal + 2 qt + 1 pt
Second container:	2 gal + 2 qt
Third container:	3 qt + 1 pt
Fourth container: +	2 qt + 1 pt
Total capacity:	6 gal + 9 qt + 3 pt

$$= 6 \text{ gal} + (4 \text{ qt} + 4 \text{ qt} + 1 \text{ qt}) + (2 \text{ pt} + 1 \text{ pt}) \text{ (rewriting)}$$
$$= 6 \text{ gal} + 1 \text{ gal} + 1 \text{ gal} + 1 \text{ qt} + 1 \text{ qt} + 1 \text{ pt (renaming)}$$
$$= 8 \text{ gal} + 1 \text{ qt} + 1 \text{ qt} + 1 \text{ pt (adding gallons)}$$
$$= 8 \text{ gal} + 2 \text{ qt} + 1 \text{ pt (adding quarts)}$$

Therefore, the total capacity of the containers is 8 gal 2 qt 1 pt.

Exercise 1.11
A runner passes the first timing station in a marathon after 35 minutes. The next section takes him 1 hr 13 min, and the final distance to the finish line takes him another 1 hr 37 min. What was his total time? (Hint: 60 min = 1 hr)

Solution 1.11

The total time can be represented as follows:

First section: 35 min
Second section: 1 hr + 13 min
Final section: +1 hr + 37 min
Total time: 2 hr + 85 min
                   = 2 hr + 60 min + 25 min (rewriting)
                   = 2 hr + 1 hr + 25 min (renaming)
                   = 3 hr + 25 min (adding hours)

Therefore, the total time for the runner was 3 hr 25 min.

1.3 SUBTRACTION OF WHOLE NUMBERS

In this section, we discuss subtraction of whole numbers and use the operation of subtraction to solve certain word problems.

The process of "taking away" is called **subtraction.** It is also the process that lets us tell how much larger one number is than another. The answer in a subtraction problem is called the **difference.**

To check a subtraction problem, add the answer (that is, the difference) to the number being subtracted; the result should be the number you started with.

Exercise 1.12

Solve each of the following:
    a) $9 - 5$
    b) $13 - 8$
    c) $19 - 11$
    d) $7 - 7$

Solution 1.12

    a) $9 - 5 = 4$, because $5 + 4 = 9$
    b) $13 - 8 = 5$, because $8 + 5 = 13$
    c) $19 - 11 = 8$, because $11 + 8 = 19$
    d) $7 - 7 = 0$, because $7 + 0 = 7$

Note that subtraction is the opposite (or inverse) operation of addition.

ORDER OF SUBTRACTION

You know that the order in which you add two whole numbers is not important. However, you must pay attention to the order in which you subtract one whole number from another.

To add 27 and 35, we can write either $\begin{array}{r}27\\+35\\\hline 62\end{array}$ or $\begin{array}{r}35\\+27\\\hline 62\end{array}$

However, 62 − 27 can be written only as $\begin{array}{r}62\\-27\\\hline 35\end{array}$ because subtraction is *not* commutative.

Subtraction of whole numbers is *not* an associative operation, either. Note that (11 − 6) − 4 = 5 − 4 = 1 but 11 − (6 − 4) = 11 − 2 = 9.

Because 1 ≠ 9 (read, "1 is not equal to 9"), (11 − 6) − 4 ≠ 11 − (6 − 4) shows that subtraction is *not* an associative operation; thus, how you group the whole numbers to be subtracted *does* matter.

To subtract a whole number from another whole number when the whole numbers have two or more digits, write the smaller number under the larger number, aligning the digits from the *right*. Then, subtract the digits of like place value.

Exercise 1.13
Subtract:
$$\begin{array}{r}9,675\\-4,534\end{array}$$

Solution 1.13
Step 1: Subtract the **ones:**
$$\begin{array}{r}9,675\\-4,534\\\hline 1\end{array}$$

Step 2: Subtract the **tens:**
$$\begin{array}{r}9,675\\-4,534\\\hline 41\end{array}$$

Step 3: Subtract the **hundreds:**
$$\begin{array}{r}9,675\\-4,534\\\hline 141\end{array}$$

Step 4: Subtract the **thousands:**
$$\begin{array}{r}9,675\\-4,534\\\hline 5.141\end{array}$$

Check:

$$\begin{array}{r}4,534\\+5,141\\\hline 9,675\ \checkmark\end{array}$$

BORROWING

Sometimes, when you subtract one number from another, the digit in a given place value is larger than the digit directly above it. In that case, you need to "borrow." Borrowing in subtraction is very similar to carrying in addition.

Exercise 1.14

Subtract:

a) $\quad\begin{array}{r}872\\-524\\\hline\end{array}$

b) $\quad\begin{array}{r}87,002\\-45,678\\\hline\end{array}$

Solution 1.14

a) $\quad\begin{array}{r}872\\-524\\\hline\end{array}$

Step 1: You cannot subtract 4 ones from 2 ones, because 4 > 2. Borrow 1 ten (that is, 10 ones) from the tens column in 872. Add the 10 ones to the 2 ones already in the ones column. Now, subtract 4 ones from 12 ones:

$$\begin{array}{r}{}^{6\,1}\\8\!\!\!/\,2\\-52\,4\\\hline 8\end{array}$$

Step 2: Subtract the tens:

$$\begin{array}{r}{}^{6\,1}\\8\!\!\!/\,2\\-52\,4\\\hline 48\end{array}$$

(You are subtracting 2 from 6, not 2 from 7, because you borrowed 1 from 7.)

Step 3: Subtract the hundreds:

$$\begin{array}{r}{}^{6\,1}\\\mathbf{8}\!\!\!/\,\mathbf{2}\\-52\,4\\\hline 348\end{array}$$

Check:

$$\begin{array}{r}524\\+348\\\hline 872\checkmark\end{array}$$

b) $\quad\begin{array}{r}6,404\\-4,878\\\hline\end{array}$

Step 1: You cannot subtract 8 ones from 4 ones, because 8 > 4. Borrow 1 ten (that is, 10 ones) from the tens column in 6,404. Subtract the ones:

$$\begin{array}{r}6,404\\-4,878\\\hline\end{array}$$

There are 0 tens in 6,404; hence, you cannot borrow 1 ten. Therefore, you move to the hundreds column to borrow 1 hundred (that is, 10 tens). The subtraction problem now becomes:

$$\begin{array}{r}{\scriptstyle 3\,1}\\6,\cancel{4}\cancel{0}\cancel{4}\\4,878\\\hline\end{array}$$

Step 2: Borrow 1 ten (10 ones) from the tens column in 6,404. Add the 10 ones to the 4 ones in the ones column. Subtract the ones:

$$\begin{array}{r}{\scriptstyle 9}\\{\scriptstyle 3\,\cancel{1}\,1}\\6,\cancel{4}\cancel{0}\cancel{4}\\-4,878\\\hline 6\end{array}$$

Step 3: Subtract the tens:

$$\begin{array}{r}{\scriptstyle 9}\\{\scriptstyle 3\,\cancel{1}\,1}\\6,\cancel{4}\cancel{0}\cancel{4}\\-4,878\\\hline 26\end{array}$$

Step 4: Subtract the hundreds. Because 8 > 3, borrow 1 thousand (10 hundreds) from the thousands column in 6,404. Add the 10 hundreds to the 3 hundreds. Subtract:

$$\begin{array}{r}{\scriptstyle 1\,9}\\{\scriptstyle 5\,\cancel{3}\,\cancel{1}\,1}\\\cancel{6},\cancel{4}\cancel{0}\cancel{4}\\-4,878\\\hline 526\end{array}$$

Step 5: Subtract the thousands:

$$\begin{array}{r}\scriptstyle 1\ 9\\ \scriptstyle 5\,\cancel{3}\,\cancel{1}1\\ \cancel{6,404}\\ -4,878\\ \hline 1,526\end{array}$$

Check:

$$\begin{array}{r}4,878\\ +1,526\\ \hline 6,404\ \checkmark\end{array}$$

c) $\begin{array}{r}87,002\\ -45,678\end{array}$

Step 1: Borrow 10 ones from the tens column; subtract the ones:

$$\begin{array}{r}87,002\\ -45,678\end{array}$$

There are 0 tens in 87,002 so you cannot borrow 1 ten. Move to the hundreds column and borrow 1 hundred (10 tens). There are 0 hundreds, so you move to the thousands column and borrow 1 thousand (10 hundreds), as indicated:

$$\begin{array}{r}\scriptstyle 6\ \ 1\\ 8\cancel{7},002\\ -45,678\end{array}$$

Step 2: Borrow 10 tens from the hundreds column:

$$\begin{array}{r}\scriptstyle 9\\ \scriptstyle 6\ \cancel{1}1\\ 8\cancel{7},\cancel{0}02\\ -45,678\end{array}$$

Step 3: Borrow 10 ones from the tens column:

$$\begin{array}{r}\scriptstyle 9\ 9\\ \scriptstyle 6\ \cancel{1}\cancel{1}\\ 8\cancel{7},\cancel{0}\cancel{0}2\\ -45,678\end{array}$$

Step 4: Subtract the ones:

$$\begin{array}{r}\scriptstyle 9\ 9\\ \scriptstyle 6\ \cancel{1}\cancel{1}1\\ 8\cancel{7},\cancel{0}\cancel{0}\cancel{2}\\ -45,678\\ \hline 4\end{array}$$

Step 5: Subtract the tens:

$$\begin{array}{r} {}^{9\,9}\\ 6\,\cancel{X}\cancel{X}1\\ 8\cancel{7},\cancel{0}\cancel{0}2\\ -45{,}678\\ \hline 24 \end{array}$$

Step 6: Subtract the hundreds:

$$\begin{array}{r} {}^{9\,9}\\ 6\,\cancel{X}\cancel{X}1\\ 8\cancel{7},\cancel{0}\cancel{0}2\\ -45{,}678\\ \hline 324 \end{array}$$

Step 7: Subtract the thousands:

$$\begin{array}{r} {}^{9\,9}\\ 6\,\cancel{X}\cancel{X}1\\ 8\cancel{7},\cancel{0}\cancel{0}2\\ -45{,}678\\ \hline 1{,}324 \end{array}$$

Step 8: Subtract the ten thousands:

$$\begin{array}{r} {}^{9\,9}\\ 6\,\cancel{X}\cancel{X}1\\ 8\cancel{7},\cancel{0}\cancel{0}2\\ -45{,}678\\ \hline 41{,}324 \end{array}$$

Check:

$$\begin{array}{r} 45{,}678\\ +41{,}324\\ \hline 87{,}002\,\checkmark \end{array}$$

Exercise 1.15

Maria can type 112 words per minute on her word processor, and Peter can type 84 words per minute on his typewriter. How many more words per minute can Maria type than Peter?

Solution 1.15

Use subtraction to solve this problem as follows:

Maria's rate of typing: 112 words per minute
Peter's rate of typing: − 84 words per minute
                                  28 words per minute

Therefore, Maria can type 28 words per minute more than Peter.

Exercise 1.16

Hattie has $536 in her savings account. If she withdraws $234, how much money will be left in her account?

Solution 1.16

Again, use subtraction to solve this problem as follows:

$536	The amount in Hattie's account
−234	The amount of withdrawal
$302	The amount left in the account

Therefore, Hattie will have $302 left in her account.

Exercise 1.17

Krystle and Brandan went fishing. Brandan caught a fish that weighed 9 lb 5 oz. Krystle caught a fish that weighed 6 lb 11 oz. How much heavier was Brandan's fish than Krystle's fish?

Solution 1.17

The problem may be set up and solved as follows:

Brandan's fish: 9 lb + 5 oz
Krystle's fish: −(6 lb + 11 oz)

Rewrite this problem by borrowing 1 lb from 9 lb, converting 1 lb to 16 oz, and adding 16 oz to 5 oz. Then, 9 lb 5 oz may be written as 8 lb 21 oz.

Brandan's fish: 8 lb + 21 oz
Krystle's fish: −(6 lb + 11 oz)
Difference: 2 lb + 10 oz

Therefore, Brandan's fish was 2 lb 10 oz heavier than Krystle's fish.

Exercise 1.18
Autumn has $408 in her checking account. If she writes checks in the amounts of $89 and $124, how much money will be left in her account?

Solution 1.18
Using subtraction, we can solve this problem as follows:

$408	The amount in the checking account
− 89	The amount of the first check
$319	The amount left after writing the first check
− 124	The amount of the second check
$195	The amount left after writing the second check

Therefore, Autumn will have $195 left in her checking account after writing the two checks.

1.4 MULTIPLICATION OF WHOLE NUMBERS

In this section, we consider the multiplication of whole numbers and the solution of certain word problems using the operation of multiplication.

LANGUAGE OF MULTIPLICATION

Multiplication may be thought of as repeated addition, with all the numbers added being the same. The symbol × is used to indicate multiplication. Thus,

$$5 \times 3 = 15$$

may be thought of as

$$5 \times 3 = 3 + 3 + 3 + 3 + 3 = 15$$

We could also write

$$5 \times 3 = 15$$

as

$$(5)(3) = 15$$

or

$$5 \cdot 3 = 15$$

In the expression $5 \cdot 3$, a dot ($\cdot$) is written in the middle of the space between the 5 and 3 that indicates multiplication. It should not be confused with a decimal point (see Chapter 4).

In a multiplication problem, the numbers being multiplied are called **factors,** and the answer is called the **product.**

Exercise 1.19
Find the product of 4 × 3.

Solution 1.19
4 × 3 = 3 + 3 + 3 + 3 = 12

Hence,

$$\begin{array}{r} 3 \\ \times 4 \\ \hline 12 \end{array} \begin{array}{l} \leftarrow \text{factor} \\ \leftarrow \text{factor} \\ \leftarrow \text{product} \end{array}$$

PROPERTY OF MULTIPLICATION

The order in which you multiply whole numbers is not important, because multiplication of whole numbers is a commutative operation. If you let a and b represent any two whole numbers, we can state the **commutative property for multiplication** of whole numbers as

$$a \times b = b \times a.$$

Exercise 1.20
Show that:
a) 2 × 5 = 5 × 2
b) 3 × 7 = 7 × 3

Solution 1.20
a) 2 × 5 = 5 + 5 = 10
5 × 2 = 2 + 2 + 2 + 2 + 2 = 10
Therefore, 2 × 5 = 5 × 2.
b) 3 × 7 = 7 + 7 + 7 = 21
7 × 3 = 3 + 3 + 3 + 3 + 3 + 3 + 3 = 21
Therefore, 3 × 7 = 7 × 3.

The product of a whole number and 1 is the whole number. This is known as the **identity property for multiplication** of whole numbers. The whole number 1 is called the **multiplicative identity** for the whole numbers. Let a represent any whole number. We can then state the identity property for multiplication of whole numbers as

$$a \times 1 = a \quad \text{and} \quad 1 \times a = a.$$

For example,
- 3 × 1 = 3
- 6 × 1 = 6
- 1 × 7 = 7

- $1 \times 9 = 9$
- $0 \times 1 = 0$
- $1 \times 0 = 0$

To multiply three whole numbers, you must group them by twos, just as you do in addition. Multiplication of whole numbers is an associative operation. If a, b, and c represent any whole numbers, we can state the **associative property for multiplication** of whole numbers as
$$(a \times b) \times c = a \times (b \times c).$$

Exercise 1.21
Show that:
a) $(2 \times 3) \times 4 = 2 \times (3 \times 4)$
b) $(1 \times 5) \times 1 = 1 \times (5 \times 1)$
c) $(4 \times 3) \times 5 = 4 \times (3 \times 5)$

Solution 1.21
a) $(2 \times 3) \times 4 = 6 \times 4 = 24$
$2 \times (3 \times 4) = 2 \times 12 = 24$
Therefore, $(2 \times 3) \times 4 = 2 \times (3 \times 4)$.
b) $(1 \times 5) \times 1 = 5 \times 1 = 5$
$1 \times (5 \times 1) = 1 \times 5 = 5$
Therefore, $(1 \times 5) \times 1 = 1 \times (5 \times 1)$.
c) $(4 \times 3) \times 5 = 12 \times 5 = 60$
$4 \times (3 \times 5) = 4 \times 15 = 60$
Therefore, $(4 \times 3) \times 5 = 4 \times (3 \times 5)$.

MULTIPLYING NUMBERS WITH MORE THAN ONE DIGIT
To multiply numbers having more than one digit, multiply, in turn, all the digits in one number by every digit in the other number. Then add all the products obtained.

Exercise 1.22
Determine the products of the following:
a) 6×29
b) 325×246
c) 399×408

Solution 1.22

a)
```
   29
 ×  6
 ────
   54   (the product of 6 and 9)
 +120   (the product of 6 and 20)
 ────
  174   (the product)
```
An alternate method of solution for the above exercise shows the following shortcut.

Step 1:

Multiply 9 ones by 6 ones. The product is 54 ones. Write the 4 in the ones column below the 6. Carry the 5 to the tens column as indicated.

```
   5
   29
  ×6
  ───
    4
```

Step 2:

Multiply 2 tens by 6 ones. The product is 12 tens. Add to this the 5 tens from the previous step. The sum is 17 tens. Write the 7 in the tens column and the 1 in the hundreds column in the answer.

```
   5
   29
 ×  6
 ────
  174
```

b)
```
  325
 ×246
```
Use the shortcut method.

Step 1:

Multiply 325 by 6 ones following the same procedure as used in solution a, above.
```
   325
  ×246
  ────
  1950
```

Step 2:

Multiply 325 by 4 tens. The answer (1,300) is lined up from the *right* under the tens column, because you are multiplying by 4 *tens*.
```
    325
   ×246
   ────
   1950
   1300
```

Step 3:

Multiply 325 by 2 hundreds. The answer (650) is lined up from the *right* under the hundreds column, because you are multiplying by 2 *hundreds*.

```
    325
   ×246
   1950
   1300
    650
```

Step 4: Add the products:

```
     325
    ×246
    1950
    1300
     650
   79,950
```

c)

```
    399
   ×408    Use the shortcut method.
```

Step 1:

Multiply 399 by 8 ones. Write the product (3,192) under 408, aligned from the *right*.

```
    399
   ×408
   3192
```

Step 2:

Multiply 399 by 0 tens. Because the product is 0, you won't write it.

```
    399
   ×408
   3192
```

Step 3:

Multiply 399 by 4 hundreds. Write the product (1,596) under 3,192, but align the numbers from the *right* under the *hundreds* column.

```
     399
    ×408
    3192
   1596
```

Step 4:
Add the products.

$$\begin{array}{r} 399 \\ \times 408 \\ \hline 3192 \\ 1596 \\ \hline 162,792 \end{array}$$

Because you can multiply in any order, the method of checking multiplication is to interchange, or reverse, the factors and multiply.

Exercise 1.23

Multiply the following numbers. Check your results.

a) 567×73

b) $2,039 \times 306$

Solution 1.23

a)
$$\begin{array}{r} 567 \\ \times 73 \\ \hline 1701 \\ 3969 \\ \hline 41,391 \end{array}$$

Check:
$$\begin{array}{r} 73 \\ \times 567 \\ \hline 511 \\ 438 \\ 365 \\ \hline 41,391\checkmark \end{array}$$

b)
$$\begin{array}{r} 2,039 \\ \times 306 \\ \hline 12234 \\ 6117 \\ \hline 623,934 \end{array}$$

Check:
$$\begin{array}{r} 306 \\ \times 2,039 \\ \hline 2754 \\ 918 \\ 612 \\ \hline 623,934\checkmark \end{array}$$

Exercise 1.24

Large grapefruits are packed 24 to a case. How many grapefruits are in 46 cases of grapefruit?

Solution 1.24

Because there are 24 grapefruits in one case and there are 46 cases of grapefruit, multiply 24 and 46 to get the answer:

```
   24    Grapefruits in one case
  ×46    Number of cases of grapefruit
  ───
  144
   96
 ─────
 1,104   Number of grapefruits in 46 cases
```

Therefore, there are 1,104 grapefruits in 46 cases of grapefruit.

Exercise 1.25

If it costs $85 per square foot to build an office building, what will be the cost of an office building that has 7,230 square feet of floor space?

Solution 1.25

Because there are 7,230 square feet of floor space and it costs $85 per square foot to build the office building, multiply to determine the cost of the building.

```
   7,230    Number of square feet
   ×$85     Cost per square foot
  ──────
   36150
   57840
  ───────
  $614,550  Cost of the building
```

Therefore, at $85 per square foot, it would cost $614,550 to build the 7,230-square-foot office building.

Exercise 1.26

The average salary paid to employees of the Acme Corporation is $46,875 per year. What is the total salary paid if there are 18 employees in the company?

Solution 1.26

Because you know the average salary paid the employees and also know the number of employees, multiply to get the answer.

```
  $46,875   Average yearly salary paid per employee
     ×18    Number of employees
  ───────
   375000
    46875
  ────────
  $843,750  Total yearly salary paid
```

Therefore, the total yearly salary paid the 18 employees would be $843,750.

1.5 DIVISION OF WHOLE NUMBERS

In this section, we discuss division of whole numbers. We also consider certain word problems that can be solved by using the operation of division.

Think of multiplication as repeated additions, with all the terms being the same. Similarly, division may be considered repeated subtractions.

LANGUAGE OF DIVISION

The symbol ÷ indicates division. For example, $15 \div 3$ means 15 divided by 3, which is equal to 5. The number being divided (in this case, 15) is called the **dividend;** the number by which the dividend is being divided (in this case, 3) is called the **divisor;** the answer (in this case, 5) is called the **quotient.**

Division is the opposite (or inverse) operation of multiplication; therefore, you use multiplication as a check for division. Thus,

$$15 \div 3 = 5$$

means that

$$3 \times 5 = 15.$$

Division is *not* a commutative operation. The order in which you divide one whole number by another whole number *is* important.

Exercise 1.27
Show that:
a) $10 \div 5 \neq 5 \div 10$
b) $16 \div 4 \neq 4 \div 16$

Solution 1.27
a) $10 \div 5 = 2$
   $5 \div 10$ is *not* a whole number.
   Therefore, $10 \div 5 \neq 5 \div 10$.
b) $16 \div 4 = 4$
   $4 \div 16$ is *not* a whole number.
   Therefore, $16 \div 4 \neq 4 \div 16$.

Division of whole numbers is also not an associative operation.

Exercise 1.28
Show that:
a) $18 \div (6 \div 3) \neq (18 \div 6) \div 3$
b) $25 \div (5 \div 5) \neq (25 \div 5) \div 5$

24 Basic Math

Solution 1.28

a) $18 \div (6 \div 3) = 18 \div 2 = 9$
$(18 \div 6) \div 3 = 3 \div 3 = 1$
Therefore, $18 \div 5 (6 \div 3) \neq (18 \div 6) \div 3$.

b) $25 \div (5 \div 5) = 25 \div 1 = 25$
$(25 \div 5) \div 5 = 5 \div 5 = 1$. Therefore, $25 \div (5 \div 5) \neq (25 \div 5) \div 5$.

Exercise 1.29

Divide:

a) 371 by 8
b) 947 by 23
c) 3,279 by 29
d) 809,762 by 316

Solution 1.29

a) Divide 371 by 8.

Step 1:

$$\begin{array}{r} 4 \\ 8{\overline{\smash{\big)}\,371}} \end{array}$$

3 is less than 8, so consider the first two digits from the *left*. How many 8s are contained in 37? The answer is 4. Because 37 represents tens, write the 4 above the 7 in the tens place in quotient.

Step 2:

$$\begin{array}{r} 4 \\ 8{\overline{\smash{\big)}\,371}} \\ -320 \\ \hline 51 \end{array}$$

Multiply 4 tens (in the quotient) by 8 ones (in the divisor). The product is 320. Write the under the 371 and subtract. The difference is 51.

Step 3:

$$\begin{array}{r} 46 \\ 8{\overline{\smash{\big)}\,371}} \\ -320 \\ \hline 51 \end{array}$$

How many 8s are contained in 51? The answer is 6. 51 represents ones so we write the 6 in the ones place in the quotient.

Step 4:

$$\begin{array}{r}46\\8{\overline{\smash{)}\,371}}\\-320\\\hline 51\\-48\\\hline 3\end{array}$$

Multiply 6 ones (in the quotient) by 8 ones (in the divisor). The product is 48. Write the 48 under the 51 already there and subtract. The difference is 3. This 3 is called the **remainder** and may be written in either of two ways, as indicated in Step 5.

Step 5:

$$\begin{array}{r}46\\8{\overline{\smash{)}\,371}}\\-320\\\hline 51\\-48\\\hline 3R\end{array} \quad \text{or} \quad \begin{array}{r}46R3\\8{\overline{\smash{)}\,371}}\\-320\\\hline 51\\-48\\\hline 3\end{array}$$

Thus, there are 46 8s contained in 371, with a remainder of 3.

Check:

$$\begin{array}{r}46 \leftarrow \text{quotient}\\\times 8 \leftarrow \text{divisor}\\\hline 368\\+3 \leftarrow \text{remainder}\\\hline 371 \leftarrow \text{dividend}\checkmark\end{array}$$

b) Divide 947 by 23.

Step 1:

$$\begin{array}{r}4\\23{\overline{\smash{)}\,947}}\end{array}$$

The divisor is a two-digit number. How many 23s are contained in 94? The answer is 4. Write the 4 in the quotient above the 4 in the dividend.

Step 2:

$$\begin{array}{r}4\\23{\overline{\smash{)}\,947}}\\-920\\\hline 27\end{array}$$

Multiply 4 tens (in the quotient) by 23 (the divisor). The product is 920. Write 920 below the 947 and subtract.

Step 3:

$$\begin{array}{r} 41 \\ 23{\overline{\smash{\big)}\,947}} \\ \underline{-920} \\ 27 \end{array}$$

How many 23s are contained in 27? The answer is 1. Write 1 in the quotient above the 7 in the dividend.

Step 4:

$$\begin{array}{r} 41 \\ 23{\overline{\smash{\big)}\,947}} \\ \underline{-920} \\ 27 \\ \underline{-23} \\ 4 \end{array}$$

Multiply 1 one (in the quotient) by 23 (the divisor). The product is 23. Write 23 below the 27 obtained in the previous step and subtract. The difference is 4.

Step 5:

$$\begin{array}{r} 41 \\ 23{\overline{\smash{\big)}\,947}} \\ \underline{-920} \\ 27 \\ \underline{-23} \\ 4R \end{array}$$

How many 23s are contained in 4? The answer in 0. We cannot subtract any more 23s from 947. The remainder is 4.

Check:

$$\begin{array}{r} 41 \leftarrow \text{quotient} \\ \times 23 \leftarrow \text{divisor} \\ \hline 123 \\ 82 \\ \hline 943 \\ +4 \leftarrow \text{remainder} \\ \hline 947 \leftarrow \text{dividend} \checkmark \end{array}$$

We shorten the division procedure in the remaining exercises.

c) Divide 3,279 by 29.

$$\begin{array}{r} 113R2 \\ 29{\overline{\smash{\big)}\,2379}} \\ \underline{-2900}\ (100 \times 29) \\ 379 \\ \underline{-290}\ (10 \times 29) \\ 89 \\ \underline{-87}\ (1 \times 29) \\ 2 \end{array}$$

Check:

$$\begin{array}{r}113 \leftarrow \text{quotient}\\ \times 29 \leftarrow \text{divisor}\\ \hline 1017\\ 226\\ \hline 3{,}277\\ +2 \leftarrow \text{remainder}\\ \hline 3{,}279 \leftarrow \text{dividend} \checkmark\end{array}$$

d) Divide 809,762 by 316.

$$\begin{array}{r}2{,}562\\ 316\overline{)809{,}762}\\ -632{,}000\ (2{,}000 \times 316)\\ \hline 177{,}762\\ -158{,}000\ (500 \times 316)\\ \hline 19{,}762\\ -18{,}960\ (60 \times 316)\\ \hline 802\\ -632\ (2 \times 316)\\ \hline 170R\end{array}$$

Check:

$$\begin{array}{r}2{,}562 \leftarrow \text{quotient}\\ \times 316 \leftarrow \text{divisor}\\ \hline 15372\\ 2562\\ 7686\\ \hline 809{,}592\\ +170 \leftarrow \text{remainder}\\ \hline 809{,}762 \leftarrow \text{dividend} \checkmark\end{array}$$

Exercise 1.30

A state lottery prize of $289,260 was shared by 36 contestants. How much did each contestant receive if they all received the same amount?

Solution 1.30

To determine how much each contestant received, divide the total prize amount ($289,260) by the number of contestants (36).

```
         $8,035
    36)$289,260
       288
       ‾‾‾
        12
         0
        ‾‾
        126
        108
        ‾‾‾
         180
         180
         ‾‾‾
           0
```

Therefore, each contestant received $8,035.

Exercise 1.31

Mr. Sullivan's estate was $252,700. He left half of the estate to his wife. The other half was divided equally among his seven children. How much did each child receive from the estate?

Solution 1.31

There are two parts to this solution.

Part I: Determine how much of the estate was divided among the children. Because half of the estate was left to his wife, divide the total amount of the estate ($252,700) by 2.

```
      $126,350
    2)$252,700
      2
      ‾
      05
       4
       ‾
       1 2
       1 2
       ‾‾‾
          07
           6
           ‾
          10
          10
          ‾‾
           00
            0
            ‾
            0
```

Hence, $126,350 was divided by the children.

Part II: Now determine how much each child received. Because there are seven children and they all received equal amounts, divide $126,350 by 7.

```
       $18,050
  7)$126,350
     7
     ─
     56
     56
     ──
      0 3
        0
        ─
        35
        35
        ──
         0
```

Therefore, each child received $18,050 from the estate.

Exercise 1.32

If your annual salary is $32,864, how much do you earn per week?

Solution 1.32

To determine how much you earn per week, divide your annual salary ($32,864) by the number of weeks in a year (52).

```
        $632
  52)$32,864
     312
     ───
      16 6
      15 6
      ────
       1 04
       1 04
       ────
          0
```

Therefore, you earn $632 per week.

1.6 ORDER OF OPERATIONS

Multiplication can be thought of as repeated addition. When both multiplication and addition must be performed, the standard convention is that multiplication is done first. The multiplications are done in the order in which they occur from left to right. Then the additions are done.

DISTRIBUTIVE PROPERTY OF MULTIPLICATION OVER ADDITION

There is another property associated with whole numbers involving both addition and multiplication: the **distributive property of multiplication over addition of whole numbers.** If a, b, and c represent any whole numbers, we can write this property as follows:
$$a \times (b + c) = (a \times b) + (a \times c).$$

Exercise 1.33
Show that:
a) $4 \times (5 + 6) = (4 \times 5) + (4 \times 6)$
b) $8 \times (5 + 3) = (8 \times 5) + (8 \times 3)$

Solution 1.33
a) $4 \times (5 + 6) = 4 \times 11 = 44$
$(4 \times 5) + (4 \times 6) = 20 + 24 = 44$
Therefore, $4 \times (5 + 6) = (4 \times 5) + (4 \times 6)$.
b) $8 \times (5 + 3) = 8 \times 8 = 64$
$(8 \times 5) + (8 \times 3) = 40 + 24 = 64$
Therefore, $8 \times (5 + 3) = (8 \times 5) + (8 \times 3)$.

MIXED OPERATIONS
Now consider finding the value of an expression using a prescribed order of operations.

Exercise 1.34
Perform the indicated operations:
a) $2 + 3 \times 5$
b) $2 \times 3 + 4 \times 5 + 6$

Solution 1.34
a) $2 + 3 \times 5$
There are two operations involved in this expression. Do the multiplication first:
$2 + \mathbf{3 \times 5} = 2 + 15$
Next, we add:
$2 + 3 \times 5 = \mathbf{2 + 15}$
$= 17$
Therefore, $2 + 3 \times 5 = 17$.
b) $2 \times 3 + 4 \times 5 + 6$
Again, there are two operations involved in this expression. Do the multiplications first:
$\mathbf{2 \times 3} + \mathbf{4 \times 5} + 6 = 6 + 20 + 6$
Next, add from left to right:
$2 \times 3 + 4 \times 5 + 6 = \mathbf{6 + 20} + 6$
$= \mathbf{26 + 6}$
$= 32$
Therefore, $2 \times 3 + 4 \times 5 + 6 = 32$.

MIXED OPERATIONS WITH PARENTHESES

When two or more operations, together with parentheses, are involved in the evaluation of an expression, use the following rule.

Rule for Performing Operations on Whole Numbers
1. First, do all operations within parentheses, if any.
2. Then, do all multiplications and divisions *in order from left to right*.
3. Finally, do all additions and subtractions *in order from left to right*.

Exercise 1.35
Evaluate:

a) $(9 - 5) \times (6 + 3) - 12 \div 6 \times 4$

b) $3 \times 5 \times 6 - 8 \div 4 + 11$

Solution 1.35

a) $(9 - 5) \times (6 + 3) - 12 \div 6 \times 4$

Step 1: First, do the work within the parentheses:
$$(\mathbf{9 - 5}) \times (\mathbf{6 + 3}) - 12 \div 6 \times 4$$
$$= 4 \times 9 - 12 \div 6 \times 4$$

Step 2: Next, do all multiplications and divisions in order from left to right:
$$\mathbf{4 \times 9} - 12 \div 6 \times 4$$
$$= 36 - \mathbf{12 \div 6} \times 4$$
$$= 36 - \mathbf{2 \times 4}$$
$$= 36 - 8$$

Step 3: Finally, do all additions and subtractions in order from left to right:
$$\mathbf{36 - 8} = 28$$

Therefore, $(9 - 5) \times (6 + 3) - 12 \div 6 \times 4 = 28$.

b) $3 \times 5 \times 6 - 8 \div 4 + 11$

Step 1: There are no parentheses, so first do all the multiplications and division in order from left to right:
$$\mathbf{3 \times 5} \times 6 - 8 + 4 + 11$$
$$= \mathbf{15 \times 6} - 8 \div 4 + 11$$
$$= 90 - \mathbf{8 \div 4} + 11$$
$$= 90 - 2 + 11$$

Step 2: Next, do all additions and subtractions in order from left to right:
$$\mathbf{90 - 2} + 11$$
$$= \mathbf{88 + 11}$$
$$= 99$$

Therefore, $3 \times 5 \times 6 - 8 - 4 + 11 = 99$.

EXPONENTS

A whole number is sometimes multiplied by itself, such as 5 × 5. The expression may also be written as 5^2, where the number 5 is written with a *superscript* 2 to the right and above the 5 as indicated. The expression 5^2 is read "5 squared" and means 5 × 5. The number being squared, 5, is called the **base**, and the superscript 2, is called the **exponent**. In a similar manner, 6^3 is read "6 cubed" and means 6 × 6 × 6.

Exercise 1.36
Evaluate:
a) 3^2
b) 4^3
c) $6^2 + 2^3$

Solution 1.36
a) In the expression 3^2, the base is 3 and the exponent is 2: $3^2 = 3 \times 3 = 9$
b) In the expression 4^3, the base is 4 and the exponent is 3: $4^3 = 4 \times 4 \times 4 = 64$
c) $6^2 + 2^3$

Step 1: First, evaluate the expressions containing exponents:
$$6^2 + 2^3 = 36 + 8$$

Step 2: Next, add:
$$6^2 + 2^3 = \mathbf{36 + 8} = 44$$
Therefore, $6^2 + 2^3 = 44$.

MIXED OPERATIONS WITH EXPONENTS

When exponents are involved in an expression, the expression is evaluated according to the previous rule for order of operations but with an additional step.

Rule for Performing Operations on Whole Numbers
1. First, do all operations within parentheses, if any.
2. Next, evaluate all expressions with exponents.
3. Then, do all multiplications and divisions *in order from left to right*.
4. Finally, do all additions and subtractions *in order from left to right*.

Exercise 1.37
Evaluate:
a) $3^2 - 2^3 \times 5 \div 8 + 2 \times (3 + 5)$
b) $(3 + 6)^2 \div (7 + 2) + 8 - 3 \times 5 + 4^2$
c) $(2 + 3) \times (8 \div 2) \div 2^2 + (3 \times 2)^2$

Solution 1.37

a) $3^2 - 2^3 \times 5 \div 8 + 2 \times (3 + 5)$

Step 1: Work within parentheses:
$3^2 - 2^3 \times 5 \div 8 + 2 \times (\mathbf{3 + 5})$
$= 3^2 - 2^3 \times 5 \div 8 + 2 \times 8$

Step 2: Evaluate all expressions with exponents:
$\mathbf{3^2} - \mathbf{2^3} \times 5 \div 8 + 2 \times 8$
$= 9 - 8 \times 5 \div 8 + 2 \times 8$

Step 3: Do all multiplications and divisions in order from left to right:
$9 - \mathbf{8 \times 5} \div 8 + 2 \times 8$
$= 9 - \mathbf{40 \div 8} + 2 \times 8$
$= 9 - 5 + \mathbf{2 \times 8}$
$= 9 - 5 + 16$

Step 4: Do all additions and subtractions in order from left to right:
$\mathbf{9 - 5} + 16$
$= \mathbf{4 + 16}$
$= 20$

Therefore, $3^2 - 2^3 \times 5 \div 8 + 2 \times (3 + 5) = 20$.

b) $(3 + 6)^2 \div (7 + 2) + 8 - 3 \times 5 + 4^2$

Step 1: Work within parentheses:
$(\mathbf{3 + 6})^2 \div (\mathbf{7 + 2}) + 8 - 3 \times 5 + 4^2$
$= 9^2 \div 9 + 8 - 3 \times 5 + 4^2$

Step 2: Evaluate all expressions with exponents:
$\mathbf{9^2} \div 9 + 8 - 3 \times 5 + \mathbf{4^2}$
$= 81 \div 9 + 8 - 3 \times 5 + 16$

Step 3: Do all multiplications and divisions in order from left to right:
$\mathbf{81 \div 9} + 8 - 3 \times 5 + 16$
$= 9 + 8 - \mathbf{3 \times 5} + 16$
$= 9 + 8 - 15 + 16$

Step 4: Do all additions and subtractions in order from left to right:
$\mathbf{9 + 8} - 15 + 16$
$= \mathbf{17 - 15} + 16$
$= \mathbf{2 + 16}$
$= 18$

Therefore, $(3 + 6)^2 \div (7 + 2) + 8 - 3 \times 5 + 4^2 = 18$.

c) $(2 + 3) \times (8 \div 2) \div 2^2 + (3 \times 2)^2$

Step 1: Work within parentheses:
$$(\mathbf{2 + 3}) \times (\mathbf{8 \div 2}) \div 2^2 + (\mathbf{3 \times 2})^2$$
$$= 5 \times 4 \div 2^2 + 6^2$$

Step 2: Evaluate all expressions with exponents:
$$5 \times 4 \div \mathbf{2^2} + \mathbf{6^2}$$
$$= 5 \times 4 \div 4 + 36$$

Step 3: Do all multiplications and divisions in order from left to right:
$$\mathbf{5 \times 4} \div 4 + 36$$
$$= \mathbf{20 \div 4} + 36$$
$$= 5 + 36$$

Step 4: Do all additions and subtractions in order from left to right:
$$\mathbf{5 + 36}$$
$$= 41$$

Therefore, $(2 + 3) \times (8 \div 2) \div 2^2 + (3 \times 2)^2 = 41$.

Exercise 1.38

If 7 is multiplied by the square of the sum of 3 and 5, what is the result?

Solution 1.38

For this exercise, translate the words into mathematics. In particular, "the square of the sum of 3 and 5" means that you first form the sum (that is, add 3 and 5), and then square it. Hence, you must evaluate:
$$7 \times (\mathbf{3 + 5})^2$$
$$= 7 \times \mathbf{8^2}$$
$$= 7 \times \mathbf{64}$$
$$= 448$$

Therefore, $7 \times (3 + 5)^2 = 448$.

Exercise 1.39

The cube of 5 is divided by the square of 5. The quotient is then subtracted from 16. What is the result?

Solution 1.39

For this exercise, you have "the cube of 5" (which means 5^3) is divided by the "square of 5" (which means 5^2). This quotient is then subtracted from 16. Hence, you must evaluate:

$$16 - (5^3 \div 5^2)$$
$$= 16 - (125 \div 25)$$
$$= 16 - 5$$
$$= 11$$

Therefore, $16 - (5^3 \div 5^2) = 11$.

1.7 OPERATIONS INVOLVING THE WHOLE NUMBER 0

In this section, we continue our discussion of evaluating expressions when more than one operation is involved. We look at operations involving the whole number 0, starting with the operations of addition and subtraction.

ADDITION AND SUBTRACTION WITH 0

In Section 1.2, we learn that the sum of 0 and any whole number is that whole number. That is, if a is any whole number, then

$$a + 0 = a \quad \text{and} \quad 0 + a = a.$$

The whole number 0 is called the **additive identity** for the whole numbers. For example:
- $4 + 0 = 4$
- $0 + 12 = 12$
- $3 + 0 = 3$
- $0 + 7 = 7$
- $15 + 0 = 15$
- $0 + 0 = 0$

Because 0 is the smallest of the whole numbers, you can't subtract a whole number (other than 0) from 0, and get a whole number, because you can't subtract a larger whole number from a smaller whole number and get a whole number. However, you can subtract 0 from any whole number. Hence, if a represents any whole number, then

$$a - 0 = a.$$

For example:
- $5 - 0 = 5$
- $7 - 0 = 7$
- $9 - 0 = 9$
- $17 - 0 = 17$
- $20 - 0 = 20$
- $0 - 0 = 0$

MULTIPLICATION AND DIVISION WITH 0

We now examine multiplication with the whole number 0. The product of 0 with any whole number is always 0. That is, if a is any whole number, then
$$a \times 0 = 0 \quad \text{and} \quad 0 \times a = 0.$$

For example:
- $8 \times 0 = 0$
- $0 \times 4 = 0$
- $11 \times 0 = 0$
- $0 \times 14 = 0$
- $19 \times 0 = 0$
- $0 \times 0 = 0$

Finally, we examine division involving the whole number 0. **You must exercise caution when 0 is involved in division.**

Division Involving 0

1. If a is a whole number and $a \neq 0$, then $0 \div a = 0$.
2. Division by 0 is not defined.

Exercise 1.40

If possible, evaluate each of the following. If not possible, indicate why not.

a) $0 \div 4$
b) $0 \div 7$
c) $3 \div 0$
d) $0 \div 20$
e) $7 \div 0$
f) $0 \div 0$
g) $0 \div 51$
h) $0 \div 63$

Solution 1.40

a) $0 \div 4 = 0$

b) $0 \div 7 = 0$

c) $3 \div 0$ is not *defined,* because there is no whole number that, when multiplied by 0, is equal to 3. (Remember, multiplication is a check for division.)

d) $0 \div 20 = 0$

e) $7 \div 0$ is *not defined,* because there is no whole number that, when multiplied by 0, is equal to 7.

f) $0 \div 0$ *is not defined*. If $0 \div 0$ were defined, then $0 \div 0 = b$, where b is a whole number. But $0 \times b = 0$ for any whole number b. Therefore, $0 \div 0$ is not defined, because the solution *is not unique.*

g) $0 \div 51 = 0$

h) $0 \div 63 = 0$

In this chapter, we discussed counting, numerals, and place value. We also discussed the ordering of whole numbers and the four basic arithmetic operations on whole numbers: addition, subtraction, multiplication, and division. Exponents were also considered. The order of operations was also discussed, together with operations involving the whole number 0. Applications involving whole numbers were also considered.

TEST YOURSELF

1) Add: 43 + 197 + 64 + 1,083
2) Add: 2,691 + 723 + 1,740 + 638
3) Add: 2,304 + 597 + 621 + 3,008
4) Subtract: 6,089 − 4,076
5) Subtract: 90,082 − 9,705
6) From 7,984, subtract 4,097.
7) Multiply: 209 × 316
8) Multiply: 9,205 × 609
9) Multiply: 8,063 × 295
10) Divide: 4,940 ÷ 52
11) Divide: 1,452,168 ÷ 486
12) Divide: 908,070 ÷ 304
13) Evaluate: $216 \div (5 \times 3 - 3) - 2^2 \times 3$
14) Evaluate: $4^2 - (5^3 \div 5^2) + 5 \times 4 - 3^2$
15) Evaluate: $3^2 - 2^3 + 4^2 \div 2^4$
16) The student enrollment in a public high school decreased by 923 in one year. If the previous enrollment was 12,041, what is the new enrollment?
17) In a recent town election for mayor, June Bright received 13,069 votes and Jim Day received 11,049 votes. How many more votes did June receive than Jim?
18) Joyce went fishing and caught five fish. Their weights were 4 lb 3 oz, 3 lb 11 oz, 5 lb 2 oz, 4 lb 13 oz, and 3 lb 9 oz. What was the total weight of her catch?
19) Jacinta starts the month with $312 in her checking account. She writes two checks for $57 each, makes a deposit of $128, and then writes two more checks for $76 and $101. What is her checking account balance at the end of these transactions, if there is also a $6 service charge?
20) Brian bought two shirts for $28 each and a tie for $13. These prices included sales tax. He gave the salesclerk $100 to pay for his purchases. How much change should he have received?
21) Dick has the following monthly expenses: $425 for rent; $163 for utilities; $259 for car payment; $113 for food; and $109 for miscellaneous expenses. What are his total monthly expenses?
22) If 1 gallon of paint will cover 450 square feet of wall space, how many square feet of wall surface will 19 gallons of paint cover?
23) A particular make of car has an EPA rating of 26 miles per gallon. If a woman drives one of these cars on a 3,302-mile trip, how many gallons of gasoline should she expect to use?
24) A manufacturer has 604 widgets in inventory. An additional 213 widgets are added to the inventory and 219 are removed. How many widgets are left in the inventory?
25) In a particular community, a special tax of $26 is levied on each adult resident. If there are 6,013 residents in the community, including 1,168 children, what will be the total tax levied?

TEST YOURSELF WITH THE HANDHELD CALCULATOR

In Exercises 1–5, add.
1) 63,409 + 1,256 + 79,489 + 52,600 + 807
2) 47,123 + 54,675 + 853 + 9,041 + 88,776
3) 7,605 + 11,762 + 49,734 + 63,999 + 10,251
4) 323,146 + 179,286 + 27,904 + 13,615 + 541,982
5) 415,623 + 546,875 + 398,398 + 710,049 + 852,146

In Exercises 6–10, subtract.
6) 72,894 − 37,097
7) 89,000 − 43,099
8) 241,005 − 87,986
9) 23,496,075 − 16,072,906
10) 341,523,106 − 116,203,407

In Exercises 11–15, multiply.
11) 4,987 × 546
12) 19,087 × 2,046
13) 861 × 23,097
14) 23,154 × 87,906
15) 399,009 × 12,305

In Exercises 16–20, divide.
16) 107,067 ÷ 89
17) 62,238 ÷ 123
18) 1,096,236 ÷ 111
19) 2,099,097 ÷ 231
20) 1,109,931 ÷ 419

In Exercises 21–25, perform the indicated operations.
21) 76,219 + 159,021 − 23,999 − 17,600
22) (72,186 + 28,499) − (23,156 + 11,469)
23) (56,239 − 39,999) + (47,087 − 28,008)
24) (109,267 − 36,765) − (31,052 − 19,569)
25) (87,269 + 72,942 − 57,036) + (59,714 − 23,459 − 12,497)

In Exercises 26–30, perform the indicated operations. (*Note:* On a basic electronic calculator, you evaluate the expression 3^2 as 3×3, 4^3 as $4 \times 4 \times 4$, and so on.)
26) $432 \div (6 \times 3 - 6) - 3 \times 4$
27) $27 \div 3 \times 5 - 16 \div 4 + 2 \times 8$
28) $5^2 - 4^3 \div 4^2 + 3 \times 7 - 5^2$
29) $156 - 3 \times 2^2 + 2^3 - 4^2 \times 3^2$
30) $36 \div 12 \times 4^2 - 20 + 3^3 \div 3 \times 2^4$

TEST YOURSELF ANSWERS

1)
```
    43
   197
    64
+1,083
─────
 1,387
```

2)
```
 2,691
   723
 1,740
  +638
─────
 5,792
```

3)
```
 2,304
   597
   621
+3,008
─────
 6,530
```

4)
```
 6,089
-4,076
─────
 2,013
```

5)
```
 90,082
 -9,705
──────
 80,377
```

6)
```
 7,984
-4,097
─────
 3,887
```

7)
```
    209
   ×316
   ────
   1254
    209
    627
  ─────
 66,044
```

8)
```
    9,205
     ×609
    ─────
    82845
    55230
  ───────
 5,605,845
```

9)
```
    8,063
     ×295
    ─────
    40315
    72567
    16126
  ───────
 2,378,585
```

10)
```
         95
     ┌─────
   52)4,940
      4 68
      ────
        260
        260
        ───
          0
```

11)
```
             2,988
       ┌──────────
    486)1,452,168
        972
        ─────
        480 1
        437 4
        ─────
         42 76
         38 88
         ─────
          3 888
          3 888
          ─────
              0
```

12)
```
            2,987
       ┌─────────
    304)908,070
        608
        ─────
        300 0
        273 6
        ─────
         26 47
         24 32
         ─────
          2 150
          2 128
          ─────
             22 R
```

The Whole Numbers

13) $216 \div (\mathbf{5 \times 3} - 3) - 2^2 \times 3$
    $= 216 \div (\mathbf{15 - 3}) - 2^2 \times 3$
    $= 216 \div 12 - \mathbf{2^2} \times 3$
    $= \mathbf{216 \div 12} - 4 \times 3$
    $= 18 - \mathbf{4 \times 3}$
    $= \mathbf{18 - 12}$
    $= 6$

14) $4^2 - (\mathbf{5^3 \div 5^2}) + 5 \times 4 - 3^2$
    $= 4^2 - (\mathbf{125 \div 25}) + 5 \times 4 - 3^2$
    $= \mathbf{4^2} - 5 + 5 \times 4 - \mathbf{3^2}$
    $= 16 - 5 + \mathbf{5 \times 4} - 9$
    $= \mathbf{16 - 5} + 20 - 9$
    $= \mathbf{11 + 20} - 9$
    $= \mathbf{31 - 9}$
    $= 22$

15) $\mathbf{3^2 - 2^3 + 4^2 \div 2^4}$
    $= 9 - 8 + \mathbf{16 \div 16}$
    $= \mathbf{9 - 8} + 1$
    $= 1 + 1$
    $= 2$

16)
12,041	Previous enrollment
923	Decrease
11,118	New enrollment

17)
13,069	Number of votes for June
−11,049	Number of votes for Jim
2,020	Number of more votes for June

18) 4 lb + 3 oz
    3 lb + 11 oz
    5 lb + 2 oz
    4 lb + 13 oz
    3 lb + 9 oz
    ─────────
    19 lb + 38 oz
    19 lb + 16 oz + 16 oz + 6 oz
    19 lb + 1 lb + 1 lb + 6 oz
    21 lb + 6 oz

19) $312   Amount in checking account
    −57   Amount of first check
    $255   New amount in checking account
    −57   Amount of second check
    $198   New amount in checking account
    +128   Amount of deposit
    $326   New amount in checking account
    −76   Amount of third check
    $250   New amount in checking account
    −101   Amount of fourth check
    $149   New amount in checking account
    −6   Service charge
    $143   New amount in checking account

20) $28   Price of one shirt
    ×2   Number of shirts
    $56   Price of two shirts
    +13   Price of tie
    $69   Total purchase price

    $100   Amount given to clerk
    −69   Amount of purchases
    $31   Amount of change

21) $425   Rent
    163   Utilities
    259   Car
    113   Food
    +109   Miscellaneous
    $1,069 Total monthly expenses

22) 450   Square feet per gallon
    ×19   Number of gallons
    4050
    450
    8,550   Number of square feet covered by 19 gallons

23) We must divide 3,302 miles by 26 gallons.

$$\begin{array}{r} 127 \\ 26\overline{)3302} \\ \underline{26} \\ 70 \\ \underline{52} \\ 182 \\ \underline{182} \\ 0 \end{array}$$

Hence, the woman should use about 127 gallons of gasoline.

24)
```
  604   Number of widgets in inventory
 +213   Number of widgets added
  817   New inventory
 -219   Number of widgets removed
  598   New inventory
```

25)
```
   6,013   Number of residents
  -1,168   Number of children
   4,845   Number of adults
     ×26   Tax per adult
   29070
    9690
$125,970   Total tax on adults
```

TEST YOURSELF ANSWERS WITH THE HANDHELD CALCULATOR

1) 63,409 + 1,256 + 79,489 + 52,600 + 807 = 197,561
2) 47,123 + 54,675 + 853 + 9,041 + 88,776 = 200,468
3) 7,605 + 11,762 + 49,734 + 63,999 + 10,251 = 143,351
4) 323,146 + 179,286 + 27,904 + 13,615 + 541,982 = 1,085,933
5) 415,623 + 546,875 + 398,398 + 710,049 + 852,146 = 2,923,091
6) 72,894 − 37,097 = 35,797
7) 89,000 − 43,099 = 45,901
8) 241,005 − 87,986 = 153,019
9) 23,496,075 − 16,072,906 = 7,423,169
10) 341,523,106 − 116,203,407 = 225,319,699
11) 4,987 × 546 = 2,722,902
12) 19,087 × 2,046 = 39,052,002
13) 861 × 23,097 = 198,865,517
14) 23,154 × 87,906 = 2,035,375,524
15) 399,009 × 12,305 = 4,909,805,745
16) 107,067 ÷ 89 = 1,203
17) 62,238 ÷ 123 = 506
18) 1,096,236 ÷ 111 = 9,876
19) 2,099,097 ÷ 231 = 9,087
20) 1,109,931 ÷ 419 = 2,649
21) 76,219 + 159,021 − 23,999 − 17,600 = 193,641
22) (72,186 + 28,499) − (23,156 + 11,469) = 66,060
23) (56,239 − 39,999) + (47,087 − 28,008) = 35,319
24) (109,267 − 36,765) − (31,052 − 19,569) = 61,019
25) (87,269 + 72,942 − 57,036) + (59,714 − 23,459 − 12,497) = 126,933
26) 432 ÷ (6 × 3 − 6) − 3 × 4 = 24
27) 27 ÷ 3 × 5 − 16 ÷ 4 + 2 × 8 = 57
28) $5^2 - 4^3 \div 4^2 + 3 \times 7 - 5^2 = 17$
29) $156 - 3 \times 2^2 + 2^3 - 4^2 \times 3^2 = 8$
30) $36 \div 12 \times 4^2 - 20 + 3^3 \div 3 \times 2^4 = 172$

CHAPTER 2

Primes, Factors, and Multiples

In this chapter, we discuss factoring of natural numbers and multiples. Prime and composite numbers are introduced. We also discuss greatest common factors and least common multiples.

2.1 FACTORS AND MULTIPLES OF NATURAL NUMBERS

You encountered factors of natural numbers in Chapter 1. For example, in the statement $3 \times 5 = 15$, the numbers 3 and 5 are called **factors,** and the number 15 is called the **product.**

FACTORED FORM

When a natural number is written as the product of two or more whole number factors, it is said to be expressed in **factored form.**

Some factored forms of the natural number 24 are:
- 1×24
- 2×12
- 3×8
- $2 \times 2 \times 6$
- $2 \times 2 \times 2 \times 3$

Some factored forms of the natural number 110 are:
- 1×110
- 2×55
- 5×22
- $2 \times 5 \times 11$

Fact 1: The natural number 1 is a factor of every natural number.

Fact 2: Every natural number is a factor of itself.

When one natural number is a factor of another natural number, we say that the first natural number **divides** the second.
- Because $15 = 3 \times 5$, 3 divides 15 and 5 divides 15.
- Because $48 = 2 \times 3 \times 8$, 2 divides 48, 3 divides 48, and 8 divides 48.
- Because $17 = 1 \times 17$, 1 divides 17 and 17 divides 17.

To determine whether the natural number a is a factor of the natural number b, divide b by a. If the remainder is 0, we say that a divides b and also that a is a factor of b. To determine all the natural number factors of a given natural number, divide the given number by all the natural numbers less than it to find how many of them divide it.

Exercise 2.1
Determine all the natural number factors of:
   a) 18
   b) 24
   c) 112
   d) 19

Solution 2.1
   a) All the natural number factors of 18 are 1, 2, 3, 6, 9, and 18.
   b) All the natural number factors of 24 are 1, 2, 3, 4, 6, 8, 12, and 24.
   c) All the natural number factors of 112 are 1, 2, 4, 7, 8, 14, 16, 28, 56, and 112.
   d) The only natural number factors of 19 are 1 and 19.

MULTIPLES

Because $3 \times 5 = 15$, we say that 3 and 5 are factors of 15. The product, 15, is called a multiple of both 3 and 5.

A natural number is called a **multiple** of another natural number if it can be written as the product of that number and a second natural number. For example, the multiples of 3 are 3, 6, 9, 12, 15, 18, 21, 24, 27, and so on because
- $3 = 3 \times 1$
- $6 = 3 \times 2$
- $9 = 3 \times 3$
- $12 = 3 \times 4$
- $15 = 3 \times 5$
- $18 = 3 \times 6$

and so on.

Exercise 2.2

List all the natural number multiples of 3 that are less than 25.

Solution 2.2

From above, we determine that all the natural number multiples of 3 that are less than 25 are: 3, 6, 9, 12, 15, 18, 21, and 24.

Exercise 2.3

Determine the smallest natural number that is a multiple of the following:
   a) 3 and 7
   b) 3, 4, and 5

Solution 2.3

   a) The natural number multiples of 3 are 3, 6, 9, 12, 15, 18, 21, 24, 27, 30, and so on. The natural number multiples of 7 are 7, 14, 21, 28, 35, 42, 49, and so on. Therefore, the smallest natural number that is a multiple of both 3 and 7 is **21.**
   b) The natural number multiples of 3 are 3, 6, 9, 12, 15, 18, 21, 24, 27, 30, 33, 36, 39, 42, 45, 48, 51, 54, 57, **60,** 63, 66, 69, and so on. The natural number multiples of 4 are 4, 8, 12, 16, 20, 24, 28, 32, 36, 40, 44, 48, 52, 56, **60,** 64, 68, 72, and so on. The natural number multiples of 5 are 5, 10, 15, 20, 25, 30, 35, 40, 45, 50, 55, **60,** 65, 70, 75, 80, and so on. Therefore, the smallest natural number multiple of 3, 4, and 5 is **60.**

Exercise 2.4

Determine the largest natural number multiple of 3, 5, and 6 that is less than 100.

Solution 2.4

We determine all the natural number multiples of 3, 5, and 6 that are less than 100.

For 3: 3, 6, 9, 12, 15, 18, 21, 24, 27, 30, 33, 36, 39, 42, 45, 48, 51, 54, 57, 60, 63, 66, 69, 72, 75, 78, 81, 84, 87, **90,** 93, 96, 99
For 5: 5, 10, 15, 20, 25, 30, 35, 40, 45, 50, 55, 60, 65, 70, 75, 80, 85, **90,** 95
For 6: 6, 12, 18, 24, 30, 36, 42, 48, 54, 60, 66, 72, 78, 84, **90,** 96

Therefore, the largest natural number multiple of 3, 5, and 6 that is less than 100 is **90.**

48 Basic Math

2.2 PRIME AND COMPOSITE NUMBERS

PRIME NUMBERS

Some natural numbers have several natural number factors (divisors). Others have only two: 1 and the number itself. Natural numbers greater than 1 that have only two factors—1 and itself—are called **prime numbers.**

Exercise 2.5
List all the prime numbers that are less than 100.

Solution 2.5
2, 3, 5, 7, 11, 13, 17, 19, 23, 29, 31, 37, 41, 43, 47, 53, 59, 61, 67, 71, 73, 79, 83, 89, and 97. Two prime numbers that differ by 2 are called **twin primes.**

Exercise 2.6
List all pairs of twin primes that are less than 100.

Solution 2.6
From an examination of the prime numbers less than 100 listed in Exercise 2.5, we have: 3 and 5; 5 and 7; 11 and 13; 17 and 19; 29 and 31; 41 and 43; 59 and 61; 71 and 73.

COMPOSITE NUMBER

A natural number greater than 1 that has more than two factors (divisors) is called a **composite number.** Every natural number, other than 1, is either a prime number or a composite number. (The natural number 1 is neither prime nor composite.) If a natural number is not divisible by 2, it is not divisible by 4, 6, 8, or any other natural number multiple of 2. If a natural number is not divisible by 3, it is not divisible by 6, 9, 12, or any other natural number multiple of 3, and so on. To determine whether a natural number is prime, it is necessary only to divide it by the primes less than the number.

Exercise 2.7
Determine whether the following numbers are prime or composite:
   a) 2,368
   b) 107
   c) 147

Solution 2.7
   a) 2,368 is an even number and is, therefore, divisible by 2. (A natural number is an even number if its numeral has a ones digit of 0, 2, 4, 6, or 8.) Because 2,368 is divisible by 2 in addition to 1 and itself, 2,368 is a composite number.

b) Divide 107 by all the primes that are less than 107 but only until the divisor is greater than the quotient.

$$2\overline{)107} \quad 3\overline{)107} \quad 5\overline{)107} \quad 7\overline{)107} \quad 11\overline{)107}$$

(quotients: 53, 35, 21, 15, 9; remainders: 1R, 2R, 2R, 2R, 8R)

Because the remainder is different from 0 for all these divisions, 107 is prime.

c) Divide 147 by all the primes less than 147 but only until the divisor is greater than the quotient.

$$2\overline{)147} \quad 3\overline{)147} \quad 5\overline{)147} \quad 7\overline{)147} \quad 11\overline{)147} \quad 13\overline{)147}$$

(quotients: 73, 49, 29, 21, 13, 11; remainders: 1R, 0R, 2R, 0R, 4R, 4R)

Because $147 = 3 \times 49$, 3 divides (or is factor of) 147. Hence, 147 is composite, because it has a divisor (or factor) other than 1 and itself.

A prime natural number can be written as the product of two natural number factors in only one way, except for the order of the factors.

- $5 = 1 \times 5$
- $11 = 1 \times 11$
- $31 = 1 \times 31$
- $43 = 1 \times 43$
- $97 = 1 \times 97$

A composite natural number can be written as the product of natural number factors in more than one way.

- $20 = 1 \times 20$
- $20 = 2 \times 10$
- $20 = 4 \times 5$
- $20 = 2 \times 2 \times 5$
- $72 = 1 \times 72$
- $72 = 2 \times 36$
- $72 = 8 \times 9$
- $72 = 2 \times 4 \times 9$
- $72 = 2 \times 2 \times 2 \times 3 \times 3$
- $240 = 1 \times 240$
- $240 = 2 \times 120$

- $240 = 4 \times 60$
- $240 = 2 \times 2 \times 3 \times 20$
- $240 = 2 \times 2 \times 2 \times 2 \times 3 \times 5$

PRIME FACTORIZATION

Every composite natural number can be written as a product of prime factors in exactly one way, except for the order of the factors. Such a factorization is called a **prime factorization.**

Exercise 2.8

Determine a prime factorization of:

a) 205

b) 160

c) 1,368

Solution 2.8

a) $205 = 5 \times 41$. Because both 5 and 41 are prime numbers, a prime factorization of 205 is $205 = 5 \times 41$.

b) $160 = 2 \times 80$ (80 is composite)
$160 = 2 \times 2 \times 40$ (40 is composite)
$160 = 2 \times 2 \times 2 \times 20$ (20 is composite)
$160 = 2 \times 2 \times 2 \times 2 \times 10$ (10 is composite)
$160 = 2 \times 2 \times 2 \times 2 \times 2 \times 5$ (5 is prime)
Therefore, a prime factorization of 160 is $160 = 2 \times 2 \times 2 \times 2 \times 2 \times 5$.

c) $1,368 = 2 \times 684$ (684 is composite)
$1,368 = 2 \times 2 \times 342$ (342 is composite)
$1,368 = 2 \times 2 \times 2 \times 171$ (171 is composite)
$1,368 = 2 \times 2 \times 2 \times 3 \times 57$ (57 is composite)
$1,368 = 2 \times 2 \times 2 \times 3 \times 3 \times 19$ (19 is prime)
Therefore, a prime factorization of 1,368 is $1,368 = 2 \times 2 \times 2 \times 3 \times 3 \times 19$.

In a prime factorization of a natural number, a prime factor sometimes occurs more than once. Exponents may then be used to write the prime factorization.

Exercise 2.9

Determine a prime factorization of the following. Write your answer using exponents.

a) 320

b) 2,360

Solution 2.9

a) $320 = 2 \times 160$ (160 is composite)
$320 = 2 \times 2 \times 80$ (80 is composite)
$320 = 2 \times 2 \times 2 \times 40$ (40 is composite)
$320 = 2 \times 2 \times 2 \times 2 \times 20$ (20 is composite)
$320 = 2 \times 2 \times 2 \times 2 \times 2 \times 10$ (10 is composite)
$320 = 2 \times 2 \times 2 \times 2 \times 2 \times 2 \times 5$ (5 is prime)
Therefore, a prime factorization of 320 is $320 = 2 \times 2 \times 2 \times 2 \times 2 \times 2 \times 5$. Using exponents, $320 = 2^6 \times 5$.

b) $2{,}360 = 2 \times 1{,}180$ (1,180 is composite)
$2{,}360 = 2 \times 2 \times 590$ (590 is composite)
$2{,}360 = 2 \times 2 \times 2 \times 295$ (295 is composite)
$2{,}360 = 2 \times 2 \times 2 \times 5 \times 59$ (59 is prime)
Therefore, a prime factorization of 2,360 is $2{,}360 = 2 \times 2 \times 2 \times 5 \times 59$. Using exponents, $2{,}360 = 2^3 \times 5 \times 59$.

2.3 TESTS FOR DIVISIBILITY

When determining a prime factorization for a natural number, it is helpful to know whether the number is divisible by small prime numbers without actually doing the division. In this section, we look at some tests for divisibility.

TEST FOR DIVISIBILITY BY 2

A natural number is divisible by 2 if the ones digit of its numeral is 0, 2, 4, 6, or 8. Such a number is said to be an **even number**.

Exercise 2.10

Determine which of the following is divisible by 2:
a) 168
b) 139
c) 1,064
d) 6,125

Solution 2.10

a) 168 *is* divisible by 2, because 168 is even.
b) 139 *is not* divisible by 2, because 139 is not even.
c) 1,064 *is* divisible by 2, because 1,064 is even.
d) 6,125 *is not* divisible by 2, because 6,125 is not even.

TEST FOR DIVISIBILITY BY 3
A natural number is divisible by 3 if the sum of the digits in its numeral is divisible by 3.

Exercise 2.11
Determine which of the following is divisible by 3:
a) 1,221
b) 2,970
c) 1,672
d) 3,469

Solution 2.11
a) 1,221 *is* divisible by 3, because $1 + 2 + 2 + 1 = 6$ is divisible by 3.
b) 2,970 *is* divisible by 3, because $2 + 9 + 7 = 0 = 18$ is divisible by 3.
c) 1,672 *is not* divisible by 3, because $1 + 6 + 7 + 2 = 16$ is not divisible by 3.
d) 3,469 *is not* divisible by 3, because $3 + 4 + 6 + 9 = 22$ is not divisible by 3.

TEST FOR DIVISIBILITY BY 4
A natural number is divisible by 4 if the number formed by using the tens and ones digits in its numeral is divisible by 4.

Exercise 2.12
Determine which of the following is divisible by 4:
a) 128
b) 14,696
c) 23,035
d) 2,699,762

Solution 2.12
a) 128 *is* divisible by 4, because 28 is divisible by 4.
b) 14,696 *is* divisible by 4, because 96 is divisible by 4.
c) 23,035 *is not* divisible by 4, because 35 is not divisible by 4.
d) 2,699,762 *is not* divisible by 4, because 62 is not divisible by 4.

TEST FOR DIVISIBILITY BY 5

A natural number is divisible by 5 if the ones digit in its numeral is either 0 or 5.

Exercise 2.13

Determine which of the following is divisible by 5:
- a) 1,055
- b) 2,130
- c) 119
- d) 237

Solution 2.13

a) 1,055 *is* divisible by 5, because the ones digit is 5.

b) 2,130 *is* divisible by 5, because the ones digit is 0.

c) 119 *is not* divisible by 5, because the ones digit is 9.

d) 237 *is not* divisible by 5, because the ones digit is 7.

TEST FOR DIVISIBILITY BY 6

A natural number is divisible by 6 if it is divisible by both 2 and 3 (that is, the number is even and the sum of its digits is divisible by 3).

Exercise 2.14

Determine which of the following is divisible by 6:
- a) 234
- b) 5,961
- c) 482,144
- d) 2,556

Solution 2.14

a) 234 *is* divisible by 6, because it is divisible by 2 (the ones digit is 4) *and* by 3 (2 + 3 + 4 = 9 is divisible by 3).

b) 5,961 *is not* divisible by 6, because it is not divisible by 2 (the ones digit is 1).

c) 482,144 *is not* divisible by 6. It is divisible by 2 (the ones digit is 4), but it is not divisible by 3 (4 + 8 + 2 + 1 + 4 + 4 = 23, and 23 is not divisible by 3).

d) 2,556 *is* divisible by 6, because it is divisible by 2 (the ones digit is 6) *and* by 3 (2 + 5 + 5 + 6 = 18 is divisible by 3).

TEST FOR DIVISIBILITY BY 7

A natural number is divisible by 7 if the following procedure is satisfied: Take the ones digit, double it, and subtract that quantity from the number named by the remaining digits. If the new number is divisible by 7, the original number is divisible by 7. The procedure may be repeated until the resulting new number is small enough to determine whether it is divisible by 7.

Exercise 2.15
Determine which of the following is divisible by 7:
- a) 301
- b) 1,866
- c) 24,192

Solution 2.15
a) 301: The ones digit is 1. Double it and subtract the result, 2, from 30: 30 − 2 = 28; 28 is divisible by 7. Therefore, 301 *is* divisible by 7.

b) 1,866: The ones digit is 6. Double it and subtract the result, 12, from 186: 186 − 12 = 174. Repeat the procedure. The ones digit is 4. Double it and subtract the result, 8, from 17: 17 − 8 = 9. Because 9 is not divisible by 7, 1,866 *is not* divisible by 7.

c) 24,192: The ones digit is 2. Double it and subtract the result, 4, from 2,419: 2,419 − 4 = 2,415. Repeat the procedure. The ones digit is 5. Double it and subtract the result, 10, from 241: 241 − 10 = 231. Repeat the procedure. The ones digit is 1. Double it and subtract the result, 2, from 23: 23 − 2 = 21. 21 is divisible by 7. Therefore, 24,192 *is* divisible by 7.

TEST FOR DIVISIBILITY BY 8

A natural number is divisible by 8 if the number formed by using the hundreds, tens, and ones digit of the given number is divisible by 8.

Exercise 2.16
Determine which of the following is divisible by 8:
- a) 2,168
- b) 27,901
- c) 13,240
- d) 163,164

Solution 2.16
a) 2,168 *is* divisible by 8, because 168 is divisible by 8.

b) 27,901 *is not* divisible by 8, because 901 is not divisible by 8.

c) 13,240 *is* divisible by 8, because 240 is divisible by 8.

d) 163,164 *is not* divisible by 8, because 164 is not divisible by 8.

TEST FOR DIVISIBILITY BY 9
A natural number is divisible by 9 if the sum of the digits in its numeral is divisible by 9.

Exercise 2.17
Determine which of the following is divisible by 9:
- a) 1,087
- b) 216
- c) 494,874
- d) 364,703

Solution 2.17
- a) 1,087 *is not* divisible by 9, because $1 + 0 + 8 + 7 = 16$ is not divisible by 9.
- b) 216 *is* divisible by 9, because $2 + 1 + 6 = 9$ is divisible by 9.
- c) 494,874 *is* divisible by 9, because $4 + 9 + 4 + 8 + 7 + 4 = 36$ is divisible by 9.
- d) 364,703 *is not* divisible by 9, because $3 + 6 + 4 + 7 + 0 + 3 = 23$ is not divisible by 9.

TEST FOR DIVISIBILITY BY 10
A natural number is divisible by 10 if the ones digit in its numeral is 0.

Exercise 2.18
Determine which of the following is divisible by 10:
- a) 1,180
- b) 23,897

Solution 2.18
- a) 1,180 *is* divisible by 10, because the ones digit is 0.
- b) 23,897 *is not* divisible by 10, because the ones digit is 7.

TEST FOR DIVISIBILITY BY 11
A natural number is divisible by 11 if the sum of the digits in the odd-numbered columns of its numeral (counting from the *right*) minus the sum of the digits in the even-numbered columns is divisible by 11.

Exercise 2.19
Determine which of the following is divisible by 11:
- a) 4,917
- b) 2,309,876
- c) 109,736

Solution 2.19

a) 4,917:
- The sum of the digits in the odd-numbered columns of its numeral (counting from the right) is 7 + 9 = 16.
- The sum of the digits in the even-numbered columns of its numeral is 1 + 4 = 5.
- 16 − 5 = 11, which is divisible by 11.

Therefore, 4,917 is divisible by 11.

b) 2,309,876:
- The sum of the digits in the odd-numbered columns of its numeral (counting from the right) is 6 + 8 + 0 + 2 = 16.
- The sum of the digits in the even-numbered columns of its numeral is 7 + 9 + 3 = 19.
- 16 − 19 *is not* a natural number. Instead, consider 19 − 16 − 3, which is not divisible by 11.

Therefore, 2,309,876 *is not* divisible by 11.

c) 109,736:
- The sum of the digits in the odd-numbered columns of its numeral (counting from the right) is 6 + 7 + 0 = 13.
- The sum of the digits in the even-numbered columns of its numeral is 3 + 9 + 1 = 13.
- 13 − 13 = 0, which *is* divisible by 11.

Therefore, 109,736 *is* divisible by 11.

2.4 GREATEST COMMON FACTOR

In this section, we discuss the greatest common factor of two or more natural numbers.

THE G.C.F OF TWO NUMBERS

The **greatest common factor** of two natural numbers a and b is the largest natural number c, such that c divides a and b. The greatest common factor of a and b is denoted by **g. c. f.** (a, b). The greatest common factor is also called the **greatest common divisor.** To determine the greatest common factor of two natural numbers, use the procedure in the following example.

Exercise 2.20

Determine the following:

a) g.c.f. (24, 36)

b) g.c.f. (110, 132)

Solution 2.20

a) g.c.f. (24, 36)

Step 1: Write the prime factorizations of 24 and 36:
$$24 = 2 \times 2 \times 2 \times 3$$
$$36 = 2 \times 2 \times 3 \times 3$$

Step 2: Write the product of all the prime factors that are common in these factorizations:
$$2 \times 3$$

Step 3: Raise each of the factors in Step 2 to its smallest power in either of the factorizations:
$$2^2 \times 3^1$$
Therefore, the g.c.f. $(24, 36) = 2^2 \times 3^1 = 12$.

b) g.c.f. (110, 132).

Step 1: Write the prime factorizations of 110 and 132:
$$110 = 2 \times 5 \times 11$$
$$132 = 2 \times 2 \times 3 \times 11$$

Step 2: Write the product of all the prime factors that are common in these factorizations:
$$2 \times 11$$

Step 3: Raise each of the factors in Step 2 to its smallest power in either of the factorizations:
$$2^1 \times 11^1$$
Therefore, the g.c.f. $(110, 132) = 2^1 \times 11^1 = 2 \times 11 = 22$.

THE G.C.F. OF MORE THAN TWO NUMBERS

The same procedure can be extended to determine the g.c.f. of three or more natural numbers.

Exercise 2.21

Determine the following:
a) g.c.f. (88, 128, 180)
b) g.c.f. (24, 48, 96)
c) g.c.f. (60, 70, 80, 90)
d) g.c.f. (8, 27)
e) g.c.f. (23, 48, 92, 210)

Solution 2.21

a) g.c.f. (88, 128, 180)

Step 1: Write the prime factorizations of 88, 128, and 180:
$$88 = 2 \times 2 \times 2 \times 11$$

$128 = 2 \times 2 \times 2 \times 2 \times 2 \times 2 \times 2$
$180 = 2 \times 2 \times 3 \times 3 \times 5$

Step 2: Write the product of all the prime factors that are common in these factorizations:
  2

Step 3: Raise each of the factors in Step 2 to its smallest power in any of the factorizations:
  2^2

Therefore, the g.c.f. $(88, 128, 180) = 2^2 = 4$.

b) g.c.f. (24, 48, 96).

Step 1: Write the prime factorizations of 24, 48, and 96:
  $24 = 2 \times 2 \times 2 \times 3$
  $48 = 2 \times 2 \times 2 \times 2 \times 3$
  $96 = 2 \times 2 \times 2 \times 2 \times 2 \times 3$

Step 2: Write the product of all the prime factors that are common in these factorizations:
  2×3

Step 3: Raise each of the factors in Step 2 to its smallest power in any of the factorizations:
  $2^3 \times 3^1$

Therefore, the g.c.f. $(24, 48, 96) = 2^3 \times 3^1 = 8 \times 3 = 24$.

c) g.c.f. (60, 70, 80, 90)

Step 1: Write the prime factorizations for 60, 70, 80, and 90:
  $60 = 2 \times 2 \times 3 \times 5$
  $70 = 2 \times 5 \times 7$
  $80 = 2 \times 2 \times 2 \times 2 \times 5$
  $90 = 2 \times 3 \times 3 \times 5$

Step 2: Write the product of all the prime factors that are common in these factorizations:
  2×5

Step 3: Raise each of the factors in Step 2 to its smallest power in any of the factorizations:
  $2^1 \times 5^1$

Therefore, the g.c.f. $(60, 70, 80, 90) = 2^1 \times 5^1 = 2 \times 5 = 10$.

d) g.c.f. (8, 27).

Step 1: Write the prime factorizations of 8 and 27:
  $8 = 2 \times 2 \times 2$
  $27 = 3 \times 3 \times 3$

Step 2: Write the product of all the prime factors that are common in these factorizations: 8 and 27 share no common prime factors. 1, which is neither prime nor composite, is the greatest common factor. (Recall that 1 is a factor of every natural number.)

Therefore, the g.c.f. (8, 27) = 1.

Note that both 8 and 27 are composite numbers. However, the g.c.f. (8, 27) = 1. Each number is thus said to be a **prime relative to the other.**

e) g.c.f. (23, 48, 92, 210).

Step 1: Write the prime factorizations for 23, 48, 92, and 210:
$$23 = 1 \times 23$$
$$48 = 2 \times 2 \times 2 \times 2 \times 3$$
$$92 = 2 \times 2 \times 23$$
$$210 = 2 \times 3 \times 5 \times 7$$

Step 2: Write the product of all the prime factors that are common in these factorizations: There are no common prime factors; 2 is common to 48, 92, and 210 but not to 23, and so on. Therefore, the g.c.f. (23, 48, 92, 210) = 1.

Exercise 2.22

Let a and b be two distinct prime numbers. Determine the g.c.f. (a, b).

Solution 2.22

Step 1: Write the prime factorization of each of the two numbers:

Because a is prime, $a = 1 \times a$.

Because b is prime, then $b = 1 \times b$.

Step 2: Write the product of all the prime factors that are common in these factorizations: There are no common prime factors for these numbers.

Therefore, the g.c.f. $(a, b) = 1$.

Exercise 2.22 shows that the greatest common factor of any two prime natural numbers is always 1.

2.5 LEAST COMMON MULTIPLE

THE L.C.M. OF TWO NUMBERS

A natural number that is a multiple of two or more natural numbers is called a **common multiple** of those numbers. The smallest natural number that is a common multiple of two or more natural numbers is called the **least common multiple,** denoted by **l.c.m.,** of those numbers. To determine the least common multiple of two natural numbers, use the procedure in the following example.

Exercise 2.23
Determine the following:
- a) l.c.m. (4, 5)
- b) l.c.m. (12, 27)

Solution 2.23
a) l.c.m. (4, 5)

Step 1: Write the prime factorizations of 4 and 5:
$$4 = 2 \times 2$$
$$5 = 1 \times 5$$

Step 2: Write the product of all the prime factors that occur in these factorizations:
$$2 \times 5$$

Step 3: Raise each of the factors in Step 2 to its largest power in any of the factorizations:
$$2^2 \times 5^1$$
Therefore, the l.c.m. $(4, 5) = 2^2 \times 5^1 = 4 \times 5 = 20$.

b) l.c.m. (12, 27).

Step 1: Write the prime factorizations of 12 and 27:
$$12 = 2 \times 2 \times 3$$
$$27 = 3 \times 3 \times 3$$

Step 2: Write the product of all the prime factors that occur in these factorizations:
$$2 \times 3$$

Step 3: Raise each of the factors in Step 2 to its largest power in either of the factorizations:
$$2^2 \times 3^3$$
Therefore, the l.c.m. $(12, 27) = 2^2 \times 3^3 = 4 \times 27 = 108$.

THE L.C.M. OF MORE THAN TWO NUMBERS
The preceding procedure can also be used to find the l.c.m. for three or more natural numbers.

Example 2.24
Determine the following:
- a) l.c.m. (21, 35, 42)
- b) l.c.m. (44, 64, 90)
- c) l.c.m. (8, 12, 15, 21)
- d) l.c.m. (36, 42, 54, 70)
- e) l.c.m. (312, 524, 628)
- f) l.c.m. (30, 35, 40, 45)

Solution 2.24

a) l.c.m. (21, 35, 42)

Step 1: Write the prime factorizations for 21, 35, and 42:
$$21 = 3 \times 7$$
$$35 = 5 \times 7$$
$$42 = 2 \times 3 \times 7$$

Step 2: Write the product of all the prime factors that occur in these factorizations:
$$2 \times 3 \times 5 \times 7$$

Step 3: Raise each of the factors in Step 2 to its largest power in any of the factorizations:
$$2^1 \times 3^1 \times 5^1 \times 7^1$$
Therefore, the l.c.m. (21, 35, 42) = $2^1 \times 3^1 \times 5^1 \times 7^1 = 210$.

b) l.c.m. (44, 64, 90)

Step 1: Write the prime factorizations of 44, 64, and 90:
$$44 = 2 \times 2 \times 11$$
$$64 = 2 \times 2 \times 2 \times 2 \times 2 \times 2$$
$$90 = 2 \times 3 \times 3 \times 5$$

Step 2: Write the product of all the prime factors that occur in these factorizations:
$$2 \times 3 \times 5 \times 11$$

Step 3: Raise each of the factors in Step 2 to its largest power in any of the factorizations:
$$2^6 \times 3^2 \times 5^1 \times 11^1$$
Therefore, the l.c.m. (44, 64, 90) = $2^6 \times 3^2 \times 5^1 \times 11^1 = 64 \times 9 \times 5 \times 11 = 31,680$.

c) l.c.m. (8, 12, 15, 21)

Step 1: Write the prime factorizations for 8, 12, 15, and 21:
$$8 = 2 \times 2 \times 2$$
$$12 = 2 \times 2 \times 3$$
$$15 = 3 \times 5$$
$$21 = 3 \times 7$$

Step 2: Write the product of all the prime factors that occur in these factorizations:
$$2 \times 3 \times 5 \times 7$$

Step 3: Raise each of the factors in Step 2 to its largest power in any of the factorizations:
$$2^3 \times 3^1 \times 5^1 \times 7^1$$
Therefore, the l.c.m. (8, 12, 15, 21) = $2^3 \times 3^1 \times 5^1 \times 7^1 = 8 \times 3 \times 5 \times 7 = 840$.

d) l.c.m. (36, 42, 54, 70)
   Step 1: Write the prime factorizations of 36, 42, 54, 70:
   $$36 = 2 \times 2 \times 3 \times 3$$
   $$42 = 2 \times 3 \times 7$$
   $$54 = 2 \times 3 \times 3 \times 3$$
   $$70 = 2 \times 5 \times 7$$

   Step 2: Write the product of all the prime factors that occur in these factorizations:
   $$2 \times 3 \times 5 \times 7$$

   Step 3: Raise each of the factors in Step 2 to its largest power in any of the factorizations:
   $$2^2 \times 3^3 \times 5^1 \times 7^1$$
   Therefore, the l.c.m. (36, 42, 54, 70) = $2^2 \times 3^3 \times 5^1 \times 7^1 = 4 \times 27 \times 5 \times 7 = 3{,}780$.

e) l.c.m. (312, 524, 628)
   Step 1: Write the prime factorizations for 312, 524, and 628:
   $$312 = 2 \times 2 \times 2 \times 3 \times 13$$
   $$524 = 2 \times 2 \times 131$$
   $$628 = 2 \times 2 \times 157$$

   Step 2: Write the product of all the prime factors that occur in these factorizations:
   $$2 \times 3 \times 13 \times 131 \times 157$$

   Step 3: Raise each of the factors in Step 2 to its largest power in any of the factorizations:
   $$2^3 \times 3^1 \times 13^1 \times 131^1 \times 157^1$$
   Therefore, the l.c.m. (312, 524, 628) = $2^3 \times 3^1 \times 13^1 \times 131^1 \times 157^1 = 8 \times 3 \times 13 \times 131 \times 157 = 6{,}416{,}904$.

f) l.c.m. (30, 35, 40, 45)
   Step 1: Write the prime factorizations for 30, 35, 40, and 45:
   $$30 = 2 \times 3 \times 5$$
   $$35 = 5 \times 7$$
   $$40 = 2 \times 2 \times 2 \times 5$$
   $$45 = 3 \times 3 \times 5$$

   Step 2: Write the product of all the prime factors that occur in these factorizations:
   $$2 \times 3 \times 5 \times 7$$

   Step 3: Raise each of the factors in Step 2 to its largest power in any of the factorizations:
   $$2^3 \times 3^2 \times 5^1 \times 7^1$$
   Therefore, the l.c.m. (30, 35, 40, 45) = $2^3 \times 3^2 \times 5^1 \times 7^1 = 8 \times 9 \times 5 \times 7 = 2{,}520$.

In this chapter, we discussed prime and composite numbers. Prime factorization of composite numbers was introduced, as were greatest common factors and least common multiples. Divisibility tests for some natural numbers were also introduced.

TEST YOURSELF

Note: Wherever possible, use your calculator to help arrive at your solutions.

1) Determine all the natural number factors of 22.
2) Determine all the natural number factors of 48.
3) Determine all the natural number factors of 174.
4) Determine all the natural number factors of 2,010.
5) Determine all the multiples of 6 that are less than 115.
6) Determine all the multiples of 13 that are less than 150.
7) Determine the smallest natural number that is a multiple of 6 and 11.
8) Determine the smallest natural number that is a multiple of 12 and 17.
9) Determine the smallest natural number that is a multiple of 7, 9, and 11.
10) Determine which of the following is divisible by 3.
    a) 467
    b) 2,406
    c) 9,257
    d) 8,721
11) Determine which of the following is divisible by 4.
    a) 964
    b) 1,239
    c) 4,620
    d) 54,919
12) Determine which of the following is divisible by 7.
    a) 602
    b) 989
    c) 1,235
    d) 24,199
13) Determine which of the following is divisible by 9.
    a) 432
    b) 2,078
    c) 593,478
    d) 5,364,307
14) Determine which of the following is divisible by 11.
    a) 1,749
    b) 23,467
    c) 497,662
    d) 2,314,609
15) Determine a prime factorization of 375.
16) Determine a prime factorization of 1,282.
17) Determine a prime factorization of 2,512.
18) Determine the g.c.f. (9, 26).
19) Determine the g.c.f. (16, 56).
20) Determine the g.c.f. (5, 8, 34).
21) Determine the g.c.f. (14, 42, 86).
22) Determine the l.c.m. (54, 81).
23) Determine the l.c.m. (87, 120).
24) Determine the l.c.m. (3, 13, 37).
25) Determine the l.c.m. (5, 11, 70, 88).

TEST YOURSELF ANSWERS

1) The natural number factors of 22 are: 1, 2, 11, and 22.
2) The natural number factors of 48 are: 1, 2, 3, 4, 6, 8, 12, 16, 24, and 48.
3) The natural number factors of 174 are: 1, 2, 3, 6, 29, 58, 87, and 174.
4) The natural number factors of 2,010 are: 1, 2, 3, 5, 6, 10, 15, 30, 67, 134, 201, 335, 402, 670, 1,005, and 2,010.
5) Multiples of 6 that are less than 115 are: 6, 12, 18, 24, 30, 36, 42, 48, 54, 60, 66, 72, 78, 84, 90, 96, 102, 108, and 114.
6) Multiples of 13 that are less than 150 are: 13, 26, 39, 52, 65, 78, 91, 104, 117, 130, and 143.
7) Multiples of 6 are 6, 12, 18, 24, 30, 36, 42, 48, 54, 60, **66**, 72, 78 . . . Multiples of 11 are 11, 22, 33, 44, 55, **66**, 77, 88 . . . Therefore, the smallest natural number that is a multiple of 6 and 11 is 66.
8) Multiples of 12 are 12, 24, 36, 48, 60, 72, 84, 96, 108, 120, 132, 144, 156, 168, 180, 192, **204**, 216, 228 . . . Multiples of 17 are 17, 34, 51, 68, 85, 102, 119, 136, 153, 170, 187, **204**, 221 . . . Therefore, the smallest natural number that is a multiple of 12 and 17 is 204.
9) The natural numbers 7, 9, and 11 are all prime. Therefore, the smallest natural number that is a multiple of all three is their product, or $7 \times 9 \times 11 = 693$.
10) a) 467 *is not* divisible by 3, because $4 + 6 + 7 = 17$ is not divisible by 3.
    b) 2,406 *is* divisible by 3, because $2 + 4 + 0 + 6 = 12$ is divisible by 3.
    c) 9,257 *is not* divisible by 3, because $9 + 2 + 5 + 7 = 23$ is not divisible by 3.
    d) 8,721 *is* divisible by 3, because $8 + 7 + 2 + 1 = 18$ is divisible by 3.
11) a) 964 *is* divisible by 4, because 64 is divisible by 4.
    b) 1,239 *is not* divisible by 4, because 39 is not divisible by 4.
    c) 4,620 *is* divisible by 4, because 20 is divisible by 4.
    d) 54,919 *is not* divisible by 4, because 19 is not divisible by 4.
12) a) 602 *is* divisible by 7, because $60 - 4 = 56$ is divisible by 7.
    b) 989 *is not* divisible by 7, because $98 - 18 = 80$ is not divisible by 7.
    c) 1,235 *is not* divisible by 7, because $123 - 10 = 113$, because $11 - 6 = 5$ is not divisible by 7.
    d) 24,199 *is* divisible by 7, because $2,419 - 18 = 2,401$ is divisible by 7, because $240 - 2 = 238$ is divisible by 7, because $23 - 16 = 7$ is divisible by 7.

13) a) 432 *is* divisible by 9, because $4 + 3 + 2 = 9$ is divisible by 9.
   b) 2,078 *is not* divisible by 9, because $2 + 0 + 7 + 8 = 17$ is not divisible by 9.
   c) 593,478 *is* divisible by 9, because $5 + 9 + 3 + 4 + 7 + 8 = 36$ is divisible by 9.
   d) 5,364,307 *is not* divisible by 9, because $5 + 3 + 6 + 4 + 3 + 0 + 7 = 28$ is not divisible by 9.

14) a) 1,749 *is* divisible by 11, because $(9 + 7) - (4 + 1) = 16 - 5 = 11$ is divisible by 11.
   b) 23,467 *is not* divisible by 11, because $(7 + 4 + 2) - (6 + 3) = 13 - 9 = 4$ is not divisible by 11.
   c) 497,662 *is* divisible by 11, because $(2 + 6 + 9) - (6 + 7 + 4) = 17 - 17 = 0$ is divisible by 11.
   d) 2,314,609 *is* divisible by 11, because $(9 + 6 + 1 + 2) - (0 + 4 + 3) = 18 - 7 = 11$ is divisible by 11.

15) $375 = 5 \times 75$ (75 is composite)
    $= 5 \times 5 \times 15$ (15 is composite)
    $= 5 \times 5 \times 5 \times 3$ (3 is prime)
    Therefore, a prime factorization of $375 = 3 \times 5 \times 5 \times 5$.

16) $1,282 = 2 \times 641$ (641 is prime)
    Therefore, a prime factorization of 1,282 is 2×641.

17) $2,512 = 2 \times 1,256$ (1,256 is composite)
    $= 2 \times 2 \times 628$ (628 is composite)
    $= 2 \times 2 \times 2 \times 314$ (314 is composite)
    $= 2 \times 2 \times 2 \times 2 \times 157$ (157 is prime)
    Therefore, a prime factorization of 2,512 is $2 \times 2 \times 2 \times 2 \times 157$.

18) $9 = 3 \times 3$
    $26 = 2 \times 13$
    g.c.f. $(9, 26) = 2^0 \times 3^0 \times 13^0$ (lowest powers of prime factors)
    $= 1 \times 1 \times 1$
    $= 1$

19) $16 = 2 \times 2 \times 2 \times 2$
    $56 = 2 \times 2 \times 2 \times 7$
    g.c.f. $(16, 56) = 2^3 \times 7^0$ (lowest powers of prime factors)
    $= 8 \times 1$
    $= 8$

20) $5 = 5 \times 1$
$8 = 2 \times 2 \times 2$
$34 = 2 \times 17$
g.c.f. $(5, 8, 34) = 2^0 \times 5^0 \times 17^0$ (lowest powers of prime factors)
$= 1 \times 1 \times 1$
$= 1$

21) $14 = 2 \times 7$
$42 = 2 \times 3 \times 7$
$86 = 2 \times 43$
g.c.f. $(14, 42, 86) = 2^1 \times 3^0 \times 7^0 \times 43^0$ (lowest powers of prime factors)
$= 2 \times 1 \times 1 \times 1$
$= 2$

22) $54 = 2 \times 3 \times 3 \times 3$
$81 = 3 \times 3 \times 3 \times 3$
l.c.m. $(54, 81) = 2^1 \times 3^4$ (highest powers of prime factors)
$= 2 \times 81$
$= 162$

23) $87 = 3 \times 29$
$120 = 2 \times 2 \times 2 \times 3 \times 5$
l.c.m. $(87, 120) = 2^3 \times 3^1 \times 5^1 \times 29^1$ (highest powers of prime factors)
$= 8 \times 3 \times 5 \times 29$
$= 3,480$

24) $3 = 3$
$13 = 13$
$37 = 37$
l.c.m. $(3, 13, 37) = 3^1 \times 13^1 \times 37^1$ (highest powers of prime factors)
$= 3 \times 13 \times 37$
$= 1,443$

25) $5 = 5$
$11 = 11$
$70 = 2 \times 5 \times 7$
$88 = 2 \times 2 \times 2 \times 11$
l.c.m. $(5, 11, 70, 88) = 2^3 \times 5^1 \times 7^1 \times 11^1$ (highest powers of prime factors)
$= 8 \times 5 \times 7 \times 11$
$= 3,080$

CHAPTER 3
Fractions

In Chapter 1, we discussed whole numbers and operations on them. You learned that we can always add or multiply two whole numbers and get a whole number for the answer. However, you also learned that is not true with the operations of subtraction and division.

Whole numbers can be used to solve many of our everyday problems. However, other types of numbers are needed to solve other problems. In this chapter, we introduce one of these types of numbers: fractions. We also consider the four basic arithmetic operations done on fractions, together with the ordering of fractions. Applications involving fractions are introduced throughout the chapter.

3.1 FRACTIONS

In this section, we write the name for a fraction and classify a fraction as being **proper** or **improper.** We also discuss the relationship between improper fractions and mixed numbers.

PROPER AND IMPROPER FRACTIONS

If the numerator of a fraction is less than its denominator, the fraction is called a **proper fraction.** If the numerator of a fraction is greater than or equal to its denominator, the fraction is called an **improper fraction.**

Exercise 3.1

Classify the following fractions as proper or improper:

a) $\dfrac{3}{5}$

b) $\dfrac{13}{17}$

c) $\dfrac{9}{4}$

d) $\dfrac{23}{15}$

e) $\dfrac{0}{21}$

f) $\dfrac{7}{0}$

g) $\dfrac{39}{11}$

Solution 3.1

a) The fraction $\dfrac{3}{5}$ is *a proper* fraction. Its numerator is 3, and its denominator is 5; $3 < 5$. The fraction $\dfrac{3}{5}$ is read, "three fifths."

b) The fraction $\dfrac{13}{17}$ is a *proper* fraction. Its numerator is 13, and its denominator is 17; $13 < 17$. The fraction $\dfrac{13}{17}$ is read, "thirteen seventeenths."

c) The fraction $\dfrac{9}{4}$ is an *improper* fraction. Its numerator is 9, and its denominator is 4; $9 > 4$. The fraction $\dfrac{9}{4}$ is read, "nine fourths."

d) The fraction $\dfrac{23}{15}$ is an *improper* fraction. Its numerator is 23, and its denominator is 15; $23 > 15$. The fraction $\dfrac{23}{15}$ is read, "twenty-three fifteenths."

e) The fraction $\dfrac{0}{21}$ is a *proper* fraction. Its numerator is 0, and its denominator is 21; $0 < 21$. The fraction $\dfrac{0}{21}$ is read, "zero twenty-ones.

f) $\dfrac{7}{0}$ is *not* a fraction, because the denominator is 0.

g) The fraction $\dfrac{39}{11}$ is an *improper* fraction. Its numerator is 39, and its denominator is 11; $39 > 11$. The fraction $\dfrac{39}{11}$ is read, "thirty-nine elevenths."

The fraction $\frac{5}{5}$ is another name for 1, because $\frac{5}{5}$ means $5 \div 5$, and $5 \div 5 = 1$. Similarly, the fractions $\frac{7}{7}, \frac{9}{9}, \frac{12}{12}$, and $\frac{17}{17}$ are also names for 1. Hence, if the numerator and denominator of a fraction are the same nonzero number, the fraction is another name for the whole number 1.

Exercise 3.2

Write a fraction to represent each of the following:
a) Five Ford cars on a lot containing thirty-two cars.
b) The number of single-digit numbers among the first seventeen counting numbers.
c) The number of Republican candidates in an election among six Republican candidates, eight Democrats, and three other candidates.
d) The number of students who passed a course in a class of seventy students if eleven students failed the course and six students withdrew before the end of the course.

Solution 3.2

a) The numerator of the fraction would be the number of Ford cars, or 5. The denominator of the fraction would be the total number of cars on the lot, or 32. The required fraction, therefore, is $\frac{5}{32}$.

b) The numerator of the fraction would be the number of single-digit counting numbers, or 9. The denominator of the fraction would be the total number of counting numbers involved, or 17. The required fraction, therefore, is $\frac{9}{17}$.

c) The numerator of the fraction would be the number of Republican candidates, or 6. The denominator of the fraction would be the total number of candidates, or $6 + 8 + 3 = 17$.

The required fraction, therefore, is $\frac{6}{17}$.

d) The numerator would be the number of students who passed the course. There were seventy students in the course. Because eleven students failed the course and six students withdrew before the end of the course, the number of students who passed the course is $70 - (11 + 6) = 70 - 17 = 53$. The denominator would be the total number of students in the course, or 70. The required fraction, therefore, is $\frac{53}{70}$.

MIXED NUMBERS

A **mixed number** is the sum of a whole number and a proper fraction. An improper fraction can be rewritten as a mixed number. In a mixed number, the whole number part and the proper fraction part are written side by side without the addition symbol.

REWRITING IMPROPER FRACTIONS AS MIXED NUMBERS

To rewrite an improper fraction as a mixed number:
- Divide the numerator of the fraction by its denominator. The quotient in this division is the whole number part of the mixed number.
- The remainder in the above division, written over the denominator of the improper fraction, is the fraction part of the mixed number.

Exercise 3.3
Rewrite the following improper fractions as mixed numbers:

a) $\dfrac{16}{5}$

b) $\dfrac{38}{9}$

c) $\dfrac{47}{6}$

d) $\dfrac{243}{28}$

e) $\dfrac{187}{13}$

f) $\dfrac{909}{11}$

Solution 3.3

a) To rewrite the improper fraction $\dfrac{16}{5}$ as a mixed number, divide 16 by 5. This gives a quotient of 3 and a remainder of 1.
- The quotient, 3, is the whole number part of the mixed number.
- The remainder, 1, written over the denominator of the improper fraction, 5, is the fraction part of the mixed number.
- Therefore, the required mixed number is $3\dfrac{1}{5}$.

b) To rewrite the improper fraction $\dfrac{38}{9}$ as a mixed number, divide 38 by 9. This gives a quotient of 4 and a remainder of 2.
- The quotient, 4, is the whole number part of the mixed number.
- The remainder, 2, written over the denominator of the improper fraction, 9, is the fraction part of the mixed number.
- Therefore, the required mixed number is $4\dfrac{2}{9}$.

c) To rewrite the improper fraction $\frac{47}{6}$ as a mixed number, divide 47 by 6. This gives a quotient of 7 and a remainder of 5.
- The quotient, 7, is the whole number part of the mixed number.
- The remainder, 5, written over the denominator of the improper fraction, 6, is the fraction part of the mixed number.
- Therefore, the required mixed number is $7\frac{5}{6}$.

d) To rewrite the improper fraction $\frac{243}{28}$ as a mixed number, divide 243 by 28. This gives a quotient of 8 and a remainder of 19.
- The quotient, 8, is the whole number part of the mixed number.
- The remainder, 19, written over the denominator of the improper fraction, 28, is the fraction part of the mixed number.
- Therefore, the required mixed number is $8\frac{19}{28}$.

e) To rewrite the improper fraction $\frac{187}{13}$ as a mixed number, divide 187 by 13. This gives a quotient of 14 and a remainder of 5.
- The quotient, 14, is the whole number part of the mixed number.
- The remainder, 5, written over the denominator of the improper fraction, 13, is the fraction part of the mixed number.
- Therefore, the required mixed number is $14\frac{5}{13}$.

f) To rewrite the improper fraction $\frac{909}{11}$ as a mixed number, divide 909 by 11. This gives a quotient of 82 and a remainder of 7.
- The quotient, 82, is the whole number part of the mixed number.
- The remainder, 7, written over the denominator of the improper fraction, 11, is the fraction part of the mixed number.
- Therefore, the required mixed number is $82\frac{7}{11}$.

REWRITING MIXED NUMBERS AS IMPROPER FRACTIONS

We now consider rewriting a mixed number as an improper fraction. To rewrite a mixed number as an improper fraction:
- Multiply the whole number part of the mixed number by the denominator of its fraction part.
- Add the numerator of the fraction part of the mixed number to the product found in the above step. This is the numerator of the improper faction.
- The denominator of the required improper fraction is the same as the denominator of the fraction part of the given mixed number.

Exercise 3.4

Rewrite the following mixed numbers as improper fractions:

a) $4\frac{1}{2}$

b) $5\frac{2}{3}$

c) $11\frac{2}{7}$

d) $15\frac{4}{7}$

e) $45\frac{1}{3}$

f) $19\frac{11}{12}$

g) $80\frac{3}{7}$

h) $151\frac{1}{9}$

Solution 3.4

a) Multiply the whole number part of the mixed number, 4, by the denominator of its fraction part, 2; the product is 8. Add the numerator of the fraction part of the mixed number, 1, to the product just found, 8; the result is 9. The denominator of the improper fraction is the same as the denominator of the fraction part of the mixed number, or 2; $4\frac{1}{2}$, then, can be rewritten as $\frac{9}{2}$.

The procedure used can be shortened as follows: $4\frac{1}{2} = \frac{(4\times 2)+1}{2} = \frac{8+1}{2} = \frac{9}{2}$. We use this shortcut procedure in the remaining parts of this exercise.

b) $5\frac{2}{3} = \frac{(5\times 3)+2}{3} = \frac{15+2}{3} = \frac{17}{3}$

c) $11\frac{2}{7} = \frac{(11\times 7)+2}{7} = \frac{77+2}{7} = \frac{79}{7}$

d) $15\frac{4}{7} = \frac{(15\times 7)+4}{7} = \frac{105+4}{7} = \frac{109}{7}$

e) $45\frac{1}{3} = \frac{(45\times 3)+1}{3} = \frac{135+1}{3} = \frac{136}{3}$

f) $19\frac{11}{12} = \frac{(19 \times 12) + 11}{12} = \frac{228 + 11}{12} = \frac{239}{12}$

g) $80\frac{3}{7} = \frac{(80 \times 7) + 3}{7} = \frac{560 + 3}{7} = \frac{563}{7}$

h) $151\frac{1}{9} = \frac{(151 \times 9) + 1}{9} = \frac{1359 + 1}{9} = \frac{1360}{9}$

3.2 EQUIVALENT FRACTIONS

In this section, we discuss equivalent fractions and the Fundamental Principle of Fractions. We also consider the simplest form of a fraction.

Definition: Equivalent fractions are fractions that have different names but are the same number.

If $\frac{a}{b}$ and $\frac{c}{d}$ are equivalent fractions, then $a \times d = b \times c$.

Exercise 3.5

Which of the following are equivalent fractions?

a) $\frac{1}{3}$ and $\frac{3}{9}$

b) $\frac{2}{5}$ and $\frac{4}{10}$

c) $\frac{1}{2}$ and $\frac{2}{3}$

d) $\frac{0}{5}$ and $\frac{0}{7}$

e) $\frac{5}{6}$ and $\frac{8}{9}$

f) $\frac{3}{7}$ and $\frac{5}{9}$

g) $\frac{6}{6}$ and $\frac{8}{8}$

Solution 3.5

a) The fractions $\frac{1}{3}$ and $\frac{3}{9}$ *are* equivalent fractions, because $1 \times 9 = 3 \times 3$.

b) The fractions $\frac{2}{5}$ and $\frac{4}{10}$ *are* equivalent fractions, because $2 \times 10 = 5 \times 4$.

c) The fractions $\frac{1}{2}$ and $\frac{2}{3}$ *are not* equivalent fractions, because $1 \times 3 \neq 2 \times 2$.

d) The fractions $\dfrac{0}{5}$ and $\dfrac{0}{7}$ *are* equivalent fractions, because $0 \times 7 = 5 \times 0$.

e) The fractions $\dfrac{5}{6}$ and $\dfrac{8}{9}$ *are not* equivalent fractions, because $5 \times 9 \neq 6 \times 8$.

f) The fractions $\dfrac{3}{7}$ and $\dfrac{5}{9}$ *are not* equivalent fractions, because $3 \times 9 \neq 7 \times 5$.

g) The fractions $\dfrac{6}{6}$ and $\dfrac{8}{8}$ *are* equivalent fractions, because $6 \times 8 = 6 \times 8$.

FUNDAMENTAL PRINCIPLE OF FRACTIONS

To obtain an equivalent fraction, you can multiply or divide **both** the numerator and the denominator of the given fraction by the same **nonzero** whole number. This is known as the **Fundamental Principle of Fractions**.

Exercise 3.6

Use the first part of the Fundamental Principle of Fractions to show that the following are equivalent fractions:

a) $\dfrac{2}{5}$ and $\dfrac{8}{20}$

b) $\dfrac{4}{7}$ and $\dfrac{12}{21}$

c) $\dfrac{8}{5}$ and $\dfrac{24}{15}$

d) $\dfrac{0}{3}$ and $\dfrac{0}{21}$

e) $\dfrac{11}{7}$ and $\dfrac{44}{28}$

Solution 3.6

a) $\dfrac{2}{5} = \dfrac{2 \times 4}{5 \times 4} = \dfrac{8}{20}$

b) $\dfrac{4}{7} = \dfrac{4 \times 3}{7 \times 3} = \dfrac{12}{21}$

c) $\dfrac{8}{5} = \dfrac{8 \times 3}{5 \times 3} = \dfrac{24}{15}$

d) $\dfrac{0}{3} = \dfrac{0 \times 7}{3 \times 7} = \dfrac{0}{21}$

e) $\dfrac{11}{7} = \dfrac{11 \times 4}{7 \times 4} = \dfrac{44}{28}$

Exercise 3.7

Use the second part of the Fundamental Principle of Fractions to show that the following are equivalent fractions:

a) $\dfrac{9}{12}$ and $\dfrac{3}{4}$

b) $\dfrac{30}{25}$ and $\dfrac{6}{5}$

c) $\dfrac{0}{7}$ and $\dfrac{0}{1}$

d) $\dfrac{42}{12}$ and $\dfrac{7}{2}$

e) $\dfrac{54}{36}$ and $\dfrac{18}{12}$

Solution 3.7

a) $\dfrac{9}{12} = \dfrac{9 \div 3}{12 \div 3} = \dfrac{3}{4}$

b) $\dfrac{30}{25} = \dfrac{30 \div 5}{25 \div 5} = \dfrac{6}{5}$

c) $\dfrac{0}{7} = \dfrac{0 \div 7}{7 \div 7} = \dfrac{0}{1}$

d) $\dfrac{42}{12} = \dfrac{42 \div 6}{12 \div 6} = \dfrac{7}{2}$

e) $\dfrac{54}{36} = \dfrac{54 \div 3}{36 \div 3} = \dfrac{18}{12}$

Exercise 3.8

What number should be substituted for the question mark so that $\dfrac{2}{7}$ and $\dfrac{?}{35}$ are equivalent fractions?

Solution 3.8

$35 = 7 \times 5$. Use the Fundamental Principle of Fractions to multiply *both* the numerator and denominator of $\dfrac{2}{7}$ by 5:

$$\dfrac{2}{7} = \dfrac{2 \times 5}{7 \times 5} = \dfrac{10}{35}$$

Thus, ? = 10.

Exercise 3.9
$\frac{4}{5}$ is equivalent to what fraction that has a denominator of 40?

Solution 3.9
$40 = 5 \times 8$. If we multiply *both* the numerator and denominator of $\frac{4}{5}$ by 8: $\frac{4}{5} = \frac{4 \times 8}{5 \times 8} = \frac{32}{40}$

Exercise 3.10
$\frac{9}{8}$ is equivalent to what fraction that has a numerator of 36?

Solution 3.10
$36 = 9 \times 4$. If we multiply both the numerator and the denominator of $\frac{9}{8}$ by 4: $\frac{9}{8} = \frac{9 \times 4}{8 \times 4} = \frac{36}{32}$

Exercise 3.11
$\frac{21}{36}$ is equivalent to what fraction that has a denominator of 12?

Solution 3.11
$12 = 36 \div 3$. If we divide both the numerator and the denominator of $\frac{21}{36}$ by 3:
$\frac{21}{36} = \frac{21 \div 3}{36 \div 3} = \frac{7}{12}$

Exercise 3.12
$\frac{9}{51}$ is equivalent to what fraction that has a numerator of 3?

Solution 3.12
$3 = 9 \div 3$. If we divide both the numerator and the denominator of $\frac{9}{51}$ by 3: $\frac{9}{51} = \frac{9 \div 3}{51 \div 3} = \frac{3}{17}$

When working with equivalent fractions, we generally work with what is called the simplest form. If the numerator and denominator of a fraction are both exactly divisible only by the whole number 1, the fraction is said to be in its **simplest form** or in **lowest terms.**

SIMPLIFYING FRACTIONS

Exercise 3.13
Write the following fractions in simplest form:

a) $\frac{3}{19}$

b) $\frac{2}{12}$

Fractions

c) $\dfrac{18}{24}$

d) $\dfrac{152}{648}$

Solution 3.13

a) The fraction $\dfrac{3}{19}$ is already in simplest form, because 3 and 19 are both prime numbers that are exactly divisible only by the whole number 1.

b) The fraction $\dfrac{2}{12}$ is not yet in simplest form. Both the numerator and denominator are even and are divisible by 2: $\dfrac{2}{12} = \dfrac{2 \div 2}{12 \div 2} = \dfrac{1}{6}$. The resulting fraction, $\dfrac{1}{6}$ is simplified.

c) The fraction $\dfrac{18}{24}$ is not yet in simplest form. Both the numerator and denominator are divisible by 2: $\dfrac{18}{24} = \dfrac{18 \div 2}{24 \div 2} = \dfrac{9}{12}$. The resulting fraction, $\dfrac{9}{12}$, is still not in simplest form; both the numerator and denominator are divisible by 3: $\dfrac{9}{12} = \dfrac{9 \div 3}{12 \div 3} = \dfrac{3}{4}$. The fraction $\dfrac{3}{4}$ is in simplest form. Hence, the simplest form of $\dfrac{18}{24}$ is $\dfrac{3}{4}$.

d) The fraction $\dfrac{152}{648}$ is not yet in its simplest form. Both 152 and 648 are divisible by 4: $\dfrac{152}{648} = \dfrac{152 \div 4}{648 \div 4} = \dfrac{38}{162}$. The resulting fraction, $\dfrac{38}{162}$, is still not in simplest form; both the numerator and the denominator are divisible by 2: $\dfrac{38}{162} = \dfrac{38 \div 2}{162 \div 2} = \dfrac{19}{81}$. The resulting fraction, $\dfrac{19}{81}$, is in simplest form.

Exercise 3.14

In a baseball league, team A won seven games out of nine games played, and team B won six games out of eight games played. Did the two teams win an equivalent number of the games played?

Solution 3.14

If the fractions $\dfrac{7}{9}$ and $\dfrac{6}{8}$ are equivalent, 7×8 must be equal to 9×6. But $7 \times 8 = 56$ and $9 \times 6 = 54$. Because $56 \neq 54$, the two fractions are not equivalent. Therefore, the two teams did not win an equivalent number of the games played.

Exercise 3.15
Stacey worked twelve hours and received $40. Mike worked sixteen hours and received $52. Did they receive an equivalent hourly rate of pay?

Solution 3.15
If the fractions $\frac{\$40}{12}$ and $\frac{\$52}{16}$ are equivalent, 40×16 must be equal to $12 \times \$52$. But $\$40 \times 16 = \$640$ and $12 \times \$52 = \$624$. Because $\$640 \neq \$624$, Stacey and Mike did not receive an equivalent hourly rate of pay.

3.3 MULTIPLICATION AND DIVISION OF FRACTIONS AND MIXED NUMBERS

In this section, we discuss multiplication and division of fractions and mixed numbers and how to express the results in simplest form. We also discuss the reciprocal of a nonzero fraction.

MULTIPLYING FRACTIONS
To multiply any two fractions:
1. Multiply the numerators of the two fractions to get the numerator of the product.
2. Multiply the denominators of the two fractions to get the denominator of the product.
3. Simplify the resulting fraction, if possible.

Exercise 3.16
Multiply the following fractions:

a) $\frac{2}{5} \times \frac{4}{7}$

b) $\frac{1}{4} \times \frac{6}{13}$

c) $\frac{7}{15} \times \frac{0}{6}$

d) $\frac{5}{8} \times \frac{3}{7}$

e) $\frac{6}{11} \times \frac{4}{5}$

Solution 3.16
a) $\frac{2}{5} \times \frac{4}{7} = \frac{2 \times 4}{5 \times 7}$

$= \frac{8}{35}$. This is simplified.

b) $\dfrac{1}{4} \times \dfrac{6}{13} = \dfrac{1 \times 6}{4 \times 13}$

$= \dfrac{6}{52}$. Now simplify:

$= \dfrac{6 \div 2}{52 \div 2}$

$= \dfrac{3}{26}$

c) $\dfrac{7}{15} \times \dfrac{0}{6} = \dfrac{7 \times 0}{15 \times 6}$

$= \dfrac{0}{90}$. Now simplify:

$= 0$

d) $\dfrac{5}{8} \times \dfrac{3}{7} = \dfrac{5 \times 3}{8 \times 7}$

$= \dfrac{15}{56}$. This is simplified.

e) $\dfrac{6}{11} \times \dfrac{4}{5} = \dfrac{6 \times 4}{11 \times 5}$

$= \dfrac{24}{55}$. This is simplified.

When multiplying fractions, first divide the numerator and the denominator of the product by any common factors. This simplifies your work.

Exercise 3.17

Multiply the following fractions:

a) $\dfrac{3}{4} \times \dfrac{8}{9}$

b) $\dfrac{5}{8} \times \dfrac{16}{15} \times \dfrac{32}{24}$

c) $\dfrac{6}{5} \times \dfrac{15}{27} \times \dfrac{9}{18}$

Solution 3.17

a) $\dfrac{3}{4} \times \dfrac{8}{9} = \dfrac{3 \times 8}{4 \times 9}$

$= \dfrac{3 \times 8}{4 \times 9}$. Dividing by 3:

$= \dfrac{1 \times 8}{4 \times 3}$. Dividing by 4:

$= \dfrac{1 \times 2}{1 \times 3}$

$= \dfrac{2}{3}$

The above procedure could be shortened as follows:

$\dfrac{3 \times 8}{4 \times 9} = \dfrac{1 \times 2}{1 \times 3}$

$= \dfrac{2}{3}$

b) $\dfrac{5}{8} \times \dfrac{16}{15} \times \dfrac{32}{24} = \dfrac{5 \times 16 \times 32}{8 \times 15 \times 24}$

$= \dfrac{\overset{1}{\cancel{5}} \times 16 \times 32}{8 \times \underset{3}{\cancel{15}} \times 24}$. Dividing by 5:

$= \dfrac{1 \times \overset{2}{\cancel{16}} \times \overset{4}{\cancel{32}}}{\underset{1}{\cancel{8}} \times 3 \times \underset{3}{\cancel{24}}}$. Dividing by 8:

$= \dfrac{1 \times 2 \times 4}{1 \times 3 \times 3}$

$= \dfrac{8}{9}$

or

$\dfrac{\overset{1}{\cancel{5}} \times \overset{2}{\cancel{16}} \times \overset{4}{\cancel{32}}}{\underset{1}{\cancel{8}} \times \underset{3}{\cancel{15}} \times \underset{3}{\cancel{24}}} = \dfrac{1 \times 2 \times 4}{1 \times 3 \times 3}$

$= \dfrac{8}{9}$

Fractions

c) $\dfrac{6}{5} \times \dfrac{15}{27} \times \dfrac{9}{18} = \dfrac{6 \times 15 \times 9}{5 \times 27 \times 18}$

$= \dfrac{\overset{1}{\cancel{6}} \times 15 \times 9}{5 \times 27 \times \underset{3}{\cancel{18}}}$. Dividing by 6:

$= \dfrac{1 \times \overset{3}{\cancel{15}} \times 9}{\underset{1}{\cancel{5}} \times 27 \times 3}$. Dividing by 5:

$= \dfrac{1 \times 3 \times \overset{1}{\cancel{9}}}{1 \times \underset{3}{\cancel{27}} \times 3}$. Dividing by 9:

$= \dfrac{1 \times \overset{1}{\cancel{3}} \times 1}{1 \times 3 \times \underset{1}{\cancel{3}}}$. Dividing by 3:

$= \dfrac{1 \times 1 \times 1}{1 \times 1 \times 3}$

$= \dfrac{1}{3}$

or

$\dfrac{\overset{1}{\cancel{6}} \times \overset{3}{\cancel{15}} \times \overset{1}{\cancel{9}}}{\underset{1}{\cancel{5}} \times \underset{3}{\cancel{27}} \times \underset{3}{\cancel{18}}} = \dfrac{1 \times 3 \times 1}{1 \times 3 \times 3}$

$\dfrac{1 \times \overset{1}{\cancel{3}} \times 1}{1 \times \underset{1}{\cancel{3}} \times 3}$

$= \dfrac{1}{3}$

MULTIPLYING MIXED NUMBERS
To multiply mixed numbers, first rewrite the mixed numbers as improper fractions.

Exercise 3.18
Multiply:

a) $2\dfrac{1}{3} \times 5\dfrac{2}{3}$

b) $6\dfrac{3}{7} \times 2\dfrac{2}{5}$

c) $7\dfrac{1}{3} \times 8\dfrac{1}{2}$

d) $6\dfrac{2}{3} \times 7\dfrac{1}{8}$

e) $8\dfrac{1}{2} \times 9\dfrac{1}{5}$

Solution 3.18

a) $2\frac{1}{3} \times 5\frac{2}{3} = \frac{7}{3} \times \frac{17}{3}$. Rewrite mixed numbers as improper fraction:

$$= \frac{7 \times 17}{3 \times 3}$$

$$= \frac{119}{9}.$$ Rewrite product as a mixed number:

$$= 13\frac{2}{9}$$

b) $6\frac{3}{7} \times 2\frac{2}{5} = \frac{45}{7} \times \frac{12}{5}$. Rewrite mixed numbers as improper fractions:

$$= \frac{45 \times 12}{7 \times 5}$$

$$= \frac{\overset{9}{\cancel{45}} \times 12}{7 \times \underset{1}{\cancel{5}}}$$

$$= \frac{9 \times 12}{7 \times 1}$$

$$= \frac{108}{7}.$$ Rewrite product as a mixed number:

$$= 15\frac{3}{7}$$

c) $7\frac{1}{3} \times 8\frac{1}{2} = \frac{22}{3} \times \frac{17}{2}$. Rewrite mixed numbers as improper fractions:

$$= \frac{\overset{11}{\cancel{22}} \times 17}{3 \times \underset{1}{\cancel{2}}}$$

$$= \frac{11 \times 17}{3 \times 1}$$

$$= \frac{187}{3}.$$ Rewrite product as a mixed number:

$$= 62\frac{1}{3}$$

d) $6\frac{2}{3} \times 7\frac{1}{8} = \frac{20}{3} \times \frac{57}{8}$. Rewrite mixed numbers as improper tractions:

$$= \frac{\overset{5}{\cancel{20}} \times \overset{19}{\cancel{57}}}{\underset{1}{\cancel{3}} \times \underset{2}{\cancel{8}}}$$

$$= \frac{5 \times 19}{1 \times 2}$$

$= \dfrac{95}{2}$. Rewrite product as a mixed number:

$$= 47\dfrac{1}{2}$$

e) $8\dfrac{1}{2} \times 9\dfrac{1}{5} = \dfrac{17}{2} \times \dfrac{46}{5}$. Rewrite mixed numbers as improper fractions:

$$= \dfrac{17 \times \cancel{46}^{23}}{\cancel{2}_{1} \times 5}$$

$$= \dfrac{17 \times 23}{1 \times 5}$$

$$= \dfrac{391}{5}.$$ Rewrite product as a mixed number:

$$= 78\dfrac{1}{5}$$

Exercise 3.19

In a class of forty-five, $\dfrac{1}{9}$ of the students failed a psychology test. How many of the students failed the test?

Solution 3.19

$\dfrac{1}{9}$ of 45 means $\dfrac{1}{9} \times 45$. Hence, we have:

$$\dfrac{1}{9} \times 45 = \dfrac{1}{9} \times \dfrac{45}{1}$$

$$\dfrac{1 \times \cancel{45}^{5}}{\cancel{9}_{1} \times 1}$$

$$= \dfrac{1 \cdot 5}{1 \cdot 1}$$

$$= \dfrac{5}{1}$$

$$= 5$$

Therefore, five students failed the psychology test.

Exercise 3.20

In a group of 240 children, $\dfrac{5}{60}$ were found to have cavities when checked by their dentists. How many children had cavities?

Solution 3.20

$\dfrac{5}{60}$ of 240 means $\dfrac{5}{60} \times 240$. Hence, we have:

$$\dfrac{5}{60} \times 240 = \dfrac{5}{\underset{1}{\cancel{60}}} \times \dfrac{\overset{4}{\cancel{240}}}{1}$$

$$= \dfrac{5 \times 4}{1 \times 1}$$

$$= \dfrac{20}{1}$$

$$= 20$$

Therefore, twenty of the children had cavities.

RECIPROCALS

Sometimes the product of two fractions is 1. Each fraction in the multiplication in this case is called the **reciprocal** of the other. Reciprocals of fractions are used in the division of fractions.

The **reciprocal** of the *nonzero* fraction $\dfrac{a}{b}$ is the fraction $\dfrac{b}{a}$.

Exercise 3.21

Determine the reciprocals of the following fractions:

a) $\dfrac{2}{7}$

b) $\dfrac{4}{9}$

c) $\dfrac{13}{5}$

d) $\dfrac{6}{1}$

e) $4\dfrac{1}{4}$

f) $7\dfrac{1}{5}$

Solution 3.21

a) The reciprocal of $\frac{2}{7}$ is $\frac{7}{2}$.

b) The reciprocal of $\frac{4}{9}$ is $\frac{9}{4}$.

c) $\frac{5}{13}$ is the reciprocal of $\frac{13}{5}$.

d) $\frac{1}{6}$ is the reciprocal of $\frac{6}{1}$.

e) The reciprocal of $4\frac{1}{4} = \frac{17}{4}$ is $\frac{4}{17}$.

f) The reciprocal of $7\frac{1}{5} = \frac{36}{5}$ is $\frac{5}{36}$.

DIVIDING FRACTIONS

We now consider dividing a fraction by a **nonzero** fraction. To divide a fraction by a *nonzero* fraction:

1. Replace the operation of division by multiplication.
2. Replace the divisor by its reciprocal.
3. Multiply.

Exercise 3.22

Divide and write your answer in simplest form:

a) $\frac{1}{3} \div \frac{1}{4}$

b) $\frac{2}{5} \div \frac{7}{8}$

c) $\frac{5}{8} \div \frac{3}{7}$

d) $\frac{0}{3} \div \frac{7}{8}$

Solution 3.22

a) $\frac{1}{3} \div \frac{1}{4} = \frac{1}{3} \times \frac{4}{1}$

Replace division by multiplication; use the reciprocal of the divisor.

$$= \frac{1 \times 4}{3 \times 1}$$

$$= \frac{4}{3}$$

b) $\frac{2}{5} \div \frac{7}{8} = \frac{2}{5} \times \frac{8}{7}$

$$= \frac{2 \times 8}{5 \times 7}$$

$$= \frac{16}{35}$$

c) $\frac{5}{8} \div \frac{3}{7} = \frac{5}{8} \times \frac{7}{3}$

$$= \frac{5 \times 7}{8 \times 3}$$

$$= \frac{35}{24}$$

d) $\frac{0}{3} \div \frac{7}{8} = \frac{0}{3} \times \frac{8}{7}$

$$= \frac{0 \times 8}{3 \times 7}$$

$$= \frac{0}{21}$$

$= 0$. This is simplified.

DIVIDING MIXED NUMBERS

When mixed numbers are involved in a division problem, change the mixed numbers to improper fractions and proceed.

Exercise 3.23

Divide and simplify your answers:

a) $2\frac{2}{3} \div 4\frac{1}{4}$

b) $3\frac{2}{7} \div 6\frac{1}{3}$

c) $5 \div 1\frac{1}{2}$

d) $5\frac{2}{3} \div 7$

e) $6\frac{1}{7} \div 12$

Solution 3.23

a) $2\dfrac{2}{3} \div 4\dfrac{1}{4} = \dfrac{8}{3} \div \dfrac{17}{4}$

$= \dfrac{8}{3} \times \dfrac{4}{17}$

$= \dfrac{8 \times 4}{3 \times 17}$

$= \dfrac{32}{51}$

b) $3\dfrac{2}{7} \div 6\dfrac{1}{3} = \dfrac{23}{7} \div \dfrac{19}{3}$

$= \dfrac{23}{7} \times \dfrac{3}{19}$

$= \dfrac{23 \times 3}{7 \times 19}$

$= \dfrac{69}{133}$

c) $5 \div 1\dfrac{1}{2} = 5 \div \dfrac{3}{2}$

$= \dfrac{5}{1} \times \dfrac{2}{3}$

$= \dfrac{5 \times 2}{1 \times 3}$

$= \dfrac{10}{3}$

$= 3\dfrac{1}{3}$

d) $5\dfrac{2}{3} \div 7 = \dfrac{17}{3} \div 7$

$= \dfrac{17}{3} \div \dfrac{7}{1}$

$= \dfrac{17}{3} \times \dfrac{1}{7}$

$= \dfrac{17 \times 1}{3 \times 7}$

$= \dfrac{17}{21}$

e) $6\dfrac{1}{7} \div 12 = \dfrac{43}{7} \div 12$

$\qquad = \dfrac{43}{7} \div \dfrac{12}{1}$

$\qquad = \dfrac{43}{7} \times \dfrac{1}{12}$

$\qquad = \dfrac{43 \times 1}{7 \times 12}$

$\qquad = \dfrac{43}{84}$

Exercise 3.24

An empty fifty-five-gallon tank is to be filled with water. If a $1\dfrac{3}{4}$-gallon bucket is to be used, how many times will the bucket have to be filled before the tank is filled?

Solution 3.24

For this exercise, we must determine how many $1\dfrac{3}{4}$s are contained in 55. Hence, we divide as follows:

$55 \div 1\dfrac{3}{4} = 55 \div \dfrac{7}{4}$

$\qquad = \dfrac{55}{1} \div \dfrac{7}{4}$

$\qquad = \dfrac{55}{1} \times \dfrac{4}{7}$

$\qquad = \dfrac{55 \times 4}{1 \times 7}$

$\qquad = \dfrac{220}{7}$

$\qquad = 31\dfrac{3}{7}$

The bucket would have to be filled $31\dfrac{3}{7}$ times. For all practical purposes, of course, the bucket would have to be filled 32 times.

Exercise 3.25

Which is the better buy, an 11-ounce bottle of soap bubbles that sells for 79 cents or a 13-ounce bottle of the same soap bubbles that sells for 95 cents?

Solution 3.25

To solve this exercise, we consider the per-ounce price of the two bottles of soap bubbles as follows:

For the 11-ounce bottle: 79 cents divided by 11 oz, or

$$79 \text{ cents} \div 11 \text{ oz}$$
$$= \frac{79 \text{ cents}}{1} \div \frac{11 \text{ oz}}{1}$$
$$= \frac{79 \text{ cents}}{1} \times \frac{1}{11 \text{ oz}}$$
$$= \frac{79 \text{ cents} \times 1}{1 \times 11 \text{ oz}}$$
$$= \frac{79 \text{ cents}}{11 \text{ oz}}$$
$$= \frac{79}{11} \text{ cents per oz}$$
$$= 7\frac{2}{11} \text{ cents per oz}$$

For the 13-ounce bottle: 95 cents divided by 13 oz, or

$$95 \text{ cents} \div 13 \text{ oz}$$
$$= \frac{95 \text{ cents}}{1} \div \frac{13 \text{ oz}}{1}$$
$$= \frac{95 \text{ cents}}{1} \times \frac{1}{13 \text{ oz}}$$
$$= \frac{95 \text{ cents} \times 1}{1 \times 13 \text{ oz}}$$
$$= \frac{95 \text{ cents}}{13 \text{ oz}}$$
$$= \frac{95}{13} \text{ cents per oz}$$
$$= 7\frac{4}{13} \text{ cents per oz}$$

We now compare $7\frac{2}{11}$ and $7\frac{4}{13}$ by rewriting the fraction parts with a common denominator of 143 as $7\frac{26}{143}$ and $7\frac{44}{143}$, respectively. Because $7\frac{26}{143}$ (or its equivalent $7\frac{2}{11}$) is less than $7\frac{44}{143}$ (or its equivalent $7\frac{4}{13}$), the 11-ounce bottle is the better buy.

3.4 ADDITION AND SUBTRACTION OF FRACTIONS AND MIXED NUMBERS

In this section, we discuss the addition and subtraction of fractions and mixed numbers. Ordering of fractions is also discussed.

LIKE TERMS

We can add only **like** fractions; that is, those with the same denominators.

Definitions: Fractions that have the same denominator are called **like fractions**. Fractions that have different denominators are called **unlike fractions**.

Exercise 3.26

Which of the following are like fractions?

a) $\frac{2}{9}$ and $\frac{5}{9}$

b) $\frac{1}{3}$ and $\frac{3}{8}$

c) $\frac{37}{15}$ and $\frac{57}{22}$

d) $\frac{1}{5}, \frac{3}{5},$ and $\frac{9}{5}$

Solution 3.26

a) The fractions $\frac{2}{9}$ and $\frac{5}{9}$ are *like* fractions, because they both have a denominator of 9.

b) The fractions $\frac{1}{3}$ and $\frac{3}{8}$ are *unlike* fractions, because they have different denominators.

c) The fractions $\frac{37}{15}$ and $\frac{57}{22}$ are *unlike* fractions, because they have different denominators.

d) The fractions $\frac{1}{5}, \frac{3}{5},$ and $\frac{9}{5}$ are *like* fractions, because they all have a denominator of 5.

ADDING LIKE FRACTIONS

Two like fractions can be added quite simply according to the following rule.

Rule for Adding Like Fractions

To add like fractions:
1. Add the numerators.
2. Write the sum over the like denominator
3. Simplify, if possible.

Exercise 3.27

Add the following like fractions:

a) $\dfrac{4}{9} + \dfrac{1}{9}$

b) $\dfrac{3}{17} + \dfrac{8}{17}$

c) $\dfrac{2}{7} + \dfrac{4}{7} + \dfrac{1}{7}$

d) $\dfrac{4}{13} + \dfrac{6}{13} + \dfrac{5}{13}$

e) $\dfrac{9}{15} + \dfrac{10}{15} + \dfrac{7}{15}$

Solution 3.27

a) $\dfrac{4}{9} + \dfrac{1}{9} = \dfrac{4+1}{9} = \dfrac{5}{9}$

b) $\dfrac{3}{17} + \dfrac{8}{17} = \dfrac{3+8}{17} = \dfrac{11}{17}$

c) $\dfrac{2}{7} + \dfrac{4}{7} + \dfrac{1}{7} = \dfrac{2+4+1}{7} = \dfrac{7}{7} = 1$

d) $\dfrac{4}{13} + \dfrac{6}{13} + \dfrac{5}{13} = \dfrac{4+6+5}{13} = \dfrac{15}{13} = 1\dfrac{2}{13}$

e) $\dfrac{9}{15} + \dfrac{10}{15} + \dfrac{7}{15} = \dfrac{9+10+7}{15} = \dfrac{26}{15} = 1\dfrac{11}{15}$

ADDING UNLIKE FRACTION

To add unlike fractions, we rewrite the two fractions in equivalent form with the same denominator.

Exercise 3.28

Add the following unlike fractions:

a) $\dfrac{1}{4} + \dfrac{3}{5}$

b) $\dfrac{3}{7} + \dfrac{4}{5}$

c) $\dfrac{3}{8} + \dfrac{2}{9}$

d) $\dfrac{11}{16} + \dfrac{13}{24}$

e) $\dfrac{7}{12} + \dfrac{8}{15}$

f) $\dfrac{1}{2} + \dfrac{2}{3} + \dfrac{3}{5}$

g) $\dfrac{3}{4} + \dfrac{1}{3} + \dfrac{3}{5} + \dfrac{1}{2}$

Solution 3.28

a) $\dfrac{1}{4} + \dfrac{3}{5}$

The l.c.m. (4, 5) = 20.

$$\frac{1}{4} = \frac{1 \times 5}{4 \times 5} = \frac{5}{20}$$

and

$$\frac{3}{5} = \frac{3 \times 4}{5 \times 4} = \frac{12}{20}$$

Hence,

$$\frac{1}{4} + \frac{3}{5} = \frac{5}{20} + \frac{12}{20}.$$ Adding like fractions:

$$= \frac{5+12}{20}$$

$$= \frac{17}{20}$$

b) $\dfrac{3}{7}+\dfrac{4}{5}$

$\dfrac{3}{7}+\dfrac{4}{5}=\dfrac{3\times 5}{7\times 5}+\dfrac{4\times 7}{5\times 7}$, because l.c.m. (5, 7) = 35.

$=\dfrac{15}{35}+\dfrac{28}{35}$. Adding like fractions:

$=\dfrac{15+28}{35}$

$=\dfrac{43}{35}$. Rewriting as a mixed number:

$=1\dfrac{8}{35}$

c) $\dfrac{3}{8}+\dfrac{2}{9}$

$\dfrac{3}{8}+\dfrac{2}{9}=\dfrac{3\times 9}{8\times 9}+\dfrac{2\times 8}{9\times 8}$, because l.c.m. (8, 9) = 72.

$=\dfrac{27}{72}+\dfrac{16}{72}$. Adding like fractions:

$=\dfrac{27+16}{72}$

$=\dfrac{43}{72}$

d) $\dfrac{11}{16}+\dfrac{13}{24}$

$\dfrac{11}{16}+\dfrac{13}{24}=\dfrac{11\times 3}{16\times 3}+\dfrac{13\times 2}{24\times 2}$, because l.c.m. (16, 24) = 48.

$=\dfrac{33}{48}+\dfrac{26}{48}$. Adding like fractions:

$=\dfrac{33+26}{48}$

$=\dfrac{59}{48}$. Rewriting as a mixed number:

$=1\dfrac{11}{48}$

e) $\dfrac{7}{12} + \dfrac{8}{15}$

$\dfrac{7}{12} + \dfrac{8}{15} = \dfrac{7 \times 5}{12 \times 5} + \dfrac{8 \times 4}{15 \times 4}$, because l.c.m. (12, 15) = 60.

$= \dfrac{35}{60} + \dfrac{32}{60}$. Adding like fractions:

$= \dfrac{35 + 32}{60}$

$= \dfrac{67}{60}$.   Rewriting as a mixed number:

$= 1\dfrac{7}{60}$

f) $\dfrac{1}{2} + \dfrac{2}{3} + \dfrac{3}{5}$

$\dfrac{1}{2} + \dfrac{2}{3} + \dfrac{3}{5} = \dfrac{1 \times 15}{2 \times 15} + \dfrac{2 \times 10}{3 \times 10} + \dfrac{3 \times 6}{5 \times 6}$, because l.c.m. (2, 3, 5) = 30.

$= \dfrac{15}{30} + \dfrac{20}{30} + \dfrac{18}{30}$. Adding like fractions:

$= \dfrac{15 + 20 + 18}{30}$

$= \dfrac{53}{30}$.

$= 1\dfrac{23}{30}$.   Rewriting as a mixed number:

g) $\dfrac{3}{4} + \dfrac{1}{3} + \dfrac{3}{5} + \dfrac{1}{2}$

$\dfrac{3}{4} + \dfrac{1}{3} + \dfrac{3}{5} + \dfrac{1}{2} = \dfrac{3 \times 14}{4 \times 15} + \dfrac{1 \times 20}{3 \times 20} + \dfrac{3 \times 12}{5 \times 12} + \dfrac{1 \times 30}{2 \times 30}$, because l.c.m. (4,3,5,2) = 60.

$= \dfrac{45}{60} + \dfrac{20}{60} + \dfrac{36}{60} + \dfrac{30}{60}$. Adding like fractions:

$= \dfrac{45 + 20 + 36 + 30}{60}$

$= \dfrac{131}{60}$.   Rewriting as a mixed number:

$= 2\dfrac{11}{60}$

Exercise 3.29

One piece of tubing measures $4\frac{1}{4}$ ft and another piece $3\frac{13}{16}$ ft. What is the combined length of the two pieces of tubing?

Solution 3.29

To determine the combined length of the two pieces of tubing, add the lengths of the two pieces. Hence,

$$4\frac{1}{4} \text{ ft} + 3\frac{13}{16} \text{ ft}$$

$$= \frac{17}{4} \text{ ft} + \frac{61}{16} \text{ ft. Rewriting mixed numbers as improper fractions:}$$

$$= \frac{17 \times 4}{4 \times 4} \text{ ft} + \frac{61}{16} \text{ ft, because l.c.m. } (4, 16) = 16.$$

$$= \frac{68}{16} \text{ ft} + \frac{61}{16} \text{ ft. Adding like fractions:}$$

$$= \left(\frac{68 + 61}{16}\right) \text{ ft}$$

$$= \frac{129}{16} \text{ ft}$$

$$= 8\frac{1}{16} \text{ ft. Rewriting as a mixed number:}$$

Therefore, the combined length of the two pieces of tubing is $8\frac{1}{16}$ ft.

Exercise 3.30

Brad spends $\frac{1}{4}$ of his monthly income for rent, $\frac{1}{7}$ of his monthly income for his truck, and $\frac{1}{5}$ of his monthly income for food. What fractional part of his monthly income does this represent?

Solution 3.30

To determine what fractional part of Brad's monthly income is spent on rent, his truck, and food, add $\frac{1}{4}$, $\frac{1}{7}$, and $\frac{1}{5}$. Hence, we have:

$$\frac{1}{4} + \frac{1}{7} + \frac{1}{5} = \frac{1 \times 35}{4 \times 35} + \frac{1 \times 20}{7 \times 20} + \frac{1 \times 28}{5 \times 28}, \text{ because l.c.m. } (4, 5, 7) = 140.$$

$$= \frac{35}{140} + \frac{20}{140} + \frac{28}{140}$$

$$= \frac{35 + 20 + 28}{140}$$

$$= \frac{83}{140}$$

Therefore, Brad spends $\frac{83}{140}$ of his monthly income on rent, his truck, and food.

Exercise 3.31
A triangle has three sides whose measures are $\frac{3}{16}$ cm, $\frac{7}{8}$ cm, and $\frac{3}{4}$ cm. What is the total length of the three sides?

Solution 3.31
To determine the total length of the three sides of the triangle, add the lengths of the three sides. Hence, we have

$$\frac{3}{16} \text{ cm} + \frac{7}{8} \text{ cm} + \frac{3}{4} \text{ cm}$$
$$= \left(\frac{3}{16} + \frac{7 \times 2}{8 \times 2} + \frac{3 \times 4}{4 \times 4}\right) \text{ cm, because l.c.m. }(8, 4, 16) = 16.$$
$$= \left(\frac{3}{16} + \frac{14}{16} + \frac{12}{16}\right) \text{ cm}$$
$$= \frac{3 + 14 + 12}{16} \text{ cm}$$
$$= \frac{29}{16} \text{ cm}$$
$$= 1\frac{13}{16} \text{ cm}$$

Therefore, the total length of the three sides of the triangle is $1\frac{13}{16}$ cm.

ORDERING OF FRACTIONS

We now consider ordering of fractions. If two fractions are like fractions, we can compare them by comparing their numerators. If two fractions are like fractions, the fraction with the greater numerator is the greater fraction.

Exercise 3.32
Order each of the following pairs of fractions by inserting the symbol < or > between them:

a) $\frac{3}{8}$ and $\frac{1}{8}$

b) $\frac{13}{7}$ and $\frac{5}{7}$

c) $\frac{4}{9}$ and $\frac{17}{9}$

d) $\frac{0}{5}$ and $\frac{3}{5}$

Solution 3.32

a) $\dfrac{3}{8}$ and $\dfrac{1}{8}$ are *like* fractions; $\dfrac{3}{8} > \dfrac{1}{8}$, because $3 > 1$.

b) $\dfrac{13}{7}$ and $\dfrac{5}{7}$ are *like* fractions; $\dfrac{13}{7} > \dfrac{5}{7}$, because $13 > 5$.

c) $\dfrac{4}{9}$ and $\dfrac{17}{9}$ are *like* fractions; $\dfrac{4}{9} < \dfrac{17}{9}$, because $4 < 17$.

d) $\dfrac{0}{5}$ and $\dfrac{3}{5}$ are *like* fractions; $\dfrac{0}{5} < \dfrac{3}{5}$, because $0 < 3$.

To compare unlike fractions, first rewrite the fractions as equivalent fractions with the same denominators.

Exercise 3.33

For each of the following, determine which fraction is smaller:

a) $\dfrac{2}{3}$ or $\dfrac{4}{5}$

b) $\dfrac{7}{9}$ or $\dfrac{9}{11}$

c) $\dfrac{8}{9}$ or $\dfrac{7}{8}$

d) $\dfrac{13}{11}$ or $\dfrac{15}{13}$

e) $\dfrac{17}{9}$ or $\dfrac{19}{12}$

Solution 3.33

a) First rewrite each of the fractions in equivalent form with a denominator of 15.

$$\frac{2}{3} = \frac{2 \times 5}{3 \times 5} = \frac{10}{15}$$

and

$$\frac{4}{5} = \frac{4 \times 3}{5 \times 3} = \frac{12}{15}$$

The fractions $\dfrac{10}{15}$ and $\dfrac{12}{15}$ are *like* fractions. $\dfrac{10}{15} < \dfrac{12}{15}$, because $10 < 12$.

Therefore, $\dfrac{2}{3} < \dfrac{4}{5}$.

b) First rewrite each of the fractions in equivalent form with a denominator of 99.

$$\frac{7}{9} = \frac{7 \times 11}{9 \times 11} = \frac{77}{99}$$

and

$$\frac{9}{11} = \frac{9 \times 9}{11 \times 9} = \frac{81}{99}$$

The fractions $\frac{77}{99}$ and $\frac{81}{99}$ are *like* fractions. $\frac{77}{99} < \frac{81}{99}$, because 77 < 81.

Therefore, $\frac{7}{9} < \frac{9}{11}$.

c) First rewrite each of the fractions in equivalent form with a denominator of 72.

$$\frac{8}{9} = \frac{8 \times 8}{9 \times 8} = \frac{64}{72}$$

and

$$\frac{7}{8} = \frac{7 \times 9}{8 \times 9} = \frac{63}{72}$$

The fractions $\frac{64}{72}$ and $\frac{63}{72}$ are *like* fractions. $\frac{64}{72} > \frac{63}{72}$, because 64 > 63.

Therefore, $\frac{7}{8} < \frac{8}{9}$.

d) First rewrite each of the fractions in equivalent form with a denominator of 143.

$$\frac{13}{11} = \frac{13 \times 13}{11 \times 13} = \frac{169}{143}$$

and

$$\frac{15}{13} = \frac{15 \times 11}{13 \times 11} = \frac{165}{143}$$

The fractions $\frac{169}{143}$ and $\frac{165}{143}$ are *like* fractions. $\frac{169}{143} > \frac{165}{143}$, because 169 > 165. Therefore, $\frac{15}{13} < \frac{13}{11}$.

e) First rewrite each of the fractions in equivalent form with a denominator of 36.

$$\frac{17}{9} = \frac{17 \times 4}{9 \times 4} = \frac{68}{36}$$

and

$$\frac{19}{12} = \frac{19 \times 3}{12 \times 3} = \frac{57}{36}$$

The fractions $\frac{68}{36}$ and $\frac{57}{36}$ are *like* fractions. $\frac{68}{36} > \frac{57}{36}$, because 68 > 57.

Therefore, $\frac{19}{12} < \frac{17}{9}$.

SUBTRACTING FRACTIONS

We now consider subtracting one fraction from another. First, we consider the subtraction $\frac{a}{b} - \frac{c}{d}$, where $\frac{a}{b} > \frac{c}{d}$ or $\frac{a}{b} = \frac{c}{d}$.

If two fractions have the same numerator but different denominators, the one with the larger denominator is the smaller fraction.

To subtract fractions:

1. Determine the l.c.m. of all of the denominators involved.
2. Rewrite all fractions involved as equivalent fractions with the l.c.m. as denominator.
3. Subtract the numerators of the equivalent fractions and write the result over the l.c.m.
4. Simplify the result, if possible.

Exercise 3.34

Subtract the following:

a) $\frac{3}{5} - \frac{1}{2}$

b) $\frac{21}{23} - \frac{7}{11}$

c) $\frac{16}{25} - \frac{4}{13}$

d) $\frac{10}{17} - \frac{5}{11}$

Solution 3.34

a) $\dfrac{3}{5} - \dfrac{1}{2}$

The l.c.m. (2, 5) = 10. Therefore,

$$\dfrac{3}{5} - \dfrac{1}{2} = \dfrac{3 \times 2}{5 \times 2} - \dfrac{1 \times 5}{2 \times 5}.$$ Equivalent fractions using the l.c.m.

$$= \dfrac{6}{10} - \dfrac{5}{10}$$

$$= \dfrac{6-5}{10}.$$ Subtract numerators:

$$= \dfrac{1}{10}$$

b) $\dfrac{21}{23} - \dfrac{7}{11}$

The l.c.m. (11, 23) = 253. Therefore,

$$\dfrac{21}{23} - \dfrac{7}{11} = \dfrac{21 \times 11}{23 \times 11} - \dfrac{7 \times 23}{11 \times 23}.$$ Equivalent fractions using the l.c.m.

$$= \dfrac{231}{253} - \dfrac{161}{253}.$$ Subtract numerators:

$$= \dfrac{231 - 161}{253}$$

$$= \dfrac{70}{253}$$

c) $\dfrac{16}{25} - \dfrac{4}{13}$

The l.c.m. (13, 25) = 325. Therefore,

$$\dfrac{16}{25} - \dfrac{4}{13} = \dfrac{16 \times 13}{25 \times 13} - \dfrac{4 \times 25}{13 \times 25}.$$ Equivalent fractions using the l.c.m.

$$= \dfrac{208}{325} - \dfrac{100}{325}.$$ Subtract numerators.

$$= \dfrac{208 - 100}{325}$$

$$= \dfrac{108}{325}$$

d) $\dfrac{10}{17} - \dfrac{5}{11}$

The l.c.m. (11, 17) = 187. Therefore,

$\dfrac{10}{17} - \dfrac{5}{11} = \dfrac{10 \times 11}{17 \times 11} - \dfrac{5 \times 17}{11 \times 17}$. Equivalent fractions using the l.c.m.

$= \dfrac{110}{187} - \dfrac{85}{187}$

$= \dfrac{110 - 85}{187}$. Subtract numerators:

$= \dfrac{25}{187}$

Exercise 3.35

Perform the indicated operations and write your result in simplest form:

a) $\left(\dfrac{1}{3} + \dfrac{1}{2}\right) - \dfrac{1}{4}$

b) $\dfrac{11}{12} - \left(\dfrac{1}{6} + \dfrac{1}{2}\right)$

Solution 3.35

a) Work inside the parentheses first; change to equivalent fractions:

$\left(\dfrac{1}{3} + \dfrac{1}{2}\right) - \dfrac{1}{4} = \left(\dfrac{2}{6} + \dfrac{3}{6}\right) - \dfrac{1}{4}$

$= \dfrac{5}{6} - \dfrac{1}{4}$. Add like fractions:

$= \dfrac{5 \times 2}{6 \times 2} - \dfrac{1 \times 3}{4 \times 3}$. Change to equivalent fractions:

$= \dfrac{10}{12} - \dfrac{3}{12}$

$= \dfrac{10 - 3}{12}$. Subtract like fractions:

$= \dfrac{7}{12}$

b) Work within parentheses first; change to equivalent fractions:

$$\frac{11}{12} - \left(\frac{1}{6} + \frac{1}{2}\right) = \frac{11}{12} - \left(\frac{1}{6} + \frac{3}{6}\right)$$

$$= \frac{11}{12} - \frac{4}{6}.\text{ Add like fractions:}$$

$$= \frac{11}{12} - \frac{4 \times 2}{6 \times 2}.\text{ Change to equivalent fractions:}$$

$$= \frac{11}{12} - \frac{8}{12}$$

$$= \frac{11 - 8}{12}.\text{ Subtract like fractions:}$$

$$= \frac{3}{12}$$

$$= \frac{1}{4}.\text{ Simplify to lowest terms.}$$

ADDING MIXED NUMBERS

We now discuss the addition of mixed numbers. To add mixed numbers:

1. Add the whole number parts.
2. Add the fractional parts. If the sum is an improper fraction, rewrite it as a mixed number and add the result to the sum in Step 1.

Exercise 3.36
Add:

a) $4\frac{1}{2} + 5\frac{1}{3}$

b) $3\frac{2}{7} + 5\frac{5}{9}$

Solution 3.36

a) A mixed number is the sum of a whole number and a fraction:

$$4\frac{1}{2} + 5\frac{1}{3} = \left(4 + \frac{1}{2}\right) + \left(5 + \frac{1}{3}\right)$$

$$= (4 + 5) + \left(\frac{1}{2} + \frac{1}{3}\right).\text{ Regroup:}$$

$$= (4 + 5) + \left(\frac{3}{6} + \frac{2}{6}\right).\text{ Equivalent fractions with l.c.m.} = 6.$$

$= 9 + \frac{5}{6}$. Add like parts:

$= 9\frac{5}{6}$. Rewrite.

b) $3\frac{2}{7} + 5\frac{5}{9} = \left(3 + \frac{2}{7}\right) + \left(5 + \frac{5}{9}\right)$

$= (3+5) + \left(\frac{2}{7} + \frac{5}{9}\right)$

$= (3+5) + \left(\frac{18}{63} + \frac{35}{63}\right)$. Equivalent fractions with l.c.m.= 63:

$= 8 + \frac{53}{63}$. Add like parts:

$= 8\frac{53}{63}$. Rewrite.

Exercise 3.37
Add:

a) $3\frac{1}{3} + 2\frac{3}{4} + 5\frac{1}{2}$

b) $13\frac{3}{4} + 17\frac{5}{8} + 15\frac{1}{2} + 16 + \frac{3}{8}$

Solution 3.37

a) $3\frac{1}{3} + 2\frac{3}{4} + 5\frac{1}{2}$

$= 3\frac{4}{12} + 2\frac{9}{12} + 5\frac{6}{12}$. Equivalent fractions; l.c.m. = 12:

$= 10 + \frac{19}{12}$. Add like parts:

$= 10 + 1 + \frac{7}{12}$

$= 11 + \frac{7}{12}$

$= 11\frac{7}{12}$

b) $13\frac{3}{4} + 17\frac{5}{8} + 15\frac{1}{2} + 16 + \frac{3}{8}$

$= 13\frac{6}{8} + 17\frac{5}{8} + 15\frac{4}{8} + 16\frac{3}{8}$

$$= 61\frac{18}{8}$$

$$= 61 + 2 + \frac{2}{8}$$

$$= 63 + \frac{1}{4}$$

$$= 63\frac{1}{4}$$

SUBTRACTING MIXED NUMBERS

To subtract mixed numbers, we subtract the whole number parts and the fractional parts.

Exercise 3.38
Subtract:

a) $9\frac{3}{4} - 6\frac{1}{6}$

b) $10\frac{1}{2} - 5\frac{7}{8}$

c) $13\frac{2}{7} - 7\frac{3}{4}$

Solution 3.38

a) $9\frac{3}{4} - 6\frac{1}{6} = \left(9 + \frac{3}{4}\right) - \left(6 + \frac{1}{6}\right)$

$\qquad = (9-6) + \left(\frac{3}{4} - \frac{1}{6}\right)$. Regroup:

$\qquad = (9-6) + \left(\frac{9}{12} - \frac{2}{12}\right)$. Equivalent fractions with l.c.m. = 12:

$\qquad = 3 + \frac{7}{12}$. Subtract like parts:

$\qquad = 3\frac{7}{12}$. Rewrite.

b) $10\dfrac{1}{2} - 5\dfrac{7}{8}$

$= 10\dfrac{4}{8} - 5\dfrac{7}{8}$. Equivalent fractions with l.c.m. = 8:

We cannot subtract $\dfrac{7}{8}$ from $\dfrac{4}{8}$, because $\dfrac{7}{8} > \dfrac{4}{8}$. However, we can borrow 1 (that is, $\dfrac{8}{8}$) from 10, and rewrite $10\dfrac{4}{8}$ as $9\dfrac{8}{8} + \dfrac{4}{8} = 9\dfrac{12}{8}$.

$= 9\dfrac{12}{8} - 5\dfrac{7}{8}$

$= 4\dfrac{5}{8}$. Subtract like parts:

c) $13\dfrac{2}{7} - 7\dfrac{3}{4}$

$= 13\dfrac{8}{28} - 7\dfrac{21}{28}$. Equivalent fractions with l.c.m. = 28:

We cannot subtract $\dfrac{21}{28}$ from $\dfrac{8}{28}$. However, we can rewrite $13\dfrac{8}{28}$ as $12\dfrac{28}{28} + \dfrac{8}{28} = 12 + \dfrac{36}{28}$:

$= 12\dfrac{36}{28} - 7\dfrac{21}{28}$

$= 5 + \dfrac{15}{28}$. Subtract like parts:

$= 5\dfrac{15}{28}$. Rewrite.

Exercise 3.39

Perform the indicated operations and write your result in simplest form:

a) $\left(2\dfrac{1}{2} + 3\dfrac{1}{3}\right) - 2\dfrac{7}{8}$

b) $5\dfrac{1}{6} + \left(8\dfrac{1}{2} - 5\dfrac{2}{7}\right)$

c) $\left(8\dfrac{2}{3} - 5\dfrac{1}{6}\right) + \left(10\dfrac{2}{7} - 8\dfrac{5}{9}\right)$

Solution 3.39

a) $\left(2\dfrac{1}{2}+3\dfrac{1}{3}\right)-2\dfrac{7}{8} = \left(2\dfrac{3}{6}+3\dfrac{2}{6}\right)-2\dfrac{7}{8}$

$= 5\dfrac{5}{6}-2\dfrac{7}{8}$

$= 5\dfrac{20}{24}-2\dfrac{21}{24}$

$= 4\dfrac{44}{24}-2\dfrac{21}{24}$

$= 2\dfrac{23}{24}$

b) $5\dfrac{1}{6}+\left(8\dfrac{1}{2}-5\dfrac{2}{7}\right) = 5\dfrac{1}{6}+\left(8\dfrac{7}{14}-5\dfrac{4}{14}\right)$

$= 5\dfrac{1}{6}+3\dfrac{3}{14}$

$= 5\dfrac{7}{42}+3\dfrac{9}{42}$

$= 8\dfrac{16}{42}$

$= 8\dfrac{8}{21}$

c) $\left(8\dfrac{2}{3}-5\dfrac{1}{6}\right)+\left(10\dfrac{2}{7}-8\dfrac{5}{9}\right) = \left(8\dfrac{4}{6}-5\dfrac{1}{6}\right)+\left(10\dfrac{18}{63}-8\dfrac{35}{63}\right)$

$= 3\dfrac{3}{6}+\left(9\dfrac{81}{63}-8\dfrac{35}{63}\right)$

$= 3\dfrac{1}{2}+1\dfrac{46}{63}$

$= 3\dfrac{63}{126}+1\dfrac{92}{126}$

$= 4\dfrac{155}{126}$

$= 5\dfrac{29}{126}$

Fractions

Exercise 3.40

Jan works part time at an insurance agency. On Monday, she worked $4\frac{1}{3}$ hours; on Tuesday, she worked $5\frac{2}{3}$ hours; on Wednesday she worked $3\frac{3}{4}$ hours; on Thursday, she worked $6\frac{1}{2}$ hours; and on Friday, she worked $7\frac{4}{5}$ hours. How many hours did Jan work that week?

Solution 3.40

To determine how many hours Jan worked that week, add the number of hours she worked each of the five days. We have:

$$4\frac{1}{3}+5\frac{2}{3}+3\frac{3}{4}+6\frac{1}{2}+7\frac{4}{5} = 4\frac{20}{60}+5\frac{40}{60}+3\frac{45}{60}+6\frac{30}{60}+7\frac{48}{60}$$

$$= (4+5+3+6+7) + \left(\frac{20}{60}+\frac{40}{60}+\frac{45}{60}+\frac{30}{60}+\frac{48}{60}\right)$$

$$= 25 + \frac{183}{60}$$

$$= 25 + \frac{180}{60} + \frac{3}{60}$$

$$= 25 + 3 + \frac{3}{60}$$

$$= 28 + \frac{1}{20}$$

$$= 28\frac{1}{20}$$

Therefore, Jan worked $28\frac{1}{20}$ hours that week.

Exercise 3.41

A nine-foot-wide roll of carpet is eighty-five yards long. A customer buys three lengths of the carpet measuring $12\frac{1}{3}$ yards, $16\frac{3}{4}$ yards, and $18\frac{1}{2}$ yards. How much of the carpet is left?

Solution 3.41

To determine how much of the carpet is left, subtract the sum of the three lengths purchased from the total length of the carpet, 85 yards. We have:

$$85 - \left(12\frac{1}{3}+16\frac{3}{4}+18\frac{1}{2}\right)$$

$$= 85 - \left(12\frac{4}{12}+16\frac{9}{12}+18\frac{6}{12}\right)$$

$$= 85 - \left[(12+16+18) + \left(\frac{4}{12} + \frac{9}{12} + \frac{6}{12} \right) \right]$$

$$= 85 - \left[46 + \frac{19}{12} \right]$$

$$= 85 - \left[47 + \frac{7}{12} \right]$$

$$= 85 - 47\frac{7}{12}$$

$$= 84\frac{12}{12} - 47\frac{7}{12}$$

$$= 37\frac{5}{12}$$

Therefore, $37\frac{5}{12}$ yards of carpet is left.

In this chapter, we introduced fractions as quotients of whole numbers divided by nonzero whole numbers. Equivalent fractions and the Fundamental Principle of Fractions were discussed. Mixed numbers were introduced, and conversions between improper fractions and mixed numbers were explained. The simplest form of a fraction and the reciprocal of a nonzero fraction were introduced. Addition, subtraction, multiplication, and division of fractions and mixed numbers were discussed.

TEST YOURSELF

In Exercises 1–3, write the fraction or mixed number for each word name given.

1) Ten sevenths
2) Two hundred and eight ninths.
3) One hundred three fifths.

In Exercises 4–5, rewrite each improper fraction as a mixed number.

4) $\dfrac{86}{11}$
5) $\dfrac{176}{39}$

In Exercises 6–7, rewrite each mixed number as an improper fraction.

6) $11\dfrac{3}{5}$
7) $23\dfrac{7}{11}$
8) Write a fraction that represents the number of female CEOs in a group of five female CEOs and eleven male CEOs.

In Exercises 9–10, determine whether each pair of fractions is equivalent.

9) $\dfrac{14}{11}$ and $\dfrac{17}{14}$
10) $\dfrac{13}{39}$ and $\dfrac{8}{24}$

In Exercises 11–13, determine the missing number, indicated by the question mark, so that the fractions in each pair are equivalent.

11) $\dfrac{3}{4}$ and $\dfrac{?}{20}$
12) $\dfrac{5}{7}$ and $\dfrac{15}{?}$
13) $\dfrac{7}{?}$ and $\dfrac{1}{17}$

In Exercises 14–15, rewrite each fraction in its simplest form.

14) $\dfrac{81}{45}$
15) $\dfrac{77}{143}$

In Exercises 16–25, perform the indicated operations and write your results in simplest form.

16) $\dfrac{3}{5}+\dfrac{1}{4}+\dfrac{4}{7}$
17) $\dfrac{11}{15}-\dfrac{3}{7}$
18) $3\dfrac{1}{7}+5\dfrac{1}{9}$
19) $23\dfrac{1}{6}-11\dfrac{7}{10}$
20) $\dfrac{2}{3}\times\left(\dfrac{1}{2}+\dfrac{1}{4}\right)$
21) $\left(\dfrac{3}{4}+\dfrac{4}{7}\right)\times\left(\dfrac{1}{9}+\dfrac{2}{5}\right)$
22) $\left(\dfrac{7}{8}-\dfrac{2}{5}\right)\div\left(\dfrac{3}{4}-\dfrac{1}{3}\right)$
23) $\dfrac{2}{3}\div\left(\dfrac{1}{5}\times\dfrac{2}{3}\right)$
24) $3\dfrac{1}{2}\times\left(5\dfrac{1}{3}+2\dfrac{1}{4}\right)$
25) $13\dfrac{1}{2}\div 3\dfrac{1}{5}$

TEST YOURSELF WITH THE HANDHELD CALCULATOR

Note: All exercises involving fractions can be worked out on your calculator. However, the answers are given in decimal notation rather than in fractional form or mixed numbers (except, possibly, for graphing calculators). For example, using your calculator, $\frac{3}{5} + \frac{4}{7}$ would be computed as $(3 \div 5) + (4 \div 7) = 1.17142857$ (which is approximately $\frac{41}{35}$ or $1\frac{6}{35}$).

In Exercises 1–7, add.

1) $\frac{5}{7} + \frac{2}{9}$
2) $\frac{9}{11} + \frac{5}{8}$
3) $\frac{7}{13} + \frac{1}{2}$
4) $\frac{1}{2} + \frac{1}{3} + \frac{1}{5}$
5) $\frac{2}{3} + \frac{4}{7} + \frac{5}{11}$
6) $\frac{1}{3} + \frac{2}{5} + \frac{5}{7} + \frac{1}{9}$
7) $\frac{2}{5} + \frac{4}{7} + \frac{6}{11} + \frac{1}{13}$

In Exercises 8–14, subtract.

8) $\frac{5}{8} - \frac{1}{3}$
9) $\frac{4}{9} - \frac{1}{6}$
10) $\frac{11}{13} - \frac{7}{9}$
11) $\frac{23}{35} - \frac{1}{2}$
12) $\frac{27}{29} - \frac{11}{13}$
13) $\frac{13}{29} - \frac{1}{4}$
14) $\frac{15}{16} - \frac{7}{8}$

In Exercises 15–18, perform the indicated operations.

15) $\left(\frac{1}{2} + \frac{2}{3}\right) - \left(\frac{1}{5} + \frac{1}{4}\right)$
16) $\left(\frac{5}{6} - \frac{2}{3}\right) + \left(\frac{7}{9} - \frac{1}{2}\right)$
17) $\left(\frac{2}{3} + \frac{3}{4} + \frac{4}{5}\right) - \left(\frac{11}{15} - \frac{7}{9}\right)$
18) $\left(\frac{5}{8} + \frac{4}{11} - \frac{2}{5}\right) - \left(\frac{10}{11} - \frac{1}{2} - \frac{1}{3}\right)$

In Exercises 19–22, multiply.

19) $\frac{3}{7} \times \frac{11}{15}$
20) $2\frac{1}{3} \times \frac{17}{19}$
21) $\frac{2}{3} \times \frac{7}{9} \times \frac{13}{16}$
22) $\frac{3}{8} \times 2\frac{1}{4} \times 3\frac{3}{4}$

In Exercises 23–26, divide.

23) $\frac{13}{14} \div \frac{5}{7}$
24) $4\frac{1}{3} \div \frac{23}{10}$
25) $16 \div 5\frac{1}{7}$
26) $47\frac{1}{2} \div 31\frac{1}{3}$

In Exercises 27–30, perform the indicated operations.

27) $\left(\dfrac{5}{8}+\dfrac{3}{7}\right)\times\left(\dfrac{17}{19}-\dfrac{1}{6}\right)$

28) $\left(3\dfrac{2}{3}\div\dfrac{1}{2}\right)+4\dfrac{1}{7}$

29) $\left(\dfrac{7}{8}+\dfrac{3}{5}\right)\div\left(\dfrac{15}{17}-\dfrac{2}{3}\right)$

30) $\left(\dfrac{5}{6}\div\dfrac{4}{7}\right)+\left(3\dfrac{1}{2}\times 1\dfrac{4}{9}\right)$

In Exercises 31–35, determine which pairs of factions are equivalent.

31) $\dfrac{119}{87}$ and $\dfrac{238}{174}$

32) $\dfrac{315}{405}$ and $\dfrac{122}{162}$

33) $\dfrac{607}{916}$ and $\dfrac{1,124}{1,382}$

34) $\dfrac{1,023}{69}$ and $\dfrac{341}{23}$

35) $\dfrac{987}{3,456}$ and $\dfrac{658}{2,304}$

In Exercises 36–40, insert the symbol <, >, or = between each pair of fractions to make a true statement.

36) $\dfrac{39}{37}\quad\dfrac{102}{150}$

37) $\dfrac{412}{341}\quad\dfrac{806}{607}$

38) $\dfrac{705}{809}\quad\dfrac{816}{905}$

39) $\dfrac{1,012}{2,121}\quad\dfrac{567}{869}$

40) $\dfrac{3,071}{4,192}\quad\dfrac{113}{237}$

TEST YOURSELF ANSWERS

1) Ten sevenths = $\dfrac{10}{7}$

2) Two hundred and eight ninths = $200 + \dfrac{8}{9} = 200\dfrac{8}{9}$

3) One hundred three fifths = $\dfrac{103}{5}$

4) $\dfrac{86}{11} = 7 + \dfrac{9}{11} = 7\dfrac{9}{11}$

5) $\dfrac{176}{39} = 4 + \dfrac{20}{39} = 4\dfrac{20}{39}$

6) $11\dfrac{3}{5} = \dfrac{(11\times 5)+3}{5} = \dfrac{58}{3}$

7) $23\dfrac{7}{11} = \dfrac{(23\times 11)+7}{11} = \dfrac{260}{11}$

8) $\dfrac{\text{Number Female}}{\text{Total Number}} = \dfrac{5}{5+11} = \dfrac{5}{16}$

9) $\dfrac{14}{11}$ and $\dfrac{17}{14}$ *are not* equivalent, because $14 \times 14 = 196 \neq 187 = 11 \times 17$.

10) $\dfrac{13}{39}$ and $\dfrac{8}{24}$ *are* equivalent, because $13 \times 24 = 312 = 39 \times 8$.

11) $\dfrac{3}{4} = \dfrac{3\times 5}{4\times 5} = \dfrac{15}{20}$; ? = 15

12) $\dfrac{5}{7} = \dfrac{5\times 3}{7\times 3} = \dfrac{15}{21}$; ? = 21

13) $\dfrac{1}{17} = \dfrac{1\times 7}{17\times 7} = \dfrac{7}{119}$; ? = 119

14) $\dfrac{81}{45} = \dfrac{81\div 9}{45\div 9} = \dfrac{9}{5}$

15) $\dfrac{77}{143} = \dfrac{77\div 11}{143\div 11} = \dfrac{7}{13}$

16) l.c.m. (5, 4, 7) = 140

$$\frac{3}{5}+\frac{1}{4}+\frac{4}{7}=\frac{3\times 28}{5\times 28}+\frac{1\times 35}{4\times 35}+\frac{4\times 20}{7\times 20}=\frac{84+35+80}{140}=\frac{199}{140}=1\frac{59}{140}$$

17) l.c.m. (15, 7) = 105

$$\frac{11}{15}-\frac{3}{7}=\frac{11\times 7}{15\times 7}-\frac{3\times 15}{7\times 15}=\frac{77-45}{105}=\frac{32}{115}$$

18) l.c.m. (7, 9) = 63

$$3\frac{1}{7}+5\frac{1}{9}=3\frac{9}{63}+5\frac{7}{63}=(3+5)+\left(\frac{9}{63}+\frac{7}{63}\right)=8+\frac{16}{63}=8\frac{16}{63}$$

19) l.c.m. (6, 10) = 30

$$23\frac{1}{6}-11\frac{7}{10}=23\frac{5}{30}-11\frac{21}{30}=22\frac{35}{30}-11\frac{21}{30}=(22-11)+\left(\frac{35}{30}-\frac{21}{30}\right)=11\frac{14}{30}=11\frac{7}{15}$$

20) $\frac{2}{3}\times\left(\frac{1}{2}+\frac{1}{4}\right)=\frac{2}{3}\times\frac{3}{4}=\frac{6}{12}=\frac{1}{2}$

21) $\left(\frac{3}{4}+\frac{4}{7}\right)\times\left(\frac{1}{9}+\frac{2}{5}\right)=\frac{(21+16)}{28}\times\frac{(5+18)}{45}=\frac{37}{28}\times\frac{23}{45}=\frac{(37\times 23)}{(28\times 45)}=\frac{851}{1,260}$

22) $\left(\frac{7}{8}-\frac{2}{5}\right)\div\left(\frac{3}{4}-\frac{1}{3}\right)=\left(\frac{35-16}{40}\right)\div\left(\frac{9-4}{12}\right)=\frac{19}{40}\div\frac{5}{12}=\frac{19}{40}\times\frac{12}{5}=\frac{228}{200}=\frac{57}{50}=1\frac{7}{50}$

23) $\frac{2}{3}\div\left(\frac{1}{5}\times\frac{2}{3}\right)=\frac{2}{3}\div\frac{2}{15}=\frac{2}{3}\times\frac{15}{2}=\frac{30}{6}=5$

24) $3\frac{1}{2}\times\left(5\frac{1}{3}+2\frac{1}{4}\right)=3\frac{1}{2}\times\left(\frac{16}{3}+\frac{9}{4}\right)=\frac{7}{2}\times\left(\frac{64+27}{12}\right)=\frac{7}{2}\times\frac{91}{12}=\frac{637}{24}=26\frac{13}{24}$

25) $13\frac{1}{2}\div 3\frac{1}{5}=\frac{27}{2}\div\frac{16}{5}=\frac{27}{2}\times\frac{5}{16}=\frac{135}{32}=4\frac{7}{32}$

TEST YOURSELF ANSWERS WITH HANDHELD CALCULATOR

1) $(5 \div 7) + (2 \div 9) = .9365079365$ (which is $\frac{59}{63}$)

2) $(9 \div 11) + (5 \div 8) = 1.443181818$ (which is $\frac{127}{88} = 1\frac{39}{88}$)

3) $(7 \div 13) + (1 \div 2) = 1.038461538$ (which is $\frac{27}{26} = 1\frac{1}{26}$)

4) $(1 \div 2) + (1 \div 3) + (1 \div 5) = 1.033333333$ (which is $\frac{31}{30} = 1\frac{1}{30}$)

5) $(2 \div 3) + (4 \div 7) + (5 \div 11) = 1.692640693$ (which is $\frac{391}{231} = 1\frac{160}{231}$)

6) $(1 \div 3) + (2 \div 5) + (5 \div 7) + (1 \div 9) = 1.558730159$ (which is $\frac{491}{315} = 1\frac{176}{315}$)

7) $(2 \div 5) + (4 \div 7) + (6 \div 11) + (1 \div 13) = 1.593806194$ (which is $\frac{7,977}{5,005} = 1\frac{2,972}{5,005}$)

8) $(5 \div 8) - (1 \div 3) = 0.291666667$ (which is $\frac{7}{24}$)

9) $(4 \div 9) - (1 \div 6) = 0.277777778$ (which is $\frac{5}{18}$)

10) $(11 \div 13) - (7 \div 9) = 0.0683760684$ (which is $\frac{8}{117}$)

11) $(23 \div 35) - (1 \div 2) = 0.1571428571$ (which is $\frac{11}{70}$)

12) $(27 \div 29) - (11 \div 13) = 0.0848806366$ (which is $\frac{32}{377}$)

13) $(13 \div 29) - (1 \div 4) = 0.1982758621$ (which is $\frac{23}{116}$)

14) $(15 \div 16) - (7 \div 8) = 0.0625$ (which is $\frac{1}{16}$)

15) $((1 \div 2) + (2 \div 3)) - ((1 \div 5) + (1 \div 4)) = 0.716666667$ (which is $\frac{43}{60}$)

16) $((5 \div 6) - (2 \div 3)) + ((7 \div 9) - (1 \div 2)) = 0.444444444$ (which is $\frac{4}{9}$)

17) $((2 \div 3) + (3 \div 4) + (4 \div 5)) - ((11 \div 15) - (7 \div 9)) = 2.261111111$ (which is $\frac{407}{180} = 2\frac{47}{180}$)

18) $((5 \div 8) + (4 \div 11) - (2 \div 5)) - ((10 \div 11) - (1 \div 2) - (1 \div 3)) = 0.51287879$ (which is $\frac{677}{1,320}$)

19) $(3 \div 7) \times (11 \div 15) = 0.3142857143$ (which is $\frac{11}{35}$)

20) $\left(2\frac{1}{3}\right) \times \frac{17}{19} = \frac{7}{3} \times \frac{17}{19} = 2.087719298$ (which is $\frac{119}{57} = 2\frac{5}{57}$)

21) $(2 \div 3) \times (7 \div 9) \times (13 \div 16) = 0.4212962963$ (which is $\frac{91}{216}$)

22) $\left(\frac{3}{8} \times 2\frac{1}{4} \times 3\frac{3}{4}\right) = \left(\frac{3}{8} \times \frac{9}{4} \times \frac{15}{4}\right) = 3.1640625$ (which is $3\frac{21}{128}$)

23) $(13 \div 14) \div (5 \div 7) = 1.3$ (which is $1\frac{3}{10}$)

24) $\left(4\frac{1}{3} \div \frac{23}{10}\right) = \left(\frac{13}{3} \div \frac{23}{10}\right) = 1.884057971$ (which is $\frac{130}{69} = 1\frac{61}{69}$)

25) $16 \div \left(5\frac{1}{7}\right) = \left(16 \div \frac{36}{7}\right) = 3.111111111$ (which is $\frac{28}{9} = 3\frac{1}{9}$)

26) $\left(47\frac{1}{2} \div 31\frac{1}{3}\right) = \left(\frac{95}{2} \div \frac{94}{3}\right) = 1.515957447$ (which is $\frac{285}{188} = 1\frac{97}{188}$)

27) $((5 \div 8) + (3 \div 7)) \times ((17 \div 19) - (1 \div 6)) = 0.7670739348$ (which is $\frac{4,897}{6,384}$)

28) $\left(3\frac{2}{3} \div \frac{1}{2}\right) + 4\frac{1}{7} = \left(\frac{11}{3} \div \frac{1}{2}\right) + \frac{29}{7} = 11.47619048$ (which is $\frac{241}{21} = 11\frac{10}{21}$)

29) $((7 \div 8) + (3 \div 5)) \div ((15 \div 17) - (2 \div 3)) = 6.838636364$ (which is $\frac{3,009}{440} = 6\frac{369}{440}$)

30) $\left(\frac{5}{6} \div \frac{4}{7}\right) + \left(3\frac{1}{2} \times 1\frac{4}{9}\right) = \left(\frac{5}{6} \div \frac{4}{7}\right) + \left(\frac{7}{2} \times \frac{13}{9}\right) = 6.513888889$ (which is $\frac{469}{72} = 6\frac{37}{72}$)

31) $\frac{119}{87}$ and $\frac{238}{174}$ *are* equivalent, because $119 \times 174 = 20{,}706 = 87 \times 238$.

32) $\frac{315}{405}$ and $\frac{122}{162}$ *are not* equivalent, because $315 \times 162 = 51{,}030 \neq 49{,}410 = 405 \times 122$.

33) $\frac{607}{916}$ and $\frac{1{,}124}{1{,}382}$ *are not* equivalent, because $607 \times 1{,}382 = 838{,}874 \neq 1{,}029{,}584 = 916 \times 1{,}124$.

34) $\frac{1{,}023}{69}$ and $\frac{341}{23}$ *are* equivalent, because $1023 \times 23 = 23{,}529 = 69 \times 341$.

35) $\frac{987}{3{,}456}$ and $\frac{658}{2{,}304}$ *are* equivalent, because $987 \times 2{,}304 = 2{,}274{,}048 = 3{,}456 \times 658$.

Note: For Exercises 36–40, use the following:

a) $\dfrac{a}{b} > \dfrac{c}{d}$ if, and only if, $ad > bc$.

b) $\dfrac{a}{b} < \dfrac{c}{d}$ if, and only if, $ad < bc$.

c) $\dfrac{a}{b} = \dfrac{c}{d}$ if, and only if, $ad = bc$.

36) $\dfrac{39}{37} > \dfrac{102}{150}$, because $39 \times 150 = 5{,}850$; $37 \times 102 = 3{,}774$ and $5{,}850 > 3{,}774$.

37) $\dfrac{412}{341} < \dfrac{806}{607}$, because $412 \times 607 = 250{,}084$; $341 \times 806 = 274{,}846$ and $250{,}084 < 274{,}846$.

38) $\dfrac{705}{809} < \dfrac{816}{905}$, because $705 \times 905 = 638{,}025$; $809 \times 816 = 660{,}144$ and $638{,}025 < 660{,}144$.

39) $\dfrac{1{,}012}{2{,}121} < \dfrac{567}{869}$, because $1{,}012 \times 869 = 879{,}428$; $2{,}121 \times 567 = 1{,}202{,}607$ and $879{,}428 < 1{,}202{,}607$.

40) $\dfrac{3{,}071}{4{,}192} > \dfrac{113}{237}$, because $3{,}071 \times 237 = 727{,}827$; $4{,}192 \times 113 = 473{,}696$ and $727{,}827 > 473{,}696$.

CHAPTER 4

Decimals

In this chapter, we extend our number system to include decimals. Performing operations on fractions is not always easy. However, if the fractions have denominators that are powers of 10, such as 10, 100, 1,000, and so on, performing these operations becomes easier. We can eliminate the need for denominators in many cases if we extend our knowledge of the place value of the digits located to the right of the ones place. The places to the right of the ones place are called **decimal places.** The ones digit and the digit immediately to its right are separated by a dot, called a **decimal point.** The new numbers formed are called **decimal numbers.**

In this chapter, we discuss the arithmetic of decimal numbers. Applications involving decimals are also introduced.

4.1 DECIMALS AS NUMBERS

In this section, you learn how to determine the place value for a particular digit in a given decimal number. You will also learn how to read and write the name of a decimal number. Comparing two decimal numbers is also discussed.

DECIMAL PLACE VALUES

The digits in a whole number have place values according to their locations. For example, in 5,347

- The digit 7 is located in the ones place and has a value of 7, because $7 \times 1 = 7$.
- The digit 4 is located in the tens place and has a value of 40, because $4 \times 10 = 40$.
- The digit 3 is located in the hundreds place and has a value of 300, because $3 \times 100 = 300$.
- The digit 5 is located in the thousands place and has a value of 5,000, because $5 \times 1,000 = 5,000$.

As you move from the left digit to the right digit in a whole number, the place value of each digit is $\frac{1}{10}$ of the place value on its left. Continuing with digits to the **right** of the ones place, this pattern continues. Place a dot, called a **decimal point,** immediately after the ones digit. The place values for the digits to the right of the decimal point are named, in order, tenths, hundredths, thousandths, ten thousandths, hundred thousandths, millionths, and so on.

Consider the numeral .7549.

- The digit 7 is located in the tenths place and has a value of $\frac{7}{10}$, because $7 \times \frac{1}{10} = \frac{7}{10}$.
- The digit 5 is located in the hundredths place and has a value of $\frac{5}{100}$, because $5 \times \frac{1}{100} = \frac{5}{100}$.
- The digit 4 is located in the thousandths place and has a value of $\frac{4}{1000}$, because $4 \times \frac{1}{1000} = \frac{4}{1000}$.
- The digit 9 is located in the ten-thousandths place and has a value of $\frac{9}{10,000}$, because $9 \times \frac{1}{10,000} = \frac{9}{10,000}$.

For place values to the **right** of the ones place, write:

- $\frac{1}{10}$ as .1 (read, "one tenth")
- $\frac{1}{100}$ as .01 (read, "one hundredth")
- $\frac{1}{1000}$ as .001 (read, "one thousandth")
- $\frac{1}{10,000}$ as .0001 (read, "one ten thousandth")

and so on. Numbers such as .1, .01, .001, and .0001 are called **decimal fractions,** or simply **decimals.** The numbers .6, 3.2, 73.69, and 8.0074 are also examples of decimals.

Exercise 4.1

What is the value of each digit in the numeral .614598?

Solution 4.1

- The digit 6 is located in the tenths place and has a value of $\frac{6}{10}$, because $6 \times \frac{1}{10} = \frac{6}{10}$.
- The digit 1 is located in the hundredths place and has a value of $\frac{1}{100}$, because $1 \times \frac{1}{100} = \frac{1}{100}$.

- The digit 4 is located in the thousandths place and has a value of $\frac{4}{1000}$, because $4 \times \frac{1}{1,000} = \frac{4}{1,000}$.
- The digit 5 is located in the ten-thousandths place and has a value of $\frac{5}{10,000}$, because $5 \times \frac{1}{10,000} = \frac{5}{10,000}$.
- The digit 9 is located in the hundred-thousandths place and has a value of $\frac{9}{100,000}$, because $9 \times \frac{1}{100,000} = \frac{9}{100,000}$.
- The digit 8 is located in the millionths place and has a value of $\frac{8}{1,000,000}$, because $8 \times \frac{1}{1,000,000} = \frac{8}{1,000,000}$.

READING DECIMAL NUMBERS

To read a decimal number,
1. Read the digits to the **left** of the decimal point, if any, as you would read a whole number.
2. Read the decimal point as "and."
3. Read the digits to the **right** of the decimal point as though they represent a whole number, but state the place value of the digit on the extreme right.

Exercise 4.2

Read each of the following decimals.
a) .67
b) .789
c) 32.87
d) 56.098
e) 314.1069
f) 89.259874

Solution 4.2

a) .67 is read, "sixty-seven hundredths."
b) .789 is read, "seven hundred eighty-nine thousandths."
c) 32.87 is read, "thirty-two and eighty-seven hundredths."
d) 56.098 is read, "fifty-six and ninety-eight thousandths."
e) 314.1069 is read, "three hundred fourteen and one thousand, sixty-nine ten thousandths."
f) 89.259874 is read, "eighty-nine and two hundred fifty-nine thousand, eight hundred seventy-four millionths."

If a decimal number is less than 1, you may find it helpful to write a 0 in the ones place. For example, .9 would be written as 0.9; .568 would be written as 0.568; .8091 would be written as 0.8091; and so on.

Also, placing zeros to the **right** of a decimal number does not change its value. For example, the decimal 0.7 can be written as 0.70, because

a) $0.7 = \dfrac{7}{10}$

b) $0.70 = \dfrac{70}{100}$

c) $\dfrac{7}{10} = \dfrac{7 \times 10}{10 \times 10} = \dfrac{70}{100}$.

Placing zeros to the right of a decimal number is sometimes helpful when comparing decimals. To do so, rewrite the decimals, if necessary, so that they have the same number of decimal places and then compare them. For example, to compare the decimals 0.43 and 0.391:

- Note that 0.43 has two decimal places and that 0.391 has three decimal places.
- Rewrite 0.43 as 0.430 (with three decimal places).
- Compare 0.430 and 0.391.
- Conclude that 0.391 < 0.43.

Exercise 4.3

Determine which is greater:

a) 0.87 or 0.563

b) 7.05 or 7.008

Solution 4.3

a) Rewrite 0.87 as 0.870 and determine that 0.870 is greater than 0.563. Hence, 0.87 > 0.563.

b) Rewrite 7.05 as 7.050 and determine that 7.050 is greater than 7.008. Hence, 7.05 > 7.008.

Exercise 4.4

Arrange the following decimals in order of size, with the smallest first:

a) 0.27; 0.6; 0.467; 0.329

b) 0.71; 0.296; 0.57; 0.9; 0.3125

Solution 4.4

a) Rewrite each of the decimals with the same number (three) of decimal places:
 0.27 = 0.270
 0.6 = 0.600

0.467 = 0.467
0.329 = 0.329

Now, determine that $0.270 < 0.329 < 0.467 < 0.600$ and that the required order is 0.27, 0.329, 0.467, 0.6.

b) Rewrite each of the decimals with the same number (four) of decimal places:
0.71 = 0.7100
0.296 = 0.2960
0.57 = 0.5700
0.9 = 0.9000
0.3125 = 0.3125

Now, determine that $0.2960 < 0.3125 < 0.5700 < 0.7100 < 0.9000$ and that the required order is 0.296, 0.3125, 0.57, 0.71, 0.9.

4.2 ADDITION AND SUBTRACTION OF DECIMALS

In this section, we discuss the addition and subtraction of decimal numbers.

Exercise 4.5
Add:
a) $2.48 + 19.2 + 21.837$
b) $1.534 + 29.09 + 8.0287 + 0.081$
c) $32.1 + 4.027 + 73 + 31.2054$

Solution 4.5

a) *Step 1:* Arrange the decimal numbers under each other, lining up their decimal points.

```
   2.48              2.480
  19.2       or     19.200
 +21.837           +21.837
```

Step 2: Add the columns to the **right** of the decimal points and place the decimal point in the sum:

```
  1 1
   2.480
  19.200
 +21.837
    .517
```

Step 3: Add the columns to the **left** of the decimal points:

```
 1 1 1
   2.480
  19.200
 +21.837
  43.517
```

b)

$$\begin{array}{r} 1.534 \\ 29.09 \\ 8.0287 \\ +0.081 \\ \hline \end{array} \quad \text{or} \quad \begin{array}{r} 1.5340 \\ 29.0900 \\ 8.0287 \\ +0.0810 \\ \hline 38.7337 \end{array}$$

c)

$$\begin{array}{r} 32.1 \\ 4.027 \\ 73. \\ +31.2054 \\ \hline \end{array} \quad \text{or} \quad \begin{array}{r} 32.1000 \\ 4.0270 \\ 73.0000 \\ +31.2054 \\ \hline 140.3324 \end{array}$$

SUBTRACTING DECIMAL NUMBERS

Subtraction of decimal numbers is done in a similar manner to addition. To subtract decimal numbers:

1. Arrange the decimal numbers under each other with the decimal points lined up directly below each other.
2. Subtract according to place value as you would with whole numbers, putting the decimal point in the answer directly below the other decimal points.

Exercise 4.6

Subtract:

a) 29.36 − 7.342

b) 503.903 − 97.06

Solution 4.6

a) *Step 1:* Arrange the decimal numbers with their decimal points lined up under each other:

$$\begin{array}{r} 29.36 \\ -7.342 \\ \hline \end{array} \quad \text{or} \quad \begin{array}{r} 29.360 \\ -7.342 \\ \hline \end{array}$$

Step 2: Subtract in the columns to the **right** of the decimal points and place the decimal point in the answer:

$$\begin{array}{r} ^{5\,1} \\ 29.360 \\ -7.342 \\ \hline .018 \end{array}$$

Step 3: Subtract in the columns to the **left** of the decimal points:

$$\begin{array}{r} ^{5\,1} \\ 29.360 \\ -7.342 \\ \hline 22.018 \end{array}$$

b)

$$\begin{array}{r}503.903\\-97.06\\\hline\end{array} \quad \text{or} \quad \begin{array}{r}503.903\\-97.060\\\hline 406.843\end{array}$$

Exercise 4.7
Subtract 19.0237 from the sum of 30.05 and 8.457.

Solution 4.7
First add 30.05 and 8.457:

$$\begin{array}{r}30.050\\+8.457\\\hline 38.507\end{array}$$

Now subtract 19.0237 from the sum:

$$\begin{array}{r}38.5070\\-19.0237\\\hline 19.4833\end{array}$$

Exercise 4.8
From the sum of 96.3 and 201.37, subtract the sum of 47.6 and 94.67.

Solution 4.8
First add 96.3 and 201.37:

$$\begin{array}{r}96.30\\+\,201.37\\\hline 297.67\end{array}$$

Next, add 47.6 and 94.67:

$$\begin{array}{r}47.60\\+\,94.67\\\hline 142.27\end{array}$$

Then subtract 142.27 (the sum of 47.6 and 94.67) from 297.67 (the sum of 96.3 and 201.37):

$$\begin{array}{r}297.67\\-142.27\\\hline 155.40\end{array}$$

4.3 ROUNDING

In this section, we discuss **rounding** of decimal numbers, a procedure used to approximate the value of a number. To round a number to a given place value, look at the digit immediately to the **right** of that place value:

1. If this digit is less than 5, change it and all digits to the right of it to zeros. Do *not* change the digit in the given place value.
2. If this digit is greater than or equal to 5, change it and all digits to the right of it zeros. **Add** 1 to the digit in the given place value.

Exercise 4.9
Round the following numbers:
a) 4,649 to the nearest ten
b) 53,169 to the nearest thousand
c) 869,753 to the nearest hundred
d) 6.3794 to the nearest hundredth
e) 7.9823 to the nearest thousandth
f) 4,361.098 to the nearest thousand

Solution 4.9
a) The digit in the tens place is 4. The digit to its right is 9, which is greater than 5. Part 2 of the rule for rounding applies, so 4,649 is rounded up to 4,650 to the nearest ten.

b) The digit in the thousands place is 3. The digit to its right is 1, which is less than 5. Part l of the rule applies, so 53,169 is rounded down to 53,000 to the nearest thousand.

c) The digit in the hundreds place is 7. The digit to its right is 5. Part 2 of the rule applies, so 869,753 is rounded up to 869,800.

d) The digit in the hundredths place is 7. The digit to its right is 9, which is greater than 5. Part 2 of the rule applies, so 6.3794 is rounded up to 6.38.

e) The digit in the thousandths place is 2. The digit to its right is 3, which is less than 5. Part l of the rule applies, so 7.9823 is rounded down to 7.982.

f) The digit in the thousands place is 4. The digit to its right is 3, which is less than 5. Part l of the rule applies, so 4,361.098 is rounded down to 4,000.

Note: You should become proficient in doing rounding. This means that you should be able to do it on sight. Practice until you are able to do this.

4.4 MULTIPLICATION OF DECIMALS
In this section, we discuss the multiplication of decimal numbers.

MULTIPLYING DECIMAL NUMBERS
Multiply decimal numbers in the same way we multiply whole numbers. The difference is that when you multiply decimal numbers, you must decide where to place the decimal point in the answer.

To multiply two decimal numbers:

1. Multiply the two decimal numbers as though they were whole numbers, disregarding their decimal points.
2. The number of decimal places in the product is equal to the **sum** of the number of decimal places in the two factors.

Exercise 4.10
Multiply:

a) 0.275 and 0.54

b) 34.12 and 7.49

c) 0.00321 and 16.4

Solution 4.10

a)

```
    0.275    (three decimal places)
  × 0.54     (two decimal places)
    1100     (multiply as whole numbers)
   1375
  0.14850    (3 + 2 = 5 decimal places)
```

b)

```
    34.12    (two decimal places)
   ×7.49     (two decimal places)
   30708     
   13648     (multiply as whole numbers)
  238 84
  255.5588   (2 + 2 = 4 decimal places)
```

c)

```
   0.00321   (five decimal places)
    ×16.4    (one decimal place)
     1284
     1926    (multiply as whole numbers)
      321
   0.052644  (5 + 1 = 6 decimal places)
```

Note, in the answer above, a 0 had to be inserted between the decimal point and the digit 5 in order to get the correct number of decimal places.

MULTIPLYING BY POWER OF TEN

We now examine multiplying decimal numbers by powers of ten, such as 10, 100, 1,000, and so on.

Exercise 4.11
Multiply:
a) 64.926×10
b) 46.576×100
c) $84.547 \times 1,000$

Solution 4.11
a)
$$\begin{array}{r} 64.926 \\ \times 10 \\ \hline 649.260 \end{array} \quad \text{or} \quad 649.26$$

b)
$$\begin{array}{r} 46.576 \\ \times 100 \\ \hline 4,657.600 \end{array} \quad \text{or} \quad 4,657.6$$

c)
$$\begin{array}{r} 84.547 \\ \times 1,000 \\ \hline 84,547.000 \end{array} \quad \text{or} \quad 84,547$$

In the preceding exercise, observe that:
- To multiply a decimal number by $10 = 10^1$, move the decimal point in the decimal number **one** place to the **right.**
- To multiply a decimal number by $100 = 10^2$, move the decimal point in the decimal number **two** places to the **right.**
- To multiply a decimal number by $1,000 = 10^3$, move the decimal point in the decimal number **three** places to the **right.**

To multiply a decimal number by 10^n (where n is a counting number), move the decimal point in the number n places to the **right.**

Exercise 4.12
Multiply:
a) 81.691×10
b) 0.9087×100
c) 0.00312×100
d) $7.8 \times 1,000$
e) $42.50670987 \times 10,000$

Solution 4.12

a) $81.691 \times 10 = 81.691 \times 10^1 = 816.91$
b) $0.9087 \times 100 = 0.9087 \times 10^2 = 90.87$
c) $0.00312 \times 100 = 0.00312 \times 10^2 = 0.312$
d) $7.8 \times 1,000 = 7.8 \times 10^3 = 7,800$ (Note that two zeros had to be placed following the 8.)
e) $42.50670987 \times 10,000 = 42.50670987 \times 10^4 = 425,067.0987$

Note: You should become proficient in doing this operation by sight. Practice until you are able to do this.

4.5 DIVISION OF DECIMALS

In this section, we discuss the division of decimal numbers. As a special case, we examine the division of decimal numbers by powers or ten.

DIVIDING DECIMAL NUMBERS

Divide decimals in the same way you divide whole numbers. The difference is that, in the division of decimal numbers, you must decide where to place the decimal point in the quotient. To divide decimal numbers:

1. If the divisor is *not* a whole number, move the decimal point all the way to the right of the number. (The divisor is now a whole number.)
2. Move the decimal point in the dividend the same number of places to the right.
3. Place the decimal point in the quotient directly above the decimal point in the dividend (as determined in Step 2) and divide as you would with whole numbers.

Exercise 4.13

Divide:

a) 336.56 by 5.6
b) 36.8245 by 2.35
c) 59.84 by 0.044

Solution 4.13

a)

$5.6 \overline{)336.56}$ (Multiply *both* the dividend and the divisor by 10.)

$5.6 \overline{)336.56}$ which becomes

```
       60.1
56 )3365.6
    336
     05
      0
      5 6
      5 6
        0
```

Therefore, $336.56 \div 5.6 = 60.1$.

Check:

```
     5.6    divisor
  × 60.1    quotient
     5 6
  3360
  336.56 ✓ dividend
```

b)

$2.35 \overline{)36.8245}$ (Multiply *both* the dividend and the divisor by 100.)

$2.35 \overline{)36.8245}$, which becomes

```
        15.67
235 )3682.45
     235
     1332
     1175
      157 4
      141 0
       16 45
       16 45
           0
```

Therefore, $36.8245 \div 2.35 = 15.67$.

Check:

$$\begin{array}{r} 2.35 \\ \times 15.67 \\ \hline 1645 \\ 1410 \\ 1175 \\ 235 \\ \hline 36.8245 \end{array}$$
divisor
quotient

✓ dividend

c)

$0.044 \overline{)59.84}$ (Multiply *both* the dividend and the divisor by 1,000.)

$0.044 \overline{)59.84}$, which becomes

$$\begin{array}{r} 1{,}360 \\ 44{\overline{\smash{\big)}\,59{,}840}} \\ \underline{44} \\ 158 \\ \underline{132} \\ 264 \\ \underline{264} \\ 0 \end{array}$$

Therefore, $59.84 \div 0.044 = 1{,}360$.

Check:

$$\begin{array}{r} 0.044 \\ \times 1{,}360 \\ \hline 2\;640 \\ 13\;2 \\ 44 \\ \hline 59.840 \end{array}$$
divisor
quotient

✓ dividend

DIVIDING BY POWERS OF TEN

We now examine dividing decimal numbers by powers of 10.

Exercise 4.14
Divide:
a) 91.7 by 10
b) 73.8 by 100
c) 31.243 by 1,000

Solution 4.14

a)
```
         9.17
    10)91.70
       90
       ─────
        1 7
        1 0
        ─────
          70
          70
          ───
           0
```

b)
```
         7.38
   100)73.800
       70 0
       ─────
        3 80
        3 00
        ─────
          800
          800
          ───
            0
```

c)
```
          0.031243
  1,000)31.243000
        30 00
        ──────
         1 243
         1 000
         ──────
           2430
           2000
           ──────
            4300
            3000
            ──────
             3000
             3000
             ──────
                0
```

In the preceding example, observe that:
- To divide a decimal number by $10 = 10^1$, move the decimal point in the decimal number **one** place to the **left**.
- To divide a decimal number by $100 = 10^2$, move the decimal point in the decimal number **two** places to the **left**.
- To divide a decimal number by $1,000 = 10^3$, move the decimal point in the decimal number **three** places to the **left**.

To divide a decimal number by 10^n (where n is a counting number), move the decimal point in the decimal number n places to the **left**.

Exercise 4.15
Divide:
a) 41.49 by 10
b) 42.106 by 100
c) 865.43 by 100
d) 0.87 by 1,000
e) 12,438.12 by 10,000

Solution 4.15
a) $41.49 \div 10 = 41.49 \div 10^1 = 4.149$
b) $42.106 \div 100 = 42.106 \div 10^2 = 0.42106$
c) $865.43 \div 100 = 865.43 \div 10^2 = 8.6543$
d) $0.87 \div 1,000 = 0.87 \div 10^3 = 0.00087$
e) $12,438.12 \div 10,000 = 12,438.12 \div 10^4 = 1.243812$

REMAINDERS

Division of whole numbers is not always exact. For example, when 35 is divided by 8, you get a remainder of 3: $\frac{35}{8} = 4\frac{3}{8}$. When dividing decimals, there may also be a remainder other than 0. If there is, place zeros to the right of the last digit in the dividend and keep dividing.

Exercise 4.16
Divide:
a) 31.2 by 5
b) 756.96 by 3.2

Solution 4.16
a)

```
      6.2                 6.24
   5)31.2              5)31.20
     30                  30
     ──                  ──
      1 2                 1 2
      1 0       or        1 0
      ──                  ──
        2                   20
                            20
                            ──
                             0
```

Check:

```
   6.24   quotient
  × 5     divisor
  31.20 ✓ dividend
```

b)

$$3.2 \overline{)756.96} \quad \text{becomes} \quad 32 \overline{)7,569.60}$$

```
          236.55
      32)7,569.60
         6 4
         1 16
           96
           209
           192
            17 6
            16 0
             1 60
             1 60
                0
```

Check:

```
    236.55    quotient
   ×  3.2     divisor
    47 310
   709 65
   756.960 ✓ dividend
```

ROUNDING AFTER DIVISION

When dividing, the division sometimes does not terminate, no matter how many zeros are placed to the right of the last digit in the dividend. By placing more zeros to the right of the last digit in the dividend, the quotient can be carried out to as many decimal places as desired.

Exercise 4.17
Determine the quotient of:
a) 32.68 ÷ 1.4 to the nearest tenth
b) 719.14 ÷ 43.9 to the nearest hundredth
c) 3.46587 ÷ 1.57 to the nearest hundredth

Solution 4.17

a) To round the quotient to the nearest tenth, it is necessary to carry the division to the hundredths place. Hence, $1.4\overline{)32.68}$ becomes

```
        23.34
    14)326.80
       28
       ‾‾
        46
        42
        ‾‾
         4 8
         4 2
         ‾‾‾
           60
           56
           ‾‾
            4
```

Therefore, to the nearest tenth, 32.68 ÷ 1.4 is 23.3. The quotient is only approximate due to rounding. We indicate this by the use of the symbol ≈. Hence, 32.68 ÷ 1.4 ≈ 23.3, which is read "32.68 divided by 1.4 *is approximately equal to* 23.3."

b) To round the quotient to the nearest hundredth, it is necessary to carry the division to thousandths. Hence, 43.9)719.14 becomes

```
           16.381
    439)7,191.400
        4 39
        ‾‾‾‾
        2 801
        2 634
        ‾‾‾‾‾
          167 4
          131 7
          ‾‾‾‾‾
           35 70
           35 12
           ‾‾‾‾‾
              580
              439
              ‾‾‾
              141
```

Therefore, 719.14 ÷ 43.9 ≈ 16.38.

c) To round the quotient to the nearest hundredth, we must carry the division to the thousandths place. 1.57)3.46587 becomes

```
          2.207
    157)346.587
        314
        ‾‾‾
         32 5
         31 4
         ‾‾‾‾
          1 18
             0
          ‾‾‾‾
          1 187
          1 099
          ‾‾‾‾‾
             88
```

Therefore, 3.46587 ÷ 1.57 ≈ 2.21.

4.6 DECIMAL CONVERSIONS

In this section, we discuss decimal conversions, including converting a decimal number to a fraction and converting a fraction to a decimal number.

DECIMAL TO FRACTION CONVERSION

To convert a decimal number to a fraction, read the decimal number and write the corresponding fraction that has the same name. Then simplify the fraction, if necessary.

Exercise 4.18

Convert the following to fractions:

a) 0.37

b) 0.582

c) 3.29

Solution 4.18

a) The decimal number 0.37 is read "thirty-seven hundredths." The corresponding fraction is $\frac{37}{100}$. $0.37 = \frac{37}{100}$

b) The decimal number 0.582 is read "five hundred eighty-two thousandths." The corresponding fraction is $\frac{582}{1,000}$.

$0.582 = \frac{582}{1,000} = \frac{291}{500}$.

c) $3.29 = 3 + 0.29$

$= \frac{300}{100} + \frac{29}{100}$ (like fractions)

$= \frac{329}{100}$

Another way to convert a decimal number to a fraction is to use the following rule:

1. The **numerator** of the fraction is the decimal number without the decimal point.
2. The **denominator** of the fraction is a 1 followed by as many zeros as there are decimal places in the decimal number.
3. Simplify the fraction, if necessary.

Exercise 4.19

Write as fractions:

a) 0.625

b) 23.42

Solution 4.19

a) The numerator is 625. The denominator is 1,000, because there are three decimal places in the decimal number. The fraction is $\dfrac{625}{1,000} = \dfrac{5}{8}$.

b) The numerator is 2,342. The denominator is 100, because there are two decimal places in the decimal number. The fraction is $\dfrac{2,342}{100} = \dfrac{1,171}{50}$.

FRACTION TO DECIMAL CONVERSION

To convert a fraction to a decimal number, divide the numerator of the fraction by its denominator.

Exercise 4.20

Convert the following to decimals:

a) $\dfrac{4}{8}$

b) $\dfrac{5}{8}$

c) $\dfrac{13}{5}$

Solution 4.20

a)

$$\frac{4}{8} = 4 \div 8 = 0.5, \text{ because}$$

```
    0.5
  ─────
8)4.0
    40
    ──
     0
```

b)

$$\frac{5}{8} = 5 \div 8 = 0.625, \text{ because}$$

```
      0.625
   ────────
 8)5.000
    4 8
    ───
     20
     16
     ──
     40
     40
     ──
      0
```

c)

$$\frac{13}{5} = 13 \div 5 = 2.6, \text{ because}$$

```
      2.6
   5)13.0
     10
      3 0
      3 0
        0
```

Exercise 4.21
Convert the following to decimals:

a) $\dfrac{19}{4}$

b) $\dfrac{83}{25}$

Solution 4.21

a) Divide 19 by 4:

```
       4.75
   4)19.00
     16
      3 0
      2 8
        20
        20
         0
```

$$\frac{19}{4} = 4.75$$

b) Divide 83 by 25:

```
         3.32
   25)83.00
      75
       8 0
       7 5
         50
         50
          0
```

$$\frac{83}{25} = 3.32$$

TERMINATING AND REPEATING DECIMALS

So far in our discussions, whenever we have converted a fraction to a decimal number, the division has had a zero remainder. The corresponding decimal numbers in such cases are called **terminating decimals.** Sometimes, however, the division will never get a zero remainder no matter how far you carry it out. For example, consider the fraction $\frac{2}{3}$. If you divide the numerator, 2, by the denominator, 3, the division does not terminate: $\frac{2}{3} = 0.6666\ldots$, where the three dots (…) indicate the repeating nature of the decimal number. Such a decimal number is called a **repeating decimal.**

In a similar manner, the fraction $\frac{7}{11}$ can be written as the repeating decimal $\frac{7}{11} = 0.636363\ldots$, with the two digits 63 repeating.

Instead of using the three dots, write a repeating decimal with a bar over the digit or digits that repeat. For example, $\frac{2}{3} = 0.\overline{6}$ (because the digit 6 repeats) and $\frac{7}{11} = 0.\overline{63}$ (because the digits 63 repeat).

When converting a fraction to a repeating decimal number, we sometimes write the quotient correct to the nearest tenth or hundredth. To do so, carry the division out to the next decimal place and round the quotient to the desired result.

Exercise 4.22

Write $\frac{9}{11}$ as a decimal number correct to the nearest hundredth.

Solution 4.22

We carry the division out three decimal places and round the quotient to the nearest hundredth:

$$
\begin{array}{r}
0.818 \\
11\overline{)9.000} \\
\underline{8\ 8} \\
20 \\
\underline{11} \\
90 \\
\underline{88} \\
2
\end{array}
$$

Therefore, $\frac{9}{11} = 0.82$ to the nearest hundredth.

4.7 APPLICATIONS INVOLVING DECIMALS

In this section, we consider the solutions to some simple word problems involving decimals.

COST PROBLEMS

A basic problem involving decimals is that of cost, which generally is given in dollars and cents. Sometimes, the cost is given in tenths of a cent. For example, 35.9 cents would be written as $.359.

Exercise 4.23

Joyce went shopping for birthday presents. She spent $89.99 for toys, $49.65 for perfume, $43.75 for books, and $23.57 for flowers. How much was the total cost of these items?

Solution 4.23

Add to find the total cost of the four items.

```
  $89.99    cost of toys
   49.65    cost of perfume
   43.75    cost of books
 + 23.57    cost of flowers
 $206.96    total cost
```

Exercise 4.24

If ground beef is on a holiday sale for $1.179 cents a pound, how much will it cost to buy 21.7 pounds for the family barbecue?

Solution 4.24

We know the cost of one pound of ground beef and wish to determine the cost of 21.7 pounds of ground beef. Therefore, we multiply.

```
   $1.179    price for one pound of ground beef
   ×21.7     number of pounds of ground beef
    8253
    1179
    2358
  $25.5843   total cost
```

To the nearest cent, the total cost of the ground beef would be $25.58.

Exercise 4.25

If one inch is equal to 2.54 centimeters, how many centimeters are there in 76.8 inches?

Solution 4.25

Solve by multiplying:

```
      2.54    number of centimeters in one inch
    ×76.8    number of inches
    ─────
    2 032
   15 24
  177 8
  ───────
  195.072   number of centimeters in 76.8 inches
```

You may round this to 195.07 centimeters to the nearest hundredth of a centimeter, 195.1 centimeters to the nearest tenth of a centimeter, or 195 centimeters to the nearest centimeter.

Exercise 4.26

A man left 0.4 of his estate to his alma mater. His alma mater received $213,184. What was the value of the total estate?

Solution 4.26

We want to find
$$0.4 \times ? = \$213,184$$

To find the value of ?, we divide $213,184 by 0.4 as follows:

$$0.4 \overline{)\$213,184}$$

Or

```
           $532,960
       4)$2,131,840
         2 0
         ───
           13
           12
           ──
            11
             8
            ──
            38
            36
            ──
             24
             24
             ──
              0
```

The man's total estate was valued at $532,960.

Check:

```
    $532,960
       ×0.4
    ────────
   $213,184.0  ✓
```

Exercise 4.27

Stacey's balance in her checking account was $307.57. She paid her telephone bill with a check for $69.09. What was her new balance?

Solution 4.27

Subtract the amount of the check from Stacey's starting balance.

$307.57	Starting balance
−69.09	Amount of check
$238.48	**New balance**

Exercise 4.28

If grapefruits sell for $3.95 a box and there are sixteen pounds of grapefruits in a box, how much will 2,112 pounds of grapefruits cost?

Solution 4.28

There are two parts to this problem. First, determine how many boxes of grapefruits are needed for 2,112 pounds. We divide to find how many 16s are contained in 2,112.

$$\begin{array}{r} 132 \\ 16\overline{)2112} \\ \underline{16} \\ 51 \\ \underline{48} \\ 32 \\ \underline{32} \\ 0 \end{array}$$

Thus, 2,112 pounds of grapefruits are equivalent to 132 boxes of grapefruits. To find the total price, we now multiply the price of one box by the total number of boxes.

$3.95	Cost for one box of grapefruits
×132	Number of boxes of grapefruits
7 90	
118 5	
395	
$521.40	**Total cost**

Hence, at the rate of $3.95 per box, 2,112 pounds of grapefruits would cost $521.40.

Exercise 4.29

Anita drove her car for 1,473 miles and used 86.7 gallons of gasoline. What was the average number of miles traveled per gallon of gasoline? Give the answer to the nearest tenth of a mile.

Solution 4.29

This problem involves division. We divide the total number of miles traveled (1,473) by the number of gallons of gasoline used (86.7). Because the answer has to be given to the nearest tenth of a mile, the division will be carried out to the hundredths place, as follows:

$$86.7 \overline{)1473}$$

Or

```
            16.98
    867)14730.00
        867
        ‾‾‾
        6060
        5202
        ‾‾‾‾
         858 0
         780 3
         ‾‾‾‾‾
          77 70
          69 36
          ‾‾‾‾‾
           8 34
```

Therefore, to the nearest tenth of a mile, Anita averaged 17.0 miles to the gallon.

Exercise 4.30

Three friends agreed to split the cost of a trip equally among them. The costs were: gasoline, $63.18; meals, $131.98; turnpike tolls, $7.70; lodging, $65; and entertainment, $29.77. How much did each person pay?

Solution 4.30

There are two parts to this problem. First, we use addition to determine the total cost of the trip, as follows:

```
  $ 63.18    gasoline
   131.98    meals
     7.70    tolls
    65.00    lodging
  + 29.77    entertainment
  ‾‾‾‾‾‾‾
  $297.63    total cost
```

Next, we divide the total cost ($297.63) by the number of friends (3) to determine how much each person paid:

```
       $99.21
    3)$297.63
       27
       ‾‾
        27
        27
        ‾‾
         0 6
           6
         ‾‾‾
          03
           3
          ‾‾
           0
```

Therefore, each person paid $99.21.

Exercise 4.31
Fern went into a fabric shop and bought 3.5 yards of cotton fabric at $1.98 per yard, two spools of thread at $.89 each, a zipper at $1.89, and 2.75 yards of lace at $2.39 per yard. What was the total amount of her purchases?

Solution 4.31
There are several parts to this problem.

1. Determine the cost of the fabric as follows:

   $1.98   cost of one yard
   ×3.5   number of yards
   ─────
   990
   594
   ─────
   $6.930
   $6.93   total cost of the fabric

2. Determine the cost of the thread as follows:

   $ .89   cost of one spool of thread
   ×2   number of spools
   ─────
   $1.78   total cost of the thread

3. Determine the cost of the lace as follows:

   $2.39   cost of one yard
   ×2.75   number of yards
   ─────
   1195
   1673
   478
   ─────
   $6.5725
   $6.58   total cost of the lace

4. Determine the cost of the four items as follows:

   $6.93   cost of the fabric
   1.78   cost of the thread
   1.89   cost of the zipper
   +6.58   cost of the lace
   ─────
   $17.18   **total cost of the purchases**

Exercise 4.32

Warren went into a hardware store and bought eighteen wood screws at 27 cents each, three hinges at $2.19 each, and 6.5 yards of patio carpet at $4.19 a yard. He gave the clerk two $20 bills. How much change should Warren have received?

Solution 4.32

Again, there are several parts to this problem.

1. Determine the cost of the screws as follows:

    ```
    $.27    cost of one screw
    ×18     number of screws
    ───
    216
    27
    ───
    $4.86   total cost of the screws
    ```

2. Determine the cost of the hinges as follows:

    ```
    $2.19   cost of one hinge
     ×3     number of hinges
    ───
    $6.57   total cost of hinges
    ```

3. Determine the cost of the carpet as follows:

    ```
    $4.19     cost of one yard of carpet
    ×6.5      number of yards
    ─────
    2095
    2514
    ─────
    $27.235
    $27.24    total cost of the carpet
    ```

4. Determine the cost of the three items as follows:

    ```
    $4.86     cost of the screws
     6.57     cost of the hinges
    +27.24    cost of the carpet
    ──────
    $38.67    total cost of the purchases
    ```

5. Determine the amount of change as follows:

    ```
    $40.00    amount of money given clerk
     38.67    amount of purchases
    ──────
    $1.33     amount of change Warren should have received
    ```

Exercise 4.33

Heather stopped at a filling station for gasoline and noted that the odometer reading on her car was 38,947.8 miles. The next time she stopped for gasoline, she noted that the odometer reading was 39,266.0 miles. If she put 14.8 gallons of gasoline into her car, what was the average number of miles per gallon that her car got between fillups?

Solution 4.33

There are two parts to this problem.

1. Determine the number of miles traveled as follows:

$$\begin{array}{r} 39,266 \\ -38,947.8 \\ \hline \end{array} \quad \text{or} \quad \begin{array}{r} 39,266.0 \\ -38,947.8 \\ \hline 318.2 \end{array} \quad \begin{array}{l} \text{odometer reading at second stop} \\ \text{odometer reading at first stop} \\ \textbf{number of miles traveled} \end{array}$$

2. Determine the average number of miles per gallon by dividing the total number of miles traveled (318.2) by the number of gallons of gasoline used (14.8), as follows:

$$14.8 \overline{)318.2}$$

or

$$\begin{array}{r} 21.5 \\ 148 \overline{)3182.0} \\ \underline{296} \\ 222 \\ \underline{148} \\ 74\,0 \\ \underline{74\,0} \\ 0 \end{array}$$

Therefore, Heather's car averaged 21.5 miles per gallon.

In this chapter, you were introduced to decimal numbers, including operations involving decimals. Conversions between fractions and decimals were examined. Applications of some simple word problems involving decimals were also discussed.

TEST YOURSELF

In Exercises 1–2, write the name of each of the decimal numbers.

1) 28.056
2) 587.682
3) Arrange the decimal numbers in order of size, with the smallest first: 0.73; 0.861; 0.7; 0.582
4) Arrange the decimal numbers in order of size, with the greatest first: 2.01; 2.1; 2.101; 2.09; 2.99

In Exercises 5–7, convert each of the decimal numbers to a fraction in simplest form.

5) 0.265
6) 1.46
7) 2.316

In Exercises 8–10, convert each of the fractions to a decimal number.

8) $\dfrac{68}{40}$
9) $\dfrac{187}{20}$
10) $\dfrac{216}{80}$
11) Round 73.69 to the nearest tenth.
12) Round 123.0341 to the nearest hundredth.
13) Round 37.98687 to the nearest thousandth.

In Exercises 14–25, perform the indicated operations.

14) 54.17 + 310.9 + 0.874
15) 21.782 − 12.87
16) 119.342 − 18.07 − 43.2
17) 77.4 + 13.069 − 43.04
18) 341.2 − (0.89 + 156.234)
19) 18.9 × 0.25
20) 4.17 + 2.03 × 0.78
21) 0.9459 ÷ 0.09
22) 231.12 ÷ 21.6 − 0.096
23) (2.56 + 0.05) × (3.14 − 1.09)
24) (3.02 − 1.9) × (4.3 + 7.19)
25) (4.16 ÷ 64) + (13.2 × 0.63)

TEST YOURSELF WITH THE HANDHELD CALCULATOR

In Exercises 1–6, write each expression without exponents.

1) 5^4
2) 4^7
3) 23^4
4) 39^5
5) 113^3
6) 207^4

In Exercises 7–12, rewrite each decimal fraction as a decimal number.

7) $\dfrac{169}{1,000}$
8) $\dfrac{207}{10}$
9) $\dfrac{765}{10,000}$
10) $\dfrac{39}{80}$
11) $\dfrac{8,056}{2,500}$
12) $\dfrac{57}{25,000}$

In Exercises 13–18, rewrite each decimal fraction as a decimal number correct to the nearest ten-thousandth.

13) $\dfrac{7}{12}$
14) $\dfrac{13}{17}$
15) $\dfrac{11}{28}$
16) $\dfrac{23}{49}$
17) $\dfrac{101}{150}$
18) $\dfrac{101}{429}$

In Exercises 19–24, add.

19) 23.016 + 39.1 + 102.201 + 3.2961
20) 0.9999 + 9.999 + 99.99 + 999.9
21) 1.111111 + 22.22222 + 333.3333 + 4444.444
22)
```
    27.9
     0.542
   106.39
     7.2
    36.17
  +462.007
```

23) 202.02
      32.887
       0.6
       7.69
       0.0084
    +319.14

24) 175.321
      32.79
    6,413.3
     287.656
       0.8543
     +16.99

In Exercises 25–30, subtract.

25) $85.07 - 32.276$

26) $243.5 - 24.357$

27) $7981.546 - 956.49$

28) 246.3
     −97.87

29) 1,234.67
     −576.5

30) 15,679.843
     −9,006.9

In Exercises 31–36, multiply.

31) 65.31×0.84

32) 71.96×97.28

33) $7.31 \times 0.96 \times 312.1$

34) $0.03 \times 0.04 \times 0.07 \times 0.09$

35) $96.142 \times 314.5 \times 0.6$

36) $23.04 \times 37.192 \times 146.2 \times 0.8$

In Exercises 37–42, divide.

37) $1.459 \div 1.03$

38) $64.422 \div 34.7$

39) $934{,}884 \div 560.0$

40) $58.379 \div 0.736$

41) $59.87 \div 0.034$

42) $283.06052 \div 8.63$

In Exercises 43–50, perform the indicated operations.

43) $0.69 \times 1.2 - 3.84 \times 0.214 + 29.92 \div 0.022$

44) $31.25 \times 69.07 - 6.479 \times 91.8 + 27.007 \div 1.13$

45) $1.23 \times 2.34 \times 3.45 - 4.56 + 5.67 \div 1.134$

46) $107.502 \div 1.9 \div 12.3 \times 0.132 + 89.679$

47) $122.304 \div 63.7 \times 56.9 - 107.1 + 56.9 \times 8.8$

48) $\dfrac{5.4 + 8.6 - 3.65 - 7.63}{32.17 - 24.6}$

49) $\dfrac{(97.6 + 23.4) - (3.21 \times 4.6)}{37.1 - 29.62}$

50) $\dfrac{3.7 + 4.12 \times 5.9}{(5.4 + 7.21) \times (9.06 - 1.009)}$

TEST YOURSELF

1) 28.056 is read, "twenty-eight and fifty-six thousandths"
2) 587.682 is read, "five hundred eighty-seven and six hundred eighty-two thousandths"
3) 0.582; 0.7; 0.73; 0.861
4) 2.99; 2.101; 2.1; 2.09; 2.01
5) $0.265 = \dfrac{265}{1000} = \dfrac{53}{200}$
6) $1.46 = \dfrac{146}{100} = \dfrac{73}{50}$
7) $2.316 = \dfrac{2316}{1000} = \dfrac{579}{250}$
8) $\dfrac{68}{40} = 1.7$
9) $\dfrac{187}{20} = 9.35$
10) $\dfrac{216}{80} = 2.7$
11) 73.7
12) 123.03
13) 37.987
14) $54.17 + 310.9 + 0.874 = 365.944$
15) $21.782 - 12.87 = 8.912$
16) $(119.342 - 18.07) - 43.2 = 101.272 - 43.2 = 58.072$
17) $(77.4 + 13.069) - 43.04 = 90.469 - 43.04 = 47.429$
18) $341.2 - (0.89 + 156.234) = 341.2 - 157.124 = 184.076$
19) $18.9 \times 0.25 = 4.725$
20) $4.17 + (2.03 \times 0.78) = 4.17 + 1.5834 = 5.7534$
21) $0.9459 \div 0.09 = 10.51$
22) $(231.2 \div 21.6) - 0.096 = 10.7 - 0.096 = 10.604$
23) $(2.56 + 0.05) \times (3.14 - 1.09) = 2.61 \times 2.05 = 5.3505$
24) $(3.02 - 1.9) \times (4.3 + 7.19) = 1.12 \times 11.49 = 12.8688$
25) $(4.16 \div 64) + (13.2 \times 0.63) = 0.065 + 8.316 = 8.381$

TEST YOURSELF ANSWERS WITH THE HANDHELD CALCULATOR

1) $5^4 = 625$
2) $4^7 = 16,384$
3) $23^4 = 279,841$
4) $39^5 = 90,224,199$
5) $113^3 = 1,442,897$
6) $207^4 = 1,836,036,801$
7) $\dfrac{169}{1,000} = 0.169$
8) $\dfrac{207}{10} = 20.7$
9) $\dfrac{765}{10,000} = 0.0765$
10) $\dfrac{39}{80} = 0.4875$
11) $\dfrac{8,056}{2,500} = 3.2224$
12) $\dfrac{57}{25,000} = 0.00228$
13) $\dfrac{7}{12} = 0.5833$
14) $\dfrac{13}{17} = 0.7647$
15) $\dfrac{11}{28} = 0.3929$
16) $\dfrac{23}{49} = 0.4694$
17) $\dfrac{101}{150} = 0.6733$
18) $\dfrac{101}{429} = 0.2354$
19) $23.016 + 39.1 + 102.201 + 3.2961 = 167.6131$
20) $0.9999 + 9.999 + 99.99 + 999.9 = 1,110.8889$
21) $1.111111 + 22.22222 + 333.3333 + 4444.444 = 4,801.110631$
22) $27.9 + 0.542 + 106.39 + 7.2 + 36.17 + 462.007 = 640.209$
23) $202.02 + 32.887 + 0.6 + 7.69 + 0.0084 + 319.14 = 562.3454$
24) $175.321 + 32.79 + 6,413.3 + 287.656 + 0.8543 + 16.99 = 6,926.9113$
25) $85.07 - 32.276 = 52.794$

26) $243.5 - 24.357 = 219.143$
27) $7,981.546 - 956.49 = 7,025.056$
28) $246.3 - 97.87 = 148.43$
29) $1,234.67 - 576.5 = 658.17$
30) $15,679.843 - 9,006.9 = 6.672.943$
31) $65.31 \times 0.84 = 54.8604$
32) $71.96 \times 97.28 = 7,000.2688$
33) $7.31 \times 0.96 \times 312.1 = 2,190.19296$
34) $0.03 \times 0.04 \times 0.07 \times 0.09 = 0.00000756$ (Your answer may look like 7.56E-6, which means that you have to move the decimal point 6 places to the **left.**)
35) $96.142 \times 314.5 \times 0.6 = 18,141.9954$
36) $23.04 \times 37.192 \times 146.2 \times 0.8 = 100,223.4544$
37) $1.459 \div 1.03 = 1.417$
38) $64.422 \div 34.7 = 1.857$
39) $934,884 \div 560.9 = 1,666.757$
40) $58.379 \div 0.736 = 79.319$
41) $59.87 \div 0.034 = 1,760.882$
42) $283.06052 \div 8.63 = 32.800$
43) $0.69 \times 1.2 - 3.84 \times 0.214 + 29.92 \div 0.022 = 1,360.00624$
44) $31.25 \times 69.07 - 6.479 \times 91.8 + 27.007 \div 1.13 = 1,587.5653$
45) $1.23 \times 2.34 \times 3.45 - 4.56 + 5.67 \div 1.134 = 10.36979$
46) $107.502 \div 1.9 \div 12.3 \times 0.132 + 89.679 = 90.2862$
47) $122.304 \div 63.7 \times 56.9 - 107.1 + 56.9 \times 8.8 = 502.868$
48) $(5.4 + 8.6 - 3.65 - 7.63) \div (32.17 - 24.6) \approx 0.3593$
49) $[(97.6 + 23.4) - (3.21 \times 4.6)] \div (37.1 - 29.62) \approx 2.1660$
50) $(3.7 + 4.12 \times 5.9) \div [(5.4 + 7.21) \times (9.06 - 1.009)] \approx 0.276$

CHAPTER 5

Ratio and Proportion

In this chapter, we discuss the meaning of a ratio. We also discuss rates, which are special types of ratios. Proportions are introduced, and applications involving ratios and proportions are discussed.

5.1 RATIOS

In this section, we discuss ratios and how to write them in lowest terms. We also solve word problems by using ratios.

EXAMPLES OF RATIOS

A **ratio** is a comparison of numbers of the same kind (units), using the operation of division. Given two numbers a and b of the same kind, the **ratio** a to b is the quotient of the two numbers and is written as $\frac{a}{b}$ or $a:b$ or a to b.

- The ratio of 7 to 4 is written $\frac{7}{4}$ or 7:4, or "7 to 4," and means that 7 is $\frac{7}{4}$ $\left(or\ 1\frac{3}{4}\right)$ times as large as 4.

- The ratio of 36 to 28 is written $\frac{36}{28}$ or 36:28, or "36 to 28," and means that 36 is $\frac{36}{28}$ $\left(or\ 1\frac{2}{7}\right)$ times as large as 28.

- The ratio of 45 to 75 is written $\frac{45}{75}$ or 45:75, or "45 to 75," and means that 45 is $\frac{45}{75}$ $\left(or\ \frac{3}{5}\right)$ as large as 75.

- The ratio of 144 to 100 can be written $\frac{144}{100}$ or 144:100. It can be simplified as the ratio 36 to 25.

- The ratio 17:68 can be simplified as the ratio 1:4.
- At one Ivy League university, 110 students were accepted to its law school from among 550 applications. The ratio of applications to acceptances, then, is $\frac{550}{110}$ or $\frac{5}{1}$, simplified. This means that there were five applications for each acceptance.

Exercise 5.1

A baseball player was at bat seventy-eight times during the season and got twenty-one hits. What was his batting average for the season?

Solution 5.1

The **batting average** for the season is the ratio of the player's number of hits to the number of times at bat and is written as a decimal number, correct to the nearest thousandth. Hence, we have:

$\frac{21}{78}$, which means $21 \div 78$, or

$$\begin{array}{r} 0.2692 \\ 78{\overline{\smash{\big)}\,21.0000}} \\ \underline{15\ 6} \\ 5\ 40 \\ \underline{4\ 68} \\ 720 \\ \underline{702} \\ 180 \\ \underline{156} \\ 24 \end{array}$$

Therefore, the baseball player's batting average for the season was approximately 0.269.

Exercise 5.2

The longest side of a triangle measures 2 feet. The shortest side of the triangle measures 15 inches. What is the ratio of the length of the shortest side of the triangle to the length of the longest side?

Solution 5.2

A ratio is a quotient of two numbers of the *same* kind. Because the measures of the sides of the triangle are given in different units, you must first convert one of the units to the other, for example, 2 feet to 24 inches. Hence, the ratio of the length of the shortest side of the triangle to the length of the longest side is:

$\frac{15\ inches}{24\ inches}$ or $\frac{15}{24}$, which simplifies to $\frac{5}{8}$.

Exercise 5.3

Express the ratio of eight days to four weeks in lowest terms.

Solution 5.3
Because the units are different, convert weeks to days as follows:
4 weeks = (4 × 7) days = 28 days.
Therefore, the ratio of 8 days to 4 weeks can now be written as the ratio of 8 days to 28 days, or

$\dfrac{8 \text{ days}}{28 \text{ days}}$ or $\dfrac{8}{28}$, which simplifies to $\dfrac{2}{7}$.

RATIOS WITH FRACTIONS
In all the preceding ratio exercises, the ratios involve numbers that are whole numbers. This need not be the case, but the numbers compared by division must be of the same kind.

Exercise 5.4
What is the ratio of $2\dfrac{1}{2}$ to $3\dfrac{3}{4}$?

Solution 5.4
The two numbers given are of the same kind; they are mixed numbers. The ratio of $2\dfrac{1}{2}$ to $3\dfrac{3}{4}$ means $2\dfrac{1}{2} \div 3\dfrac{3}{4}$

$$2\dfrac{1}{2} \div 3\dfrac{3}{4} = \dfrac{5}{2} \div \dfrac{15}{4}$$
$$= \dfrac{5}{2} \times \dfrac{4}{15}$$
$$= \dfrac{20}{30}$$
$$= \dfrac{2}{3}$$

Therefore, the ratio of $2\dfrac{1}{2}$ to $3\dfrac{3}{4}$, simplified, is $\dfrac{2}{3}$.

Exercise 5.5
A sweater is on sale at $45.75. The original price of the sweater was $75. What is the ratio of the sale price to the original price? Give your answer in lowest terms.

Solution 5.5
The ratio of the sale price to the original price is $\dfrac{\$45.75}{\$75}$, which is equivalent to

$$\dfrac{45.75}{75} = \dfrac{4575}{7500} = \dfrac{183}{300} = \dfrac{61}{100}.$$

Therefore, the ratio of the sale price to the original price, in lowest terms, is $\dfrac{61}{100}$.

154 Basic Math

Remember that, when using a ratio to compare two quantities, always compare *like* quantities. The following exercise illustrates that you can convert either quantity to the other and get the same result.

Exercise 5.6
Determine the ratio of 29 centimeters (cm) to 4 meters (m).

Solution 5.6
Step 1: Change both units to centimeters, using the conversion that 100 cm = 1 m. Hence, 4 meters = 400 cm. The ratio of 29 cm to 4 m is $\frac{29cm}{400cm}$ or 29 to 400.

Step 2: Change both units to meters, using the conversion that 1 cm = $\frac{1}{100}$ m. Hence, 29 cm = $\frac{29}{100}$ m. The ratio of 29 cm to 4 m is $\left(\frac{29}{100}m\right) \div (4m)$ or $\frac{29}{400}$ or 29 to 400.

In both conversions, the resulting ratios are the same.

5.2 RATES

In the previous section, we discussed a ratio as a comparison, by division, of two numbers or measures of the same kind. When a ratio is used to compare two **different** kinds of measure, we call it a **rate**. A **rate** is a ratio comparing two **different** kinds of measure.

When working with rates, the word *per* is often used to express the ratio. Each of the following ratios is a rate:

- 376 miles per 13 gallons of gasoline.
- 17 students per faculty member.
- 56 feet per second.
- 250 kilometers in 4 hours.
- 1 centimeter on a map representing 12 miles.

Exercise 5.7
Express each of the following ratios as a rate in lowest terms:
 a) 36 inches to 3 feet
 b) 364 miles to 13 gallons
 c) 161 clients to 7 attorneys
 d) 1,089 students to 66 faculty
 e) 1,375 gallons to 25 barrels
 f) 375 grams to 3 tablets
 g) 11 pounds to 5,000 grams
 h) $152 to 32 hours
 i) $4.32 to 27 ounces

Solution 5.7

a) The ratio is given as inches to feet. The rate, then, will be *inches per foot*. Hence,

$$\frac{\text{inches}}{\text{feet}} = \frac{36}{3}$$
$$= \frac{12}{1},$$

which means a rate of 12 inches per foot.

b) The ratio is given as miles to gallons. The rate, then, will be given as *miles per gallon*. Hence,

$$\frac{\text{miles}}{\text{gallon}} = \frac{364}{13}$$
$$= \frac{28}{1},$$

which means a rate of 28 miles per gallon.

c) The ratio is given as clients to attorneys. The rate, then, will be given as *clients per attorney*. Hence,

$$\frac{\text{clients}}{\text{attorney}} = \frac{161}{7}$$
$$= \frac{23}{1},$$

which means a rate of 23 clients per attorney.

d) The ratio is given as students to faculty. The rate, then, will be given as *students per faculty*. Hence,

$$\frac{\text{students}}{\text{faculty}} = \frac{1089}{66}$$
$$= \frac{33}{2},$$

which means a rate of 33 students for every 2 faculty members or 16.5 students per each faculty member.

e) The ratio is given as gallons to barrels. The rate, then, will be given as *gallons per barrel*. Hence,

$$\frac{\text{gallons}}{\text{barrel}} = \frac{1,375}{25}$$
$$= \frac{55}{1},$$

which means a rate of 55 gallons per barrel.

f) The ratio is given as grams to tablets. The rate, then, will be given as *grams per tablet*. Hence,

$$\frac{\text{grams}}{\text{tablet}} = \frac{375}{3}$$
$$= \frac{125}{1},$$

which means a rate of 125 grams per tablet.

g) The ratio is given as pounds to grams. The rate, then, will be given as *pounds per gram*. Hence,

$$\frac{\text{pounds}}{\text{gram}} = \frac{11}{5,000}$$
$$= \frac{2.2}{1,000},$$

which means 2.2 pounds per 1,000 grams. However, this rate is usually given as 2.2 pounds per kilogram, because 1,000 grams is equal to 1 kilogram.

h) The ratio is given as dollars to hours. The rate, then, is given as *dollars per hour*. Hence,

$$\frac{\text{dollars}}{\text{hours}} = \frac{152}{32}$$
$$= \frac{4.75}{1},$$

which means a rate of $4.75 per hour.

i) The ratio is given as dollars to ounces. The rate, then, is given as *dollars per ounce*. Hence,

$$\frac{\text{dollars}}{\text{ounce}} = \frac{4.32}{27}$$
$$= \frac{432}{2,700}$$
$$= \frac{108}{675}$$
$$= \frac{4}{25}$$
$$= \frac{0.16}{1},$$

which means a rate of $.16 per ounce, or 16 cents per ounce.

5.3 PROPORTIONS

In this section, we introduce and discuss proportions and their solutions.

EQUALITY OF TWO RATIOS

If $a{:}b$ and $c{:}d$ are two **equal** ratios, the equation $\dfrac{a}{b} = \dfrac{c}{d}$ is called a **proportion**. This proportion is read "*a* is to *b* as *c* is to *d*" and can be written as $a{:}b{::}c{:}d$.

Exercise 5.8

Which of the following represent proportions?
a) 2:6 as 14:42
b) 3:7 as 9:21
c) $\dfrac{5}{8} = \dfrac{35}{56}$
d) 6:24 as 8:36
e) $\dfrac{105}{108} = \dfrac{35}{36}$
f) 60:35 as 12:7
g) 144:56 as 80:35

Solution 5.8

a) 2 is to 6 as 14 is to 42 *is* a proportion, because the fractions $\dfrac{2}{6}$ and $\dfrac{14}{42}$ are equivalent fractions. The proportion can be written as $\dfrac{2}{6} = \dfrac{14}{42}$ or as 2:6::14:42.

b) 3:7 as 9:21 *is* a proportion, because the fractions $\dfrac{3}{7}$ and $\dfrac{9}{21}$ are equivalent fractions. The proportion can be written as $\dfrac{3}{7} = \dfrac{9}{21}$ or as 3:7::9:21.

c) $\dfrac{5}{8} = \dfrac{35}{56}$ is a proportion, because the two fractions are equivalent. The proportion may also be written as 5:8::35:56.

d) 6:24 as 8:36 *is not* a proportion, because the fractions $\dfrac{6}{24}\left(=\dfrac{1}{4}\right)$ and $\dfrac{8}{36}\left(=\dfrac{2}{9}\right)$ are not equivalent.

e) $\dfrac{105}{108} = \dfrac{35}{36}$ *is* a proportion, because the two fractions are equivalent. The proportion may also be written as 105:108::35:36.

f) 60 is to 35 as 12 is to 7 *is* a proportion, because the fractions $\dfrac{60}{35}$ and $\dfrac{12}{7}$ are equivalent fractions. The proportion may be written as $\dfrac{60}{35} = \dfrac{12}{7}$ or as 60:35::12:7.

g) 144:56 as 80:35 *is not* a proportion, because the fractions $\frac{144}{56}\left(=\frac{18}{7}\right)$ and $\frac{80}{35}\left(=\frac{16}{7}\right)$ are not equivalent.

BASIC PROPERTY FOR PROPORTIONS

In the proportion $\frac{a}{b}=\frac{c}{d}$, the quantities a and d are called the **extremes**. The quantities b and c are called the **means**. The product of the extremes, ad, is equal to the product of the means, bc. That is, $ad = bc$.

Exercise 5.9

What are the extremes and means of the following proportions? Verify that each is a proportion.

a) $\frac{3}{7}=\frac{9}{21}$

b) $\frac{7}{9}=\frac{77}{99}$

c) 2:9::6:27

Solution 5.9

a) In the proportion $\frac{3}{7}=\frac{9}{21}$, 3 and 21 are the extremes; 7 and 9 are the means. We have $(3)(21) = (7)(9)$, or $63 = 63$.

b) In the proportion $\frac{7}{9}=\frac{77}{99}$, 7 and 99 are the extremes; the means are 9 and 77. We have $(7)(99) = (9)(77)$, or $693 = 693$.

c) In the proportion 2:9::6:27, the extremes are 2 and 27; the means are 9 and 6. We have $(2)(27) = (9)(6)$, or $54 = 54$.

Exercise 5.10

What are the extremes and means of each of the following proportions? Solve each for the variable.

a) 3:5::y:25

b) $\frac{x}{5}=\frac{8}{20}$

c) 9:u::3:21

Solution 5.10

a) In the proportion 3:5::y:25, the extremes are 3 and 25; the means are 5 and y. We have
$(3)(25) = (5)(y)$
$75 = 5y$
$15 = y$

b) In the proportion $\dfrac{x}{5} = \dfrac{8}{20}$, x and 20 are the extremes; the means are 5 and 8. We have
$(x)(20) = (5)(8)$
$20x = 40$ (20x means 20 times x)
$x = 2$

c) In the proportion 9:u::3:21, the extremes are 9 and 21; the means are u and 3. We have
$(9)(21) = (u)(3)$
$189 = 3u$
$63 = u$

Exercise 5.11

Determine whether the following ratios form a proportion:

a) $\dfrac{5}{8} ? \dfrac{35}{56}$

b) $\dfrac{5}{8} ? \dfrac{9}{12}$

c) $\dfrac{22}{8} ? \dfrac{16.5}{6}$

Solution 5.11

a) The ratios form a proportion if the product of the extremes equals the product of the means. The product of the extremes is $(5)(56) = 280$; the product of the means is $(8)(35) = 280$. Because the products are equal, the ratios *do* form a proportion.

b) The product of the extremes is $(5)(12) = 60$; the product of the means is $(8)(9) = 72$. Because the products are not equal, the ratios *do not* form a proportion.

c) The product of the extremes is $(22)(6) = 132$; the product of the means is $(8)(16.5) = 132$. Because the products are equal, the ratios *do* form a proportion.

Exercise 5.12

Solve the following proportions:

a) $\dfrac{x}{4} = \dfrac{12}{20}$ for x.

b) $\dfrac{9}{y} = \dfrac{18}{5}$ for y.

c) $\dfrac{4}{11} = \dfrac{t}{3.1}$ for t.

d) $\dfrac{1.2}{6} = \dfrac{0.8}{p}$ for p.

Solution 5.12

a) Because you have a proportion, the product of the extremes, $(x)(20)$, must equal the product of the means, $(4)(12)$.

$$(x)(20) = (4)(12)$$
$$20x = 48$$
$$x = \dfrac{48}{20}$$
$$x = 2.4$$

b) Because you have a proportion, the product of the extremes, $(9)(5)$, must equal the product of the means, $(y)(18)$.

$$(9)(5) = (y)(18)$$
$$45 = 18y$$
$$\dfrac{45}{18} = y$$
$$\dfrac{5}{2} = y$$

c) Because you have a proportion, the product of the extremes, $(4)(3.1)$, must equal the product of the means, $(11)(t)$

$$(4)(3.1) = (11)(t)$$
$$12.4 = 11t$$
$$\dfrac{12.4}{11} = t$$
$$\dfrac{124}{110} = t$$
$$\dfrac{62}{55} = t$$

d) Because you have a proportion, the product of the extremes, $(1.2)(p)$, must equal the product of the means, $(6)(0.8)$.

$$(1.2)(p) = (6)(0.8)$$
$$1.2p = 4.8$$
$$12p = 48$$
$$p = 4$$

5.4 APPLICATIONS OF RATIOS AND PROPORTIONS

In this section, we use ratios and proportions to solve certain types of problems. If $\frac{a}{b}$ is a ratio and x is a **nonzero** number, $\frac{a}{b} = \frac{ax}{bx}$ and x is called the **ratio multiplier.**

RATIO MULTIPLIER

In the first two sections of this chapter, we sometimes write ratios in lowest terms. If $\frac{2}{3}$ is the simplified form of a ratio, what was the given ratio? It could have been $\frac{4}{6}$ or $\frac{20}{30}$ or $\frac{400}{600}$ and so on.

Actually, you can multiply both the numerator and denominator of a ratio by the same nonzero number and get an equivalent ratio.

Exercise 5.13

Two numbers are in the ratio 3:4. If their sum is 42, what are the numbers?

Solution 5.13

Using the ratio multiplier x, let $3x$ = one number and $4x$ = the other number.

The ratio of the two numbers, then, is $\frac{3x}{4x}$, which is equivalent to the given ratio $\frac{3}{4}$. Also, the sum of the two numbers is 42. Hence, we have:

$$3x + 4x = 42$$
$$7x = 42$$
$$x = 6.$$

Therefore, the two numbers are $3x = (3)(6) = 18$ and $4x = (4).(6) = 24$. We check and determine that $18 + 24 = 42$.

Exercise 5.14

The ratio of two numbers is 5:3. If their difference is 24, what are the numbers?

Solution 5.14

Using the ratio multiplier x,

let $5x$ = the larger number
and $3x$ = the smaller number.

Because their difference is 24, we have:

$$5x - 3x = 24$$
$$2x = 24$$
$$x = 12.$$

Therefore, the numbers are $5x = (5)(12) = 60$ and $3x = (3)(12) = 36$. You check and determine that $60 - 36 = 24$.

RATIOS WITH THREE QUANTITIES

You sometimes have three quantities in the ratio $a:b:c$, which means that $a:b$, $b:c$, and $a:c$.

Exercise 5.15

Three numbers are in the ratio 2:3:5. If their sum is 70, what are the numbers?

Solution 5.15

Using the ratio multiplier x,

$$\text{let } 2x = \text{the first number}$$
$$3x = \text{the second number}$$
$$5x = \text{the third number}$$

Because the sum of the three numbers is 70, we have:

$$2x + 3x + 5x = 70$$
$$10x = 70$$
$$x = 7$$

Therefore, $2x = (2)(7) = 14$, $3x = (3)(7) = 21$, and $5x = (5)(7) = 35$, so the numbers are 14, 21, and 35. Their sum is 70.

Exercise 5.16

The ratio of the lengths of the three sides of a triangle is 3:4:5. If the perimeter of the triangle is 72 centimeters (cm), what is the length of each side of the triangle?

Solution 5.16

Using the ratio multiplier x, let

$$3x = \text{the length of one side (in cm)}$$
$$4x = \text{the length of a second side (in cm)}$$
$$5x = \text{the length of the third side (in cm)}$$

The perimeter of a triangle is the sum of the lengths of the three sides of the triangle, all measured in the same units. We have

$$3x + 4x + 5x = 72$$
$$12x = 72$$
$$x = 6$$

Therefore, $3x = (3)(6) = 18$, $4x = (4)(6) = 24$, and $5x = (5)(6) = 30$. The sides of the triangle have lengths 18 cm, 24 cm, and 30 cm. The perimeter is 18 cm + 24 cm + 30 cm = 72 cm.

We now examine some applications involving proportions.

Exercise 5.17

If oranges sell at $2.89 for 8 lb, how much will 96 lb of oranges cost, at the same rate?

Solution 5.17

Let x represent the cost of the 96 lb of oranges (in dollars and cents). Then the ratio of the 96 lb of oranges to x dollars and cents is the same as the ratio of the 8 lb of oranges to $2.89. We have, then,

$$\frac{96}{x} = \frac{8}{\$2.89}$$

We now have

$$(96)(\$2.89) = (x)(8)$$
$$\$277.44 = 8x$$
$$\frac{\$277.44}{8} = x$$
$$\$34.68 = x$$

Therefore, at the same rate, 96 lb of oranges cost $34.68.

Exercise 5.18

A distance of 300 meters is represented on a map by 2 cm. How many centimeters on the map would represent a distance of 1,800 meters?

Solution 5.18

Let y represent the required measure (in cm). Then

$$\frac{300 \text{ meters}}{1,800 \text{ meters}} = \frac{2 \text{ cm}}{y \text{ cm}}$$

or $\dfrac{300}{1,800} = \dfrac{2}{y}$

Therefore, $(300)(y) = (1,800)(2)$
$$300y = 3,600$$
$$y = \frac{3,600}{300}$$
$$y = 12$$

Therefore, 1,800 meters on the map is represented by 12 centimeters.

Exercise 5.19

The property tax on an $84,000 condominium is $1,200. What should the property tax be on a $77,000 condominium in the same area?

Solution 5.19

Let p = the property tax on the $77,000 condominium (in dollars). We then have the proportion

$$\frac{\$77,000}{p} = \frac{\$84,000}{\$1,200}$$

$$\frac{\$77,000}{p} = \frac{840}{12}$$

Therefore,

$$(\$77,000)(12) = (p)(840)$$

$$\$924,000 = 840p$$

$$\frac{\$924,000}{840} = p$$

$$\$1,100 = p$$

Therefore, the property tax on the $77,000 condominium should be $1,100.

Exercise 5.20

One mile is equal to 1.6 km. If you travel 318 miles, how many kilometers did you travel?

Solution 5.20

Let d = the distance traveled (in kilometers). Then

$$\frac{1 \text{ mi}}{1.6 \text{ km}} = \frac{318 \text{ mi}}{d \text{ km}}$$

$$(1 \text{ mi})(d \text{ km}) = (1.6 \text{ km})(318 \text{ mi})$$

or

$$d = (1.6)(318)$$

$$d = 508.8$$

Therefore, you have traveled 508.8 km.

Exercise 5.21

If 100 pounds of grass seed are used to seed 3,200 square feet of land, how many pounds of grass seed are needed to seed 5,600 square feet of land?

Solution 5.21

Let p = the number of pounds of grass seed required. Then we have:

$$\frac{100 \text{ lb}}{3,200 \text{ ft}^2} = \frac{p \text{ lb}}{5,00 \text{ ft}^2}$$

or
$$\frac{100}{3,200} = \frac{p}{5,600}$$
$$(100)(5,600) = (3,200)(p)$$
$$560,000 = 3,200p$$
$$\frac{560,000}{3,200} = p$$
$$175 = p$$

Therefore, 175 pounds of grass seed are needed.

Exercise 5.22

The shape of a rectangle having a ratio of its length to its width of 8 to 5 is supposed to be most pleasing to the eye. If so, what should the length of a rectangle be if its width is 24 inches?

Solution 5.22

Let m = the measure of the length (in inches). Then
$$\frac{8}{5} = \frac{m \text{ in}}{24 \text{ in}}$$
$$\text{or } (8)(24 \text{ in}) = (5)(m \text{ in})$$
$$192 \text{ in} = 5m \text{ in}$$
$$\frac{192 \text{ in}}{5 \text{ in}}$$
$$38.4 = m$$

Therefore, the length of the rectangle should be 38.4 inches.

In this chapter, we discussed ratios, rates, and proportions. The basic property of proportions was introduced and used. Applications of ratios and proportions were also examined.

TEST YOURSELF

In Exercises 1–5, express each of the given statements as a ratio.

1) The number of hits to the number of times at bat if a player gets eleven hits out of thirty-four times at bat.
2) The number of hourly workers to the number of salaried workers, when 96 workers among a 217 member work force are salaried and the rest are hourly.
3) The number of feet in a mile to the number of centimeters in a meter.
4) The number of students bused to the number of students not bused in a school district with 37,062 enrolled students and 21,699 students not bused.
5) The number of girls to the number of boys if 19 girls are in a group of 37 boys and girls.

In Exercises 6–10, simplify each of the given ratios.

6) 14 to 18
7) 85:17
8) 20 to 75
9) 40 to 25
10) 119:63

In Exercises 11–15, identify the extremes and the means in each of the given proportions.

11) 4 is to 7 as 16 is to 28
12) 9 is to 8 as 63 is to 56
13) 4 is to n as 9 is to $(n + 1)$
14) 5:11::30:66
15) 16:6::8:3

In Exercises 16–21, solve the given proportions for n.

16) $\dfrac{7}{n}$ as $\dfrac{21}{33}$
17) $7:6::2n:18$
18) $n:4::9:36$
19) $\dfrac{9}{17} = \dfrac{27}{n}$
20) $6:n::42:84$
21) $\dfrac{n}{18} = \dfrac{15}{54}$

22) If $10.00 in Canadian money is discounted to $8.65 in U.S. money, how much will $35.00 in Canadian money be discounted to in U.S. money?
23) A manufacturer wants to produce rotary mowers and reel mowers in a ratio of 7 to 9. How many of each should be produced for a total production of 1,280 units?
24) In a class of college students, 6 out of 11 students receives financial aid. At that rate, how many of the 5,016 enrolled students at that college receive financial aid?
25) If a recipe calls for 1 cup of Splenda for every quart of liquid, how many cups of Splenda are needed for 1.5 gallons of the liquid?

TEST YOURSELF WITH THE HANDHELD CALCULATOR

In Exercises 1–10, solve each of the following proportions.

1) $\dfrac{n}{7} = \dfrac{104}{728}$

2) $\dfrac{n}{30} = \dfrac{45}{225}$

3) $\dfrac{3}{16} = \dfrac{627}{n}$

4) $\dfrac{23}{n} = \dfrac{2{,}323}{2{,}727}$

5) $\dfrac{4.7}{7.5} = \dfrac{192.7}{n}$

6) $\dfrac{9.2}{2.7} = \dfrac{n}{97.2}$

7) $\dfrac{5224}{3{,}016} = \dfrac{1{,}306}{n}$

8) $\dfrac{8{,}181}{5{,}463} = \dfrac{909}{n}$

9) $\dfrac{3.5}{4.2} = \dfrac{n}{27.3}$

10) $\dfrac{11.2}{5.5} = \dfrac{184.8}{n}$

11) If 3.2 centimeters on a map represent 4,148 miles on the ground, how many miles on the ground are represented by 9.6 centimeters on the map?

12) At the rate of 3 pounds of apples for 97 cents, how much will 42 pounds of apples cost?

13) If a $180,000 house is taxed at $1,350, how much should a $150,000 house in the same neighborhood be taxed?

14) If 2 gallons of paint cover 550 square feet of wall space, how many gallons of the same paint will be needed to cover 1,925 square feet of wall space?

15) Baribor works in a blouse factory, sewing on collars. She completed 2,100 collars in an eight-hour day. At that rate, how many collars should she sew on in a 40-hour work week?

TEST YOURSELF ANSWERS

1) $\dfrac{11}{34}$

2) $\dfrac{217-96}{96} = \dfrac{121}{96}$

3) $\dfrac{5,280}{100}$ or, $\dfrac{264}{5}$ (simplified)

4) $\dfrac{37,062-21,699}{21,699} = \dfrac{15,363}{21,699}$ or, $\dfrac{1,707}{2,411}$ (simplified)

5) $\dfrac{19}{37-19} = \dfrac{19}{18}$

6) $\dfrac{14}{18} = \dfrac{7}{9}$ or, 7 to 9

7) $85:17 = \dfrac{85}{17} = \dfrac{5}{1}$ or, 5:1

8) $\dfrac{20}{75} = \dfrac{4}{15}$ or, 4 to 15

9) $\dfrac{40}{25} = \dfrac{8}{5}$ or, 8 to 5

10) $119:63 = \dfrac{119}{63} = \dfrac{17}{9}$ or, 17:9

11) The extremes are 4 and 28; the means are 7 and 16.

12) The extremes are 9 and 56; the means are 8 and 63.

13) The extremes are 4 and $(n+1)$; the means are n and 9.

14) The extremes are 5 and 66; the means are 11 and 30.

15) The extremes are 16 and 3; the means are 6 and 8.

16) $\dfrac{21}{33} = \dfrac{21 \div 3}{33 \div 3} = \dfrac{7}{11}$; $n = 11$

17) $\dfrac{7}{6} = \dfrac{7 \times 3}{6 \times 3} = \dfrac{21}{18}$; $2n = 21$; $n = 10.5$

18) $\dfrac{9}{36} = \dfrac{9 \div 9}{36 \div 9} = \dfrac{1}{4}$; $n = 1$

Ratio and Proportion

19) $\dfrac{9}{17} = \dfrac{9 \times 3}{17 \times 3} = \dfrac{27}{51}$; $n = 51$

20) $\dfrac{42}{84} = \dfrac{42 \div 7}{84 \div 7} = \dfrac{6}{12}$; $n = 12$

21) $\dfrac{15}{54} = \dfrac{15 \div 3}{54 \div 3} = \dfrac{5}{18}$; $n = 5$

22) $\dfrac{\text{Canadian}}{\text{U.S.}} = \dfrac{\$10}{\$8.65} = \dfrac{\$10 \times 3.5}{\$8.65 \times 3.5} = \dfrac{\$35}{\$30.275}$; $n = \$30.275$ or, $30.28 rounded to the nearest cent.

23) $\dfrac{\text{Rotary}}{\text{Total}}$; $\dfrac{7}{9} = \dfrac{n}{1280}$; $\dfrac{7}{9} = \dfrac{7 \times 80}{16 \times 80} = \dfrac{560}{1280}$; $n = 560$. Therefore, 560 rotary mowers and (1,280 − 560 = 720) reel mowers should be produced.

24) $\dfrac{\text{\#Financial Aid}}{\text{Total \#Students}}$; $\dfrac{6}{11} = \dfrac{n}{5016}$; $\dfrac{6}{11} = \dfrac{6 \times 456}{11 \times 456} = \dfrac{2736}{5016}$; $n = 2{,}736$ (number of students with financial aid)

25) $\dfrac{\text{Splenda}}{\text{Liquid}}$; $\dfrac{\text{cups}}{\text{qts}}$; $\dfrac{1}{1} = \dfrac{n}{1.5\,\text{gal}} = \dfrac{n}{6\,\text{qt}}$; $\dfrac{1}{1} = \dfrac{1 \times 6}{1 \times 6} = \dfrac{6\,\text{cups}}{6\,\text{qt}}$; $n = 6$

TEST YOURSELF ANSWERS WITH THE HANDHELD CALCULATOR

1) $\dfrac{104}{728} = \dfrac{104 \div 104}{728 \div 104} = \dfrac{1}{7}$; $n = 7$

2) $\dfrac{45}{225} = \dfrac{45 \div 7.5}{225 \div 7.5} = \dfrac{6}{30}$; $n = 6$

3) $\dfrac{3}{16} = \dfrac{3 \times 209}{16 \times 209} = \dfrac{627}{3,344}$; $n = 3,344$

4) $\dfrac{2,323}{2,727} = \dfrac{2,323 \div 101}{2,727 \div 101} = \dfrac{23}{27}$; $n = 27$

5) $\dfrac{4.7}{7.5} = \dfrac{4.7 \times 41}{7.5 \times 41} = \dfrac{192.7}{307.5}$; $n = 307.5$

6) $\dfrac{9.2}{2.7} = \dfrac{9.2 \times 36}{2.7 \times 36} = \dfrac{331.2}{97.2}$; $n = 331.2$

7) $\dfrac{5,224}{3,016} = \dfrac{5,224 \div 4}{3,016 \div 4} = \dfrac{1306}{754}$; $n = 754$

8) $\dfrac{8,181}{5,463} = \dfrac{8,181 \div 9}{5,463 \div 9} = \dfrac{909}{607}$; $n = 607$

9) $\dfrac{3.5}{4.2} = \dfrac{3.5 \times 6.5}{4.2 \times 6.5} = \dfrac{22.75}{27.3}$; $n = 22.75$

10) $\dfrac{11.2}{5.5} = \dfrac{11.2 \times 16.5}{5.5 \times 16.5} = \dfrac{184.8}{90.75}$; $n = 90.75$

11) $\dfrac{\text{cm}}{\text{mi}}$; $\dfrac{3.2}{4,148} = \dfrac{9.6}{n}$; $3.2n = (4,148)(9.6)$; $n = 12,444$ miles

12) $\dfrac{\text{lb}}{\text{cents}}$; $\dfrac{3}{97} = \dfrac{42}{n}$; $3n = (97)(42)$; $n = 1,358$ cents or, $13.58

13) $\dfrac{\text{House}}{\text{Tax}}$; $\dfrac{\$180,000}{\$1,350} = \dfrac{\$150,000}{n}$; $\$180,000n = (\$1,350)(\$150,000)$; $n = \$1,125$

14) $\dfrac{\text{gal}}{\text{sq ft}}$; $\dfrac{2}{550} = \dfrac{n}{1,925}$; $2(1,925) = 550n$; $n = 7$ gal

15) $\dfrac{\text{Collars}}{\text{Hours}}$; $\dfrac{2,100}{8} = \dfrac{n}{40}$; $(2,100(40) = 8n$; $n = 10,500$ collars

CHAPTER 6

Percent

In this chapter, we introduce the concept of percent and the relationships among percent, decimals, and fractions. Applications of percent is also discussed.

6.1 PERCENT

In this section, we introduce the concept of percent. We also discuss the relationships among percent, fractions, and decimals. Conversions between percent and fractions and between percent and decimals is examined.

PERCENT AND FRACTIONS

A **percent** is a fraction with a denominator of 100.

Exercise 6.1

Write the following fractions as percents:

a) $\dfrac{35}{100}$

b) $\dfrac{237}{100}$

c) $\dfrac{23}{10}$

d) $\dfrac{195}{1000}$

e) $\dfrac{1234}{1000}$

Solution 6.1

a) 35 percent

b) 237 percent

c) The fraction $\dfrac{23}{10} = \dfrac{230}{100}$, which can be written as 230 percent.

d) The fraction $\dfrac{195}{1000} = \dfrac{19.5}{100}$, which can be written as 19.5 percent.

e) The fraction $\dfrac{1234}{1000} = \dfrac{123.4}{100}$, which can be written as 123.4 percent.

We use the symbol % to indicate percent. Hence, we write 35 percent as 35%, 237 percent as 237%, 19.5 percent as 19.5%, and so on.

Fractions that have denominators other than 10, 100, or 1,000 can also be written as percents.

Exercise 6.2

Write the following fractions as percents:

a) $\dfrac{1}{4}$

b) $\dfrac{2}{3}$

c) $\dfrac{2}{5}$

d) $\dfrac{9}{20}$

e) $\dfrac{3}{8}$

Solution 6.2

a) $\dfrac{1}{4} = \dfrac{1 \times 25}{4 \times 25} = \dfrac{25}{100} = 25\%$

b) $\dfrac{2}{3} = \dfrac{2 \times 100}{3 \times 100} = \dfrac{200}{3} \times \dfrac{1}{100} = 66\dfrac{2}{3} \times \dfrac{1}{100} = 66\dfrac{2}{3}\%$

c) $\dfrac{2}{5} = \dfrac{2 \times 20}{5 \times 20} = \dfrac{40}{100} = 40\%$

d) $\dfrac{9}{20} = \dfrac{9 \times 5}{20 \times 5} = \dfrac{45}{100} = 45\%$

e) $\dfrac{3}{8} = \dfrac{3 \times 125}{8 \times 125} = \dfrac{375}{1,000} = \dfrac{37.5}{100} = 37.5\%$

The fractions in Exercise 6.2 have a value less than 1. In turn, they convert to percents less than 100 percent. In the next exercise, note that the whole number 1 converts to 100 percent.

Exercise 6.3
Write the whole number 1 as a percent.

Solution 6.3
$$1 = \frac{1}{1} = \frac{1 \times 100}{1 \times 100} = \frac{100}{100} = 100\%$$

The fractions in Exercises 6.4 represent numbers greater than 1. These are written as percents greater than 100 percent.

Exercise 6.4
Write the following numbers as percents:

a) $\dfrac{9}{4}$

b) 6

c) $2\dfrac{1}{4}$

d) $3\dfrac{1}{2}$

Solution 6.4
a) $\dfrac{9}{4} = \dfrac{9 \times 25}{4 \times 25} = \dfrac{225}{100} = 225\%$

b) $6 = \dfrac{6}{1} = \dfrac{6 \times 100}{1 \times 100} = \dfrac{600}{100} = 600\%$

c) $2\dfrac{1}{4} = \dfrac{9}{4} = \dfrac{9 \times 25}{4 \times 25} = \dfrac{225}{100} = 225\%$

Notice that both $\dfrac{9}{4}$ and $2\dfrac{1}{4}$ convert to 225%.

d) $3\dfrac{1}{2} = \dfrac{7}{2} = \dfrac{7 \times 50}{2 \times 50} = \dfrac{350}{100} = 350\%$

The procedure illustrated in the previous exercises for rewriting a fraction as a percent can be shortened by using the following rule.

- If the number is a mixed number, convert it to an improper fraction.
- Divide the numerator of the fraction by its denominator and multiply the quotient obtained by 100.
- Add the % symbol after the result obtained in the preceding step.

Exercise 6.5

Write the following numbers as percents:

a) $\dfrac{5}{8}$

b) $3\dfrac{2}{5}$

c) $\dfrac{7}{16}$

d) 8

e) $\dfrac{1}{3}$

Solution 6.5

a) $5 \div 8 = 0.625$

$0.625 \times 100 = 62.5$

Add the % symbol. $\dfrac{5}{8} = 62.5\%$

b) $3\dfrac{2}{5} = \dfrac{17}{5} = 17 \div 5 = 3.4$

$3.4 \times 100 = 340$

Add the % symbol. $3\dfrac{2}{5} = 340\%$

c) $7 \div 16 = 0.4375$

$0.4375 \times 100 = 43.75$

Add the % symbol. $\dfrac{7}{16} = 43.75\%$

d) No division is needed. $(8 \div 1 = 8)$

$8 \times 100 = 800$

Add the % symbol. $8 = 800\%$

e) $1 \div 3 = 0.3\overline{3}$

$0.3\overline{3} \times 100 = 33.\overline{3}\% = 33\dfrac{1}{3}$

Add the % symbol. $\dfrac{1}{3} = 33\dfrac{1}{3}\%$

CONVERTING PERCENTS TO FRACTIONS

We can also convert percents to fractions by reversing the above.

Exercise 6.6

Write the following percents as fractions:
a) 25%
b) 148%
c) 23.4%
d) $83\frac{1}{3}\%$

Solution 6.6

a) $25\% = \dfrac{25}{100} = \dfrac{25 \div 25}{100 \div 25} = \dfrac{1}{4}$

b) $148\% = \dfrac{148}{100} = \dfrac{148 \div 4}{100 \div 4} = \dfrac{37}{25}$

c) $23.4\% = \dfrac{23.4}{100} = \dfrac{23.4 \times 10}{100 \times 10} = \dfrac{234}{1,000} = \dfrac{234 \div 2}{1,000 \div 2} = \dfrac{117}{500}$

d) $83\dfrac{1}{3}\% = \dfrac{83\frac{1}{3}}{100} = 83\dfrac{1}{3} \div \dfrac{100}{1} = \dfrac{250}{3} \times \dfrac{1}{100} = \dfrac{250}{300} = \dfrac{250 \div 50}{300 \div 50} = \dfrac{5}{6}$

To rewrite a percent as a fraction, delete the % symbol after the number and write the number over 100. Simplify the result, if necessary.

Exercise 6.7

Rewrite the following percents as fractions:
a) 70%
b) 105%
c) 275%
d) 18.6%

Solution 6.7

a) Delete the % symbol and write the number over 100.
$$70\% = \dfrac{70}{100} = \dfrac{70 \div 10}{100 \div 10} = \dfrac{7}{10}$$

b) Delete the % symbol and write the number over 100.
$$105\% = \dfrac{105}{100} = \dfrac{105 \div 5}{100 \div 5} = \dfrac{21}{20}$$

c) Delete the % symbol and write the number over 100.
$$275\% = \dfrac{275}{100} = \dfrac{275 \div 25}{100 \div 25} = \dfrac{11}{4}$$

d) Delete the % symbol and write the number over 100.

$$18.6\% = \frac{18.6}{100} = \frac{18.6 \times 10}{100 \times 10} = \frac{186}{1,000} = \frac{186 \div 2}{1,000 \div 2} = \frac{93}{500}$$

PERCENT AND DECIMALS

We now examine the conversion of percents to decimals.

Exercise 6.8

Rewrite the following percents as decimals:

a) 45%

b) 59%

c) 112%

d) 434%

e) 5.3%

f) 17.8%

g) 0.8%

h) $66\frac{2}{3}\%$

Solution 6.8

a) $45\% = \frac{45}{100}$, because $\frac{45}{100} = 0.45$, $45\% = 0.45$

b) $59\% = \frac{59}{100}$, because $\frac{59}{100} = 0.59$, $59\% = 0.59$

c) $112\% = \frac{112}{100}$, because $\frac{112}{100} = 1.12$, $112\% = 1.12$

d) $434\% = \frac{434}{100}$, because $\frac{434}{100} = 4.34$, $434\% = 4.34$

e) $5.3\% = \frac{5.3}{100} = \frac{5.3 \times 10}{100 \times 10} = \frac{53}{1,000} = 0.053$, $5.3\% = 0.053$

f) $17.8\% = \frac{17.8}{100} = \frac{17.8 \times 10}{100 \times 10} = \frac{178}{1,000} = 0.178$, $17.8\% = 0.178$

g) $0.8\% = \frac{.8}{100} = \frac{.8 \times 10}{100 \times 10} = \frac{8}{1,000} = 0.008$, $0.8\% = 0.008$

h) $66\frac{2}{3}\% = \frac{66.\overline{6}}{100} = \frac{66.\overline{6} \times 10}{100 \times 10} = \frac{666.\overline{6}}{1,000} = .66\overline{6} \approx .6$, $66\frac{2}{3}\% \approx .6$

To rewrite a percent as a decimal, remove the % symbol and move the decimal point **two** places to the **left.**

Exercise 6.9

Rewrite the following percents as decimals:

a) 37%
b) 220%
c) 7.6%
d) 0.9%
e) 0.06%
f) 0.001%
g) 23.4%
h) 187.69%
i) 4%

Solution 6.9

a) 37% = 0.37
b) 220% = 2.2
c) 7.6% = 07.6% = 0.076. (Note that a 0 had to be placed between the decimal point and the digit 7.)
d) 0.9% = 00.9% = 0.009
e) 0.06% = 00.06% = 0.0006
f) 0.001% = 00.001% = 0.00001
g) 23.4% = 0.234
h) 187.69% = 1.8769
i) 4% = 04% = 0.04

We now convert decimals to percents.

Exercise 6.10

Rewrite the following decimals as percents:

a) 0.29
b) 0.56
c) 2.14
d) 7.85
e) 0.08
f) 0.02
g) 0.068

Solution 6.10

a) $0.29 = \dfrac{29}{100}$. Because $\dfrac{29}{100} = 29\%$, $0.29 = 29\%$.

b) $0.56 = \dfrac{56}{100}$. Because $\dfrac{56}{100} = 56\%$, $0.56 = 56\%$.

c) $2.14 = \dfrac{214}{100}$. Because $\dfrac{214}{100} = 214\%$, $2.14 = 214\%$.

d) $7.85 = \dfrac{785}{100}$. Because $\dfrac{785}{100} = 785\%$, $7.85 = 785\%$.

e) $0.08 = \dfrac{8}{100}$. Because $\dfrac{8}{100} = 8\%$, $0.08 = 8\%$.

f) $0.02 = \dfrac{2}{100}$. Because $\dfrac{2}{100} = 2\%$, $0.02 = 2\%$.

e) $0.068 = \dfrac{68}{1,000} = \dfrac{68 \div 10}{1,000 \div 10} = \dfrac{6.8}{100} = 6.8\%$

Thus, $0.068 = 6.8\%$.

To rewrite a decimal as a percent, move the decimal point **two** places to the **right** and place the % symbol after the resulting number.

Exercise 6.11

Rewrite the following decimals as percents:

a) 0.37
b) 0.6
c) 3.4
d) 6.79
e) 0.08
f) 37.2
g) 0.087

Solution 6.11

a) $0.37 = 37\%$
b) $0.6 = 0.60 = 60\%$ (a 0 had to be placed after the 6 in order to move the decimal point two places to the right)
c) $3.4 = 3.40 = 340\%$
d) $6.79 = 679\%$
e) $0.08 = 8\%$
f) $37.2 = 37.20 = 3,720\%$
g) $0.087 = 8.7\%$

6.2 PROBLEMS INVOLVING PERCENTS

In this section, we examine some basic problems involving percents. All of the percent problems given in this section can be solved by using the relationship

$$A = R \times B$$

Where
- A is the **amount** or part of the total.
- R is the **rate,** given as a percent.
- B is the **base,** or the total number, and usually follows the word "of."

SOLVING WORD PROBLEMS

Exercise 6.12
What is 30% of 80?

Solution 6.12
Let A = what, R = 30%, and B = 80.

$$\text{What is 30\% of 80?}$$
$$A = R \times B$$
$$A = 30\% \times 80$$
$$A = 0.30 \times 80 \text{ (Converting } 30\% = 0.30)$$
$$= 24$$

30% of 80 is 24.

Exercise 6.13
24 is what percent of 40?

Solution 6.13
Let A = 24, R = what percent, and B = 40.

$$\text{24 is what percent of 40?}$$
$$A = R \times B$$
$$24 = R \times 40$$
$$R = \frac{24}{40}$$
$$= 0.60$$
$$= 60\%$$

24 is 60% of 40.

Exercise 6.14
100 is 50% of what number?

Solution 6.14
Let $A = 100$, $R = 50\%$, and $B =$ what number.

$$100 \text{ is } 50\% \text{ of what number?}$$
$$A = R \times B$$
$$100 = 50\% \times B$$
$$100 = 0.50 \times B$$
$$B = \frac{100}{0.50}$$
$$= 200$$

100 is 50% of 200.

Exercise 6.15
What is 55.5% of 187.2?

Solution 6.15
Let $A =$ what, $R = 55.5\%$, and $B = 187.2$.

$$\text{What is } 55.5\% \text{ of } 187.2?$$
$$A = R \times B$$
$$A = 55.5\% \times 187.2$$
$$A = 0.555 \times 187.2$$
$$= 103.896$$

103.896 is 55.5% of 187.2

Exercise 6.16
84 is 35% of what number?

Solution 6.16
Let $A = 84$, $R = 35\%$, and $B =$ what number.

$$84 \text{ is } 35\% \text{ of what number?}$$
$$A = R \times B$$
$$84 = 0.35 \times B$$
$$B = \frac{84}{0.35}$$
$$= 240$$

84 is 35% of 240.

Exercise 6.17
60 is what percent of 30?

Solution 6.17
Let $A = 60$, $R =$ what percent, and $B = 30$.

$$60 \text{ is what percent of } 30?$$
$$A = R \times B$$
$$60 = R \times 30$$
$$R = \frac{60}{30}$$
$$= 2$$
$$= 200\%$$

60 is 200% of 30.

Exercise 6.18
During an inventory clearance sale, The Wicker Shop offered all its merchandise at 65% of the regular price (35% off). If a wicker sofa regularly sold for $699, what is the sale price?

Solution 6.18
Sale price is 65% of the regular price.

$$\text{Sale price} = 65\% \times \$699$$
$$= 0.65 \times \$699$$
$$= \$454.35$$

The sale price is $454.35.

Exercise 6.19
Stacey borrowed $150 from her friend Mike. She repaid $105 of the loan within a month. What percent of the loan did Stacey repay?

Solution 6.19
$105 is what percent of $150?
$$\$105 = R \times \$150$$
$$R = \frac{\$105}{\$150}$$
$$= \frac{105}{150}$$
$$= 0.7$$
$$= 70\%$$

Stacey repaid 70% of the loan.

Exercise 6.20

Vince leases a car for $285 a month. That is 15% of his monthly income. What is his monthly income?

Solution 6.20

$285 is 15% of what?

$$\$285 = 0.15 \times B$$
$$B = \frac{\$285}{0.15}$$
$$= \$1,900$$

Vince's monthly income is $1,900.

Exercise 6.21

In an English class, sixteen of the thirty-six students are from a minority group. What percent of the class is minority?

Solution 6.21

16 is what percent of 36?

$$16 = R \times 36$$
$$R = \frac{16}{36}$$
$$= 0.44$$
$$= 44.4\%$$

Approximately 44.4% of the class is minority.

Exercise 6.22

Colin completed 76% of the homework problems assigned in his mathematics class. If there were twenty-five problems assigned, how many did Colin complete?

Solution 6.22

What is 76% of 25?

$$A = 76\% \times 25$$
$$= 0.76 \times 25$$
$$= 19$$

Colin completed nineteen of the assigned problems.

Exercise 6.23

In a recent school bond election, 1,152 people voted in favor of the school bond. If this number represents 60% of those who voted, how many people voted in the election?

Solution 6.23

1,152 is 60% of what?

$$1,152 = 60\% \times B$$
$$B = \frac{1,152}{0.60}$$
$$= 1,920$$

1,920 people voted in the election.

6.3 PERCENT INCREASE OR DECREASE

In this section, we discuss percent increase and percent decrease. We determine a new number when an old number is increased or decreased by a certain percent. We also determine the percent of increase or decrease of a number.

THE AMOUNT OF INCREASE OR DECREASE

To determine the amount of increase or decrease, multiply the original increase or amount by the rate:

Amount of Increase or Decrease = Original Amount × Rate.

Exercise 6.24

Due to a severe frost, the price of orange juice increased by 80%. If orange juice sold for $1.55 per half gallon before the increase, how much was the increase for a half gallon of orange juice?

Solution 6.24

Amount of increase = Original Price × Rate
= $1.55 × 80%
= $1.55 × 0.80
= $1.24

The amount of increase for a half gallon of orange juice was $1.24.

Exercise 6.25

The price of long-stem roses decreased by 30%. If a dozen long-stem roses sold for $39.95 before the decrease, what was the amount of decrease?

Solution 6.25

Amount of Decrease = Original Price × Rate
= $39.95 × 30%
= $39.95 × 0.30
= $11.985

The amount of decrease for a dozen long-stem roses was approximately $11.98.

Exercise 6.26

Jay's salary increased by 9.5%. If his old salary was $39,200 per year, what is his new salary?

Solution 6.26

To determine Jay's new salary, you must do two things.

Step 1: Determine the amount of increase in Jay's salary, as follows:

Amount of Increase = Old Salary × Rate
= $39,200 × 9.5%
= ($39,200)(0.095)
= $3,724

Step 2: Add the amount of increase to the old salary to obtain the new salary:

$39,200   old salary
+ 3,724   amount of increase
$42,924   new salary

Jay's new salary is $42,924.

Exercise 6.27

Due to energy conservation practices, Larry's fuel bill decreased by 18%. If his average monthly fuel bill was $167.50, what is his new average monthly bill?

Solution 6.27

To determine Larry's new average monthly fuel bill, we must do two things:

Step 1: Determine the amount of decrease (or loss) in the average monthly fuel bill, as follows:

Amount of Decrease = Original amount × Rate
= $167.50 × 18%
= ($167.50 × 0.18)
= $30.15

Step 2: Subtract the amount of decrease from the original monthly average:

$167.50   original monthly average
−30.15   amount of decrease
$137.35   new monthly average

Larry's new average monthly fuel bill is $137.35.

RATE OF INCREASE OR DECREASE

If you know the original amount and the new amount, you can determine the rate of increase or decrease according to the following rule.

$$\text{Rate of Increase or Decrease} = \frac{\text{Amount of increase or decrease}}{\text{Original amount}}$$

Exercise 6.28

The price of a liter of cola increased from $1.09 to $1.29. What was the percent of increase?

Solution 6.28

To solve this problem, we must do two things.

Step 1: Determine the *amount* of increase, as follows:

$1.29   new price per liter
−1.09   old price per liter
$.20   *amount* of increase per liter

Step 2: Determine the rate of increase, as follows:

$$\text{Amount of Decrease} = \frac{\text{Amount of decrease}}{\text{Original price}}$$

$$= \frac{\$.20}{\$1.09}$$

$$= \frac{20}{109}$$

$$= 18.3\% \text{ (approximately)}$$

The percent of increase was approximately 18.3%.

Exercise 6.29

A new boat sold for $29,800. One year later, its value was $22,350. What was the percent of decrease in the value of the boat for the one year?

Solution 6.29

Determine the amount of decrease, as follows:

$29,800   new boat value
−22,350   value of boat one year later
$7,450   *amount* of decrease in value

Determine the rate of decrease, as follows:

$$\text{Amount of Decrease} = \frac{\text{Amount of decrease}}{\text{Original price}}$$

$$= \frac{\$7450}{\$29,800}$$

$$= \frac{1}{4}$$
$$= 25\%$$

There was a 25% decrease in value of the boat for the one year.

Exercise 6.30

At Big University, the number of tenured faculty increased from 160 to 304 over a period of ten years. What was the percent of increase?

Solution 6.30

Determine the amount of increase, as follows:

    304   number of tenured faculty at the end of the period
  −160   number of tenured faculty at beginning of period
    144   *amount* of increase

Determine the rate of increase, as follows:

$$\text{Rate of Increase} = \frac{\text{Amount of increase}}{\text{Original number}}$$

$$= \frac{144}{160}$$

$$= \frac{9}{10}$$

$$= 90\%$$

Therefore, the percent of increase was 90%.

Exercise 6.31

In 2003, Krystle paid $1,096 in state income taxes. In 2005, she paid $1,401. What was the percent increase in income taxes paid from 2003 to 2005? (Give the answer to the nearest tenth of a percent.)

Solution 6.31

Determine the amount of increase, as follows:

$$\begin{array}{rl} \$1{,}401 & \text{taxes paid in 2005} \\ -1{,}096 & \text{taxes paid in 2003} \\ \hline \$305 & \textit{amount of increase} \end{array}$$

Determine the rate of increase, as follows:

$$\text{Amount of increase} = \frac{\text{Amount of increase}}{\text{Original number}}$$

$$= \frac{\$305}{\$1{,}096}$$

$$= 0.278 \text{ (to the nearest thousandth)}$$

The percent of increase was 27.8%, to the nearest tenth percent.

In this chapter, we discussed percents, decimals, and fractions. We examined the relationships between percents and fractions and between percents and decimals. We also introduced basic problems involving percents. Percent increase and percent decrease were also discussed.

TEST YOURSELF

In Exercises 1–5, rewrite each of the given numbers as a percent.

1) $\dfrac{7}{4}$
2) 47
3) $6\dfrac{1}{5}$
4) 23.4
5) 172.3

In Exercises 6–10, rewrite each of the given percents as equivalent fractions in simplest form.

6) 12%
7) 235%
8) 8.5%
9) 0.85%
10) 17.65%

In Exercises 11–15, rewrite each of the following as a percent.

11) 0.4
12) 11
13) $\dfrac{7}{5}$
14) 22.3
15) 0.38
16) 0.9 is what percent of 36?
17) 0.03 is 6% of what number?
18) A worker's salary is increased by 7.6%. If the old salary was $21,350 per year, what is the worker's new yearly salary?
19) The price of gasoline decreased from $2.199 per gallon to $1.899 per gallon. What was the percent of decrease? (Give your result to the nearest tenth of a percent.)
20) State grants to City A were increased from $785,000 to $1,020,350. What was the percent of increase in the amount of the grants? (Give your result to the nearest percent.)
21) Determine the total price of an item that sells for $963 if the sales tax rate is 6.5%.
22) Mercy purchased a sweater for $23.95, a blouse for $17.99, a pair of shoes for $37.50, and a skirt for $29.95. If the sales tax rate was 6.5%, what was the total amount of her purchases including tax?
23) A furniture item is advertised as 60% off the marked price of $1,195. What is the sale price?
24) Becky works for a commission of sales. Her commission is 4% of the first $5,000 in sales, 5.5% of the next $6,000 of sales, and 7% of all sales over $11,000. If her total net sales amounted to $18,750, what was her commission?
25) Interest is compounded semi-annually. Determine the amount of interest for $3,000 at 6.5% per year for 2 years.

TEST YOURSELF WITH THE HANDHELD CALCULATOR

In Exercises 1–5, rewrite the fractions as percents, correct to the nearest tenth of 1%.

1) $\dfrac{7}{11}$

2) $\dfrac{5}{37}$

3) $\dfrac{111}{29}$

4) $\dfrac{467}{235}$

5) $\dfrac{107}{2,005}$

In Exercises 6–12, determine the value of each expression. Give your results to the nearest hundredth.

6) 20.9% of 32.7
7) 37% of 2.07
8) 29.3% of 396.2
9) 78.55% of 211.14
10) 138% of 138.5
11) 207.6% of 151.4
12) 0.94% of 0.94

In Exercises 13–20, write your results to the nearest tenth of 1%.

13) 8 is what percent of 315?
14) 50 is what percent of 20?
15) 39.4 is what percent of 102.7?
16) 900.5 is what percent of 210?
17) What percent of 270 is 17?
18) What percent of 69.7 is 13.4?
19) What percent of 0.06 is 0.035?
20) What percent of 540 is 18?
21) University professors received a salary increase of 8.5%. If Professor Bright's monthly salary before the increase was $5,208, what is his new annual salary (correct to the nearest dollar)?
22) The unemployment rate in a particular county decreased by 3.72% during a six-month period. If 27,692 people were unemployed at the start of the six-month period, how many people were unemployed at the end of the period? (Give your result to the nearest whole number.)
23) Kandice bought a new pickup truck on sale for $22,750 plus sales tax of 7%. If she financed the total amount over a period of 60 months, what will be her average monthly payment? (Disregard tags, shipping, and other fees.)
24) Cindy bought a new car, at the end of the season, for $17,890. If this was a 22% discount, what was the price of the car before the sale?
25) The marked price of an item of furniture was $2,165.99. The discount is $762. What is the rate of discount?

TEST YOURSELF ANSWERS

1) $\dfrac{7}{4} = \dfrac{7 \times 25}{4 \times 25} = \dfrac{175}{100} = 175\%$

2) $47 = \dfrac{47}{1} = \dfrac{47 \times 100}{100} = \dfrac{4700}{100} = 4,700\%$

3) $6\dfrac{1}{5} = 6.2 = \dfrac{6.2 \times 100}{100} = \dfrac{620}{100} = 620\%$

4) $23.4 = \dfrac{23.4 \times 100}{100} = \dfrac{2340}{100} = 2,340\%$

5) $172.3 = \dfrac{172.3 \times 100}{100} = \dfrac{17230}{100} = 17,230\%$

6) $12\% = \dfrac{12}{100} = \dfrac{12 \div 4}{100 \div 4} = \dfrac{3}{25}$

7) $235\% = \dfrac{235}{100} = \dfrac{235 \div 5}{100 \div 5} = \dfrac{47}{20}$

8) $8.5\% = \dfrac{8.5}{100} = \dfrac{85}{1,000} = \dfrac{85 \div 5}{1,000 \div 5} = \dfrac{17}{200}$

9) $0.85\% = \dfrac{0.85}{100} = \dfrac{85}{10,000} = \dfrac{85 \div 5}{10,000 \div 5} = \dfrac{17}{2,000}$

10) $17.65\% = \dfrac{17.65}{100} = \dfrac{1765}{10,000} = \dfrac{1765 \div 5}{10,000 \div 5} = \dfrac{353}{2,000}$

11) $0.4 = 0.4 \times 100 = 40\%$

12) $11 = 11 \times 100 = 1,100\%$

13) $\dfrac{7}{5} = 1.4 = 1.4 \; 100 = 140\%$

14) $22.3 = 22.3 \times 100 = 2,230\%$

15) $0.38 = 0.38 \times 100 = 38\%$

16) $A = R \times B$

$0.9 = R \times 36$

$R = \dfrac{0.9}{36} = \dfrac{9}{360} = \dfrac{1}{40} = 0.025 = 2.5\%$

17) $A = R \times B$

$0.03 = 6\% \times B$

$0.03 = 0.06 \times B$

$B = \dfrac{0.03}{0.06} = \dfrac{3}{6} = \dfrac{1}{2} = 0.5$

18)

    $21,350   (old salary)
   ×0.076   (percent increase)
  $ 1,622.60   (amount of increase)
+$21,350.00
  $23,972.60   (new salary)

19)

  $2.199   (old price)
 −1.899   (new price)
  $ .30   (*price* decrease)

Percent of decrease = $\dfrac{\text{price decrease}}{\text{old price}} = \dfrac{\$.30}{\$2.199} = 0.136 = 13.6\%$ (to the nearest tenth of 1%)

20)

  $1,020,350   (new amount)
   −785,000   (old amount)
  $ 235,350   (*amount* of increase)

Percent of increase = $\dfrac{\text{amount of increase}}{\text{old amount}} = \dfrac{\$235,350}{\$785,000} = 30\%$

21)

      $963   (price of item)
    ×.065   (sales tax rate)
   $62.595   (sales tax)
 +963.00   (price of item)
$1,025.595   (total amount)

or

$1,025.60

22)

  $23.95   (sweater)
   17.99   (blouse)
   37.50   (shoes)
+ 29.95   (skirt)
$109.39   (total)
  ×.065   (sales tax rate)
  $7.11   (sales tax)
+109.39
$116.50   (total amount)

23)

$1,195	(marked price)	$1,195	(marked price)
×.6	(% off)	−717	(discount)
$717	(discount)	$478	(sales price)

24)

$5,000 (1st $5,000)
×.04 (rate)
$200 (commission)

$6,000 (next $6,000)
×.055 (rate)
$330 (commission)

$18,750 (total sales)
−11,000
$7,750 (amount over $11,000)
×.07 (rate)
$542.50 (commission)

$200.00
330.00
+542.50
$1,072.50 (total commission)

25)

$3,000 (initial amount)
×.0325 (interest rate for 6 mos)
$97.50 (interest after 6 mos)
+3,000.00
$3,097.50 (amount after 6 mos)
×.0325 (interest rate for 6 mos)
$100.67 (interest after next 6 mos)
+3,097.50
$3,198.17 (amount after 1 yr)
×.0325 (interest rate for 6 mos)
$103.94 (interest after next 6 mos)
+3,198.17
$3,302.11 (amount after 18 mos)
×.0325 (interest rate for 6 mos)
$107.32 (interest after next 6 mos)
+3,302.11
$3,409.43 (amount after 2 yrs)

TEST YOURSELF ANSWERS WITH THE HANDHELD CALCULATOR

1) $\dfrac{7}{11} = 0.6363 = 63.6\%$

2) $\dfrac{5}{37} = 0.1351 = 13.5\%$

3) $\dfrac{111}{29} = 3.8275 = 382.8\%$

4) $\dfrac{467}{235} = 1.9872 = 198.7\%$

5) $\dfrac{107}{2,005} = 0.0533 = 5.3\%$

6) 20.9% of 32.7 = ?
$0.209 \times 32.7 = 6.83$

7) 37% of 2.07 is ?
$0.37 \times 2.07 = 0.77$

8) 29.3% of 396.2 is ?
$0.293 \times 396.2 = 116.09$

9) 78.55% of 211.14 is ?
$0.7855 \times 211.14 = 165.85$

10) 138% of 138.5 is ?
$1.38 \times 138.5 = 191.13$

11) 207.6% of 151.4 is ?
$2.076 \times 151.4 = 314.31$

12) 0.94% of 0.94 is ?
$0.0094 \times 0.94 = 0.01$

13) $A = R \times B$
$8 = R \times 315$
$R = \dfrac{8}{315} = 0.0253 = 2.5\%$

14) $A = R \times B$
$50 = R \times 20$
$R = \dfrac{50}{20} = 2.5 = 250\%$

15) $A = R \times B$
$39.4 = R \times 102.7$
$R = \dfrac{39.4}{102.7} = 0.3836 = 38.4\%$

16) $A = R \times B$
$900.5 = R \times 210$
$R = \dfrac{900.5}{210} = 4.2880 = 428.8\%$

17) $A = R \times B$
$17 = R \times 270$
$R = \dfrac{17}{270} = 0.0629 = 6.3\%$

18) $A = R \times B$
$13.4 = R \times 69.7$
$R = \dfrac{13.4}{69.7} = 0.1922 = 19.2\%$

19) $A = R \times B$
$0.035 = R \times 0.06$
$R = \dfrac{0.035}{0.06} = 0.5833 = 58.3\%$

20) $A = R \times B$
$18 = R \times 540$
$R = \dfrac{18}{540} = 0.0333 = 3.3\%$

21)

$5,208	(old monthly salary)
×0.085	(percent increase)
$442.68	(amount of increase)
+5,208.00	
$5,650.68	(new monthly salary)
×12	(12 months = 1 yr)
$67,808.16	(new annual salary)

or

$67,808 (rounded to the nearest dollar)

22)

	27,692	(number unemployed at start of period)
×	0.0372	(percent decrease)
	1,030	(amount of decrease, to the nearest whole number)

	27,692	(number unemployed at start of period)
−	1,030	(amount of decrease)
	26,662	(number unemployed at end of period)

23)

	$22,750	(purchase price)
×	0.07	(sales tax rate)
	$ 1,592.50	(sales tax)
+	22,750.00	
	$24,342.50	(total price)

$24,342.50 ÷ 60 = $405.71 (monthly payment)

24) If the discount was 22%, the sale price was (100 − 22)% = 78% of the original price.

Sale price = 78% of Original price

$17,890 = 0.78 × Original price

Original price = $\frac{\$17{,}890}{0.78}$ = $22,935.90

25) Rate of discount = $\frac{\text{Discount}}{\text{Marked Price}} = \frac{\$762}{\$2165.99} = 0.3518 = 35.2\%$ (to the nearest tenth of 1%)

CHAPTER 7

Measurement

In this chapter, we introduce the concept of measurement. Both the English system and the metric system are discussed. Conversions within and between the systems are examined.

7.1 UNITS OF LENGTH

In this section, we introduce units of length using both the English and the metric systems. Conversion of units between and within the systems are discussed.

UNITS

Measurements are stated in generally agreed upon terms called **units.** The following are examples of measurements together with their units.

- 58 feet; the measure is 58 and the unit is feet.
- 83 kilometers; the measure is 83 and the unit is kilometers.
- 16 hours; the measure is 16 and the unit is hours.
- 5.6 meters; the measure is 5.6 and the unit is meters.
- 57.8 centimeters; the measure is 57.8 and the unit is centimeters.
- 62.5 grams; the measure is 62.5 and the unit is grams.
- 409 miles; the measure is 409 and the unit is miles.
- 36 pounds; the measure is 36 and the unit is pounds.
- 6.4 liters; the measure is 6.4 and the unit is liters.

The two major systems of units used throughout the world today are the **English system** and the **metric system.** The metric system, which involves only powers of ten, is more convenient to use.

LENGTH

Listed in Table 7.1 are some of the standard units of measure for length. These are given in both the English and the metric systems. Conversions between them are also given.

Table 7.1

Measure	English System	Metric System	Conversion
Length	1 ft = 12 in.	1 cm = 10 mm	1 in. = 2.54 cm
	1 yd = 3 ft	1 dm = 10 cm	1 m = 39.4 in.
	1 mi = 5,280 ft	1 m = 10 dm	1 mi = 1.61 km
		1 km = 1,000 m	1 km = 0.6 mi

Legend: ft (feet); in. (inches); yd (yard); mi (mile); cm (centimeter); mm (millimeter); m (meter); dm (decimeter); km (kilometer)

CONVERSIONS WITHIN THE ENGLISH SYSTEM

Using the entries in Table 7.1, we illustrate conversions within the English System of measurement.

Exercise 7.1

Convert each of the following as indicated:
a) 39 feet (ft) to yards (yd)
b) 10.5 yards (yd) to inches (in.)
c) 69 inches (in.) to feet (ft)
d) 0.6 miles (mi) to feet (ft)

Solution 7.1

a) From Table 7.1, we have 3 ft = 1 yd. Hence,

$$1 \text{ ft} = \frac{1}{3} \text{ yd}$$

$$39 \text{ ft} = 39\left(\frac{1}{3}\right) \text{ yd} = 13 \text{ yd}$$

Therefore, 39 feet = 13 yards.

b) From Table 7.1, we have 1 yd = 36 in. Hence,
10.5 yd = 10.5(36 in.) = 378 in.
Therefore, 10.5 yards = 378 inches.

c) From Table 7.1, we have 12 in. = 1 ft. Hence,

$$1 \text{ in.} = \left(\frac{1}{12}\right) \text{ ft}$$

$$69 \text{ inches} = 69\left(\frac{1}{12}\right) \text{ft.} = 5\frac{3}{4} \text{ ft}$$

Therefore, 69 inches = $5\frac{3}{4}$ ft.

d) From Table 7.1, we have 1 mi = 5,280 ft. Then
0.6 mi = 0.6(5,280) ft = 3,168 ft
Therefore, 0.6 miles = 3,168 feet.

In the previous exercise, each of the conversions involved only two units of measure. When more than two units are involved, we use a different procedure to help us keep track of all the different units of measurement.

Exercise 7.2
Convert 1.6 miles (mi) to inches (in.)

Solution 7.2
From Table 7.1, we have 1.mi = 5,280 ft and 1 ft = 12 in. So, changing miles to feet and changing feet to inches:

$$1.6 \text{ mi} = \frac{1.6 \text{ mi}}{1} \times \frac{5,280}{1 \text{ mi}} \times \frac{12 \text{ in}}{1 \text{ ft}}$$

$$= \frac{1.6 \times 5,280 \times 12 \text{ in}}{1 \times 1 \times 1}$$

$$= \frac{101,376 \text{ in}}{1}$$

$$= 101,376 \text{ in.}$$

Therefore, 1.6 miles = 101,376 inches.

Exercise 7.3
Convert 440 yards (yd) to miles (mi).

Solution 7.3
From Table 7.1, we have 1 mi = 5,280 ft and 1 yd = 3 ft. So, changing yards to feet and changing feet to miles:

$$440 \text{ yd} = \frac{\overset{1}{\cancel{440}} \text{ yd}}{1} \times \frac{3 \text{ ft}}{1 \text{ yd}} \times \frac{1 \text{ mi}}{\underset{12}{\cancel{5,280}} \text{ ft}}$$

$$= \frac{1 \times \overset{1}{\cancel{3}} \times 1 \text{ mi}}{1 \times 1 \times \underset{4}{\cancel{12}}}$$

$$= \frac{1 \times 1 \times 1 \text{ mi}}{1 \times 1 \times 4}$$

$$= \frac{1 \text{ mi}}{4}$$

$$= \frac{1}{4} \text{ mi}$$

Therefore, 440 yards = $\frac{1}{4}$ mile.

USING POWERS OF TEN IN THE METRIC SYSTEM MEASUREMENT

In the metric system, we convert between units by using powers of ten. For each measure, there is a basic unit. Using prefixes as indicated in Table 7.2, we convert all the other units in terms of the basic unit.

Table 7.2

Prefix	Symbol	Value
milli	m	$\frac{1}{1000}$ times basic unit
centi	c	$\frac{1}{100}$ times basic unit
deci	d	$\frac{1}{10}$ times basic unit
Basic unit		
deka	dk	10 times basic unit
hecto	h	100 times basic unit
kilo	k	1,000 times basic unit

Note that in Table 7.2 we have the symbol m, which means milli, and in Table 7.1 we also have the same symbol, which means meter. If the symbol *m* is used alone, it always represents meters. However, if a symbol has two letters and the first is *m*, that *m* represents the prefix milli.

If the basic unit is meters (m), then:

- 5 m means 5 meters
- 125 mm means 125 millimeters
- 2 km means 2 kilometers
- 11 dm means 11 decimeters
- 24 mg means 24 milligrams

CONVERSIONS WITHIN THE METRIC SYSTEM

Exercise 7.4
Convert each of the following as indicated:
a) 12 meters (m) to dekameters (dkm)
b) 360 decimeters (dm) to meters (m)

Solution 7.4
a) 10 m = 1 dkm

$$1 m = \frac{1}{10} \text{ dkm}$$

$$12 \text{ m} = 12 \left(\frac{1}{10}\right) \text{ dkm} = 1.2 \text{ dkm}$$

Therefore, 12 meters = 1.2 dekameters.

b) From Table 7.1, we have 10 dm = 1 m. Hence, 1 dm = $\frac{1}{10}$ m. So,

$$360 \text{ dm} = 360 \left(\frac{1}{10}\right) \text{ m} = 36 \text{ m}$$

Therefore, 360 decimeters = 36 meters.

Exercise 7.5
Convert 23 kilometers (km) to decimeters (dm).

Solution 7.5
From Table 7.1, we have 1 km = 1,000 m and 1 m = 10 dm. So, changing km to m and changing m to dm:

$$23 \text{ km} = \frac{23 \text{ km}}{1} \times \frac{1,000 \text{ m}}{1 \text{ km}} \times \frac{10 \text{ dm}}{1 \text{ m}}$$

$$= \frac{230,000 \text{ dm}}{1} = 230,000 \text{ dm}$$

Therefore, 23 kilometers = 230,000 decimeters.

CONVERSIONS BETWEEN THE ENGLISH AND METRIC SYSTEMS

Exercise 7.6
Convert 457 miles (mi) to hectometers (hm).

Solution 7.6
From Table 7.1, we have 1 mi = 1.61 km and 1 km = 1,000 m. From Table 7.2, we have 1 hm = 100 m. So, changing mi to km (1 mi = 1.61 km), changing km to meters (1 km = 1,000 m), and changing meters to hectometers (1 hm = 100 m):

$$457 \text{ miles} = \frac{457 \text{ mi}}{1} \times \frac{1.6 \text{ km}}{1 \text{ mi}} \times \frac{1,000 \text{ m}}{1 \text{ km}} \times \frac{1 \text{ hm}}{100 \text{ m}}$$

$$= \frac{457 \times 1.61 \times 1,000 \times 1 \text{ hm}}{1 \times 1 \times 1 \times 100}$$

$$= \frac{735,770 \text{ hm}}{100}$$

$$= 7,357.7 \text{ hm}$$

Therefore, 457 miles = 7,357.7 hectometers.

7.2 UNITS OF WEIGHT

In this section, we introduce units of weight using both the English and the metric systems. Conversion of units between and within the systems are discussed.

WEIGHT

In Table 7.3 are listed some of the standard units of measure for weight.

These are given in both the English and the metric systems. Conversions between them are also given.

Table 7.3

Measure	English System	Metric System	Conversion
Weight	1 lb = 16 oz	1 g = 1,000 mg	1 lb = 454 g
	1 T = 2,000 lb	1 kg = 1,000 g	1 kg = 2.2 lb

Legend: lb (pound); oz (ounce); T (ton); g (gram); mg (milligram); kg (kilogram)

Using the entries in Table 7.3, we illustrate conversions within and between the systems of measurement.

Exercise 7.7
Convert each of the following as indicated:
a) 27 tons (T) to pounds (lb)
b) 3,750 grams (g) to kilograms (kg)
c) 3.5 pounds (lb) to ounces (oz)
d) 72 ounces (oz) to pounds (lb)
e) 3 pounds (lb) to grams (g)

Solution 7.7
a) From Table 7.3, we have 1 T = 2,000 lb. Hence, 27 T = 27(2,000) lb = 54,000 lb
Therefore, 27 tons = 54,000 pounds.

b) Because 1 kg = 1,000 g, we have
$$1 \text{ g} = \frac{1}{1,000} \text{ kg}$$
Hence, 3,750 grams = 3,750 $(\frac{1}{1,000}$ kg$)$ = 3.75 kilograms.
Therefore, 3,750 grams = 3.75 kilograms.

c) From Table 7.3, we have 1 lb = 16 oz. Hence, 3.5 lb = 3.5(16) oz = 56 oz.
Therefore, 3.5 pounds = 56 ounces.

d) 16 oz = 1 lb
$$1 \text{ oz} = \left(\frac{1}{16}\right) \text{lb}$$
$$72 \text{ oz} = 72\left(\frac{1}{16}\right) \text{lb}$$
$$= \left(\frac{72}{16}\right) \text{lb}$$
$$= \frac{9}{2} \text{ lb}$$
$$= 4\frac{1}{2} \text{ lb}$$
Therefore, 72 ounces = $4\frac{1}{2}$ pounds.

e) From Table 7.3, we have 1 lb = 454 g
Hence, 3 lb = 3(454) g = 1,362 g. Therefore, 3 pounds = 1,362 grams.

Exercise 7.8
Convert 0.38 ton (T) to kilograms (kg).

Solution 7.8
From Table 7.3, we have 1 T = 2,000 lb and 1 kg = 2.20 lb. So, changing tons to lb and changing pounds to kilograms:

$$0.38 \text{ ton} = \frac{0.38}{1} \times \frac{2,000 \text{ lb}}{1 \text{ ton}} \times \frac{1 \text{ kg}}{2.20 \text{ lb}}$$

$$= \frac{(0.38)(2,000) \text{ kg}}{2.20}$$

$$= 345.5 \text{ kg (rounded to the nearest tenth)}$$

Therefore, 0.38 tons equal 345.5 kilograms, rounded to the nearest tenth.

Exercise 7.9
Convert 2 hectograms (hg) to centigrams (cg).

Solution 7.9
We have that 1 hg = 100 g and 1 g = 100 cg. So, changing hg to g and changing g to cg:

$$2 \text{ hg} = \frac{2 \text{ hg}}{1} \cdot \frac{100 \text{ g}}{1 \text{ hg}} \cdot \frac{100 \text{ cg}}{1 \text{ g}}$$

$$= \frac{2 \times 100 \times 100 \text{ cg}}{1 \times 1 \times 1}$$

$$= \frac{20,000 \text{ cg}}{1}$$

$$= 20,000 \text{ cg}$$

Therefore, 2 hectograms = 20,000 centigrams.

Exercise 7.10
Convert 13 pounds (lb) to kilograms (kg).

Solution 7.10
From Table 7.3, we have 1 lb = 454 g and 1 kg = 1,000 g. So, changing lb to g and changing g to kg

$$13 \text{ lb} = \frac{13 \text{ lb}}{1} \times \frac{\overset{227}{\cancel{454}} \text{ g}}{1 \text{ lb}} \times \frac{1 \text{ kg}}{\underset{500}{\cancel{1,000}} \text{ g}}$$

$$= \frac{13 \times 227 \times 1 \text{ kg}}{1 \times 1 \times 500}$$

$$= \frac{2951 \text{kg}}{500}$$

$$= 5.902 \text{ kg}$$

Therefore, 13 pounds = 5.9 kilograms (to the nearest tenth).

Exercise 7.11
Convert 13 pounds (lb) to kilograms (kg).

Solution 7.11
From Table 7.3, we have 2.2 lb = 1 kg. Hence,

$$1 \text{ lb} = \left(\frac{1}{2.2}\right) \text{kg}$$

$$\text{Then } 13 \text{ lb} = 13\left(\frac{1}{2.2}\right) \text{kg}$$

$$= \frac{13}{2.2} \text{ kg}$$

$$= 5.9 \text{ kg (to the nearest tenth)}$$

Therefore, we have that 13 pounds = 5.9 kilograms (to the nearest tenth).

7.3 UNITS OF LIQUID CAPACITY

In this section, we introduce units of liquid capacity using both the English and metric systems. Conversion of units between and within the systems are discussed. In Table 7.4 are listed some of the standard units of measure for liquid capacity. These are given in both the English and the metric systems. Conversions between them are also given.

Table 7.4

Measure	English System	Metric System	Conversion
Liquid	1 qt = 2 pt	1 L = 1,000 ml	1 L = 1.06 qt
Capacity	1 gal = 4 qt	1 kl = 1,000 L	1 qt = 946 ml
	1 qt = 32 oz		

Legend: qt (quart); pt (pint); gal (gallon); oz (ounce); L (liter); ml (milliliter); kl (kiloliter)

Using the entries in Table 7.4, we illustrate conversions within and between the systems of measurement.

Exercise 7.12
Convert 19 gallons (gal) to pints (pt).

Solution 7.12
From Table 7.4, we have 1 gallon = 4 quarts and 1 quart = 2 pints. Hence, changing gal to qt and changing qt to pt:

$$19\,\text{gal} = \frac{19\,\text{gal}}{1} \times \frac{4\,\text{qt}}{1\,\text{gal}} \times \frac{2\,\text{pt}}{1\,\text{qt}}$$
$$= \frac{19 \times 4 \times 2\,\text{pt}}{1 \times 1 \times 1}$$
$$= \frac{152\,\text{pt}}{1}$$
$$= 152\,\text{pt}$$

Therefore, 19 gallons = 152 pints.

Exercise 7.13
Convert 36 kiloliters (kl) to milliliters (ml).

Solution 7.13
Because 1 kl = 1,000 L, we have

$$36\,\text{kl} = \frac{36\,\text{kL}}{1} \times \frac{1000\,\text{L}}{1\,\text{kl}}$$

Also, 1 L = 1,000 ml. Hence, changing kl to L (1 kl=1,000 L) and changing liters to (1 L=1,000 ml):

$$36\,\text{kl} = \frac{36\,\text{kl}}{1} \times \frac{1,000\,\text{L}}{1\,\text{kl}} \times \frac{1,000\,\text{ml}}{\text{L}}$$
$$= \frac{36\,\text{kl}}{1} \times \frac{1,000\,\text{L}}{1\,\text{kl}} \times \frac{1,000\,\text{ml}}{1\,\text{L}}$$
$$= \frac{36,00,000\,\text{ml}}{1} = 36,000,000\,\text{ml}$$

Therefore, 36 kiloliters = 36,000,000 milliliters.

Exercise 7.14
Convert 37 pints (pt) to liters (L).

Solution 7.14

From Table 7.4, we have 1 qt = 2 pt and 1 L = 1.06 qt. So, changing pt to qt and changing qt to liters:

$$37\,\text{pt} = \frac{37\,\text{pt}}{1} \times \frac{1\,\text{qt}}{2\,\text{pt}} \times \frac{1\,\text{L}}{1.06\,\text{qt}}$$

$$= \frac{37\,\text{L}}{2 \times 1.06}$$

$$= \frac{37\,\text{L}}{2.12}$$

$$= 17.5\,\text{L (rounded to the nearest tenth)}$$

Therefore, 37 pints = 17.5 liters, rounded to the nearest tenth.

Exercise 7.15

Convert 21.5 pints (pt) to dekaliters (dkl).

Solution 7.15

From Table 7.4, we have 1 qt = 946 ml. Therefore, 1 pt = 473 ml.

Also, 1,000 ml = 1 L and 1 dkl = 10 L. So, changing pt to ml, changing ml to liters and changing liters to dekaliters:

$$21.5\,\text{pt} = \frac{21.5\,\text{pt}}{1} \times \frac{473\,\text{ml}}{1\,\text{pt}} \times \frac{1\,\text{l}}{1{,}000\,\text{ml}} \times \frac{1\,\text{dkl}}{10\,\text{l}}$$

$$= \frac{21.5 \times 473 \times 1 \times 1\,\text{dkl}}{1 \times 1 \times 1{,}000 \times 10}$$

$$= \frac{10{,}169.5\,\text{dkl}}{10{,}000}$$

$$= 1.01695\,\text{dkl}$$

Therefore, 21.5 pints is approximately equal to 1.02 dekaliters (to the nearest hundredth dekaliter).

Exercise 7.16

Convert 24 ounces (oz) to pints (pt).

Solution 7.16

From Table 7.4, we have 32 oz = 1 qt and 1 qt = 2 pt. So, changing oz to qt and changing qt to pt:

$$24\,oz = \frac{24\,oz}{1} \times \frac{1\,qt}{32\,oz} \times \frac{2\,pt}{1\,qt}$$

$$= \frac{\overset{3}{\cancel{24}} \times 1 \times 2\,pt}{1 \times \underset{4}{\cancel{32}} \times 1}$$

$$= \frac{3 \times 1 \times 2\,pt}{1 \times 4 \times 1}$$

$$= \frac{3\,pt}{2}$$

$$= 1\frac{1}{2}\,pt$$

Therefore, 24 ounces = 1.5 pints.

Exercise 7.17

Convert 38 gallons to pints.

Solution 7.17

From Table 7.4, we have 1 gal = 4 qt and 1 qt = 2 pt. So, changing gal to qt and changing qt to pt:

$$38\,gal = \frac{38\,gal}{1} \times \frac{4\,qt}{1\,gal} \times \frac{2\,pt}{1\,qt}$$

$$= \frac{38 \times 4 \times 2\,pt}{1 \times 1 \times 1}$$

$$= \frac{304\,pt}{1}$$

$$= 304\,pt$$

Therefore, 38 gallons = 304 pints.

Exercise 7.18

Convert 64 ounces to gallons.

Solution 7.18

From Table 7.4, we have 32 oz = 1 qt and 4 qt = 1 gal. So, changing oz to qt and changing qt to gal:

$$64\,oz = \frac{64\,oz}{1} \times \frac{1\,qt}{32\,oz} \times \frac{1\,gal}{4\,qt}$$

$$= \frac{\overset{2}{\cancel{64}} \times 1 \times 1\,gal}{1 \times \underset{1}{\cancel{32}} \times 4}$$

$$= \frac{\overset{1}{\cancel{2}} \times 1 \times 1\,gal}{1 \times 1 \times \underset{2}{\cancel{4}}}$$

$$= \frac{1 \times 1 \times 1\,gal}{1 \times 1 \times 2}$$

$$= \frac{1}{2}\,gal$$

Therefore, 64 ounces = $\frac{1}{2}$ gallon

Exercise 7.19

Convert 212 quarts to liters.

Solution 7.19

From Table 7.4, we note that 1 qt = 946 ml. We also know that 1 L = 1,000 ml. So, changing qt to ml and changing ml to L:

$$212\,qt = \frac{212\,qt}{1} \times \frac{946\,qt}{1\,qt} \times \frac{1\,L}{1,000\,mL}$$

$$= \frac{\overset{53}{\cancel{212}} \times 946 \times 1\,L}{1 \times 1 \times \underset{250}{\cancel{1,000}}}$$

$$= \frac{53 \times \overset{473}{\cancel{946}} \times 1\,L}{1 \times 1 \times \underset{125}{\cancel{250}}}$$

$$= \frac{53 \times 473 \times 1\,L}{1 \times 1 \times 125}$$

$$= \frac{25,069\,L}{125}$$

$$= 200.6\,L \text{ (to the nearest tenth)}$$

Therefore 212 quarts = 200.6 liters (to the nearest tenth).

7.4 UNITS OF TIME

In this section, we introduce units of time using both the English and the metric systems. Note, however, that the units are the same in both systems.

In Table 7.5 are listed some of the standard units of measure for time.

Table 7.5

Measure	English System	Metric System	Conversion
Time	1 min = 60 sec	1 min = 60 sec	Units are the same in both systems
	1 hr = 60 min	1 hr = 60 min	
	1 day = 24 min	1 day = 24 hr	

Legend: min (minute); sec (second); hr (hour)

Using the entries in Table 7.5, we illustrate conversions.

Exercise 7.20
Convert seventeen days to hours.

Solution 7.20
From Table 7.5, we have 1 day = 24 hr. Hence,
17 days = 17(24) hr = 408 hr
Therefore, 17 days = 408 hours.

Exercise 7.21
Convert twenty-seven hours to seconds.

Solution 7.21
From Table 7.5, we have 1 hr = 60 min and 1 min = 60 sec. Hence, changing hr to min and changing min to sec:

$$27 \text{ hr} = \frac{27 \text{ hr}}{1} \times \frac{60 \text{ min}}{1 \text{ hr}} \times \frac{60 \text{ sec}}{1 \text{ min}}$$

$$= \frac{27 \times 60 \times 60 \text{ sec}}{1 \times 1 \times 1}$$

$$= \frac{97,200 \text{ sec}}{1}$$

$$= 97,200 \text{ sec}$$

Therefore, 27 hours = 97,200 seconds.

Exercise 7.22
Convert sixty hours to days.

Solution 7.22

From Table 7.5, we have 24 hr = 1 day. Hence, changing hr to days:

$$60\,\text{hr} = \frac{60\,\text{hr}}{1} \times \frac{1\,\text{day}}{24\,\text{hr}}$$

$$= \frac{60 \times 1\,\text{day}}{1 \times 24}$$

$$= \frac{5 \times 1\,\text{day}}{1 \times 2}$$

$$= \frac{5}{2}$$

Therefore, 60 hours = 2.5 days.

Exercise 7.23

How many seconds are there in the month of July?

Solution 7.23

Because there are thirty-one days in the month of July and each day has twenty-four hours, we have 1 day = 24 hours and 31 days = 31(24 hours) = 744 hours. Because 1 hr = 60 min, we have: 744 hr = 744(60 min) = 44,640 min. Because 1 min = 60 sec, 44,640 min = 44,640(60 sec) = 2,678,400 sec.

Therefore, the month of July (31 days) has 2,678,400 seconds.

Exercise 7.24

Convert 12,000 seconds to hours.

Solution 7.24

From Table 7.5, we have 60 sec = 1 min and 60 min = 1 hr. Hence, changing sec to min and changing min to hr:

$$12,000\,\text{sec} = \frac{12,000\,\text{sec}}{1} \times \frac{1\,\text{min}}{60\,\text{sec}} \times \frac{1\,\text{hr}}{60\,\text{min}}$$

$$= \frac{\cancel{12,000}^{200} \times 1 \times 1\,\text{hr}}{1 \times \cancel{60}_{1} \times 60}$$

$$= \frac{\cancel{200}^{10} \times 1 \times 1\,\text{hr}}{1 \times 1 \times \cancel{60}_{3}}$$

$$= \frac{10 \times 1 \times 1\,\text{hr}}{1 \times 1 \times 3}$$

$$= \frac{10\,\text{hr}}{3}$$

Therefore, 12,000 seconds = $3\frac{1}{3}$ hours.

Exercise 7.25
How many seconds are there in five hours, twenty-seven minutes, and seventeen seconds?

Solution 7.25
There are three parts to this exercise.

Step 1: Determine the number of seconds in five hours as follows:

$$5\,\text{hr} = \frac{5\,\text{hr}}{1} \times \frac{60\,\text{min}}{1\,\text{hr}} \times \frac{60\,\text{sec}}{1\,\text{min}}$$

$$= \frac{5 \times 60 \times 60\,\text{sec}}{1 \times 1 \times 1}$$

$$= \frac{18,000\,\text{sec}}{1}$$

$$= 18,000\,\text{sec}$$

Therefore, there are 18,000 seconds in five hours.

Step 2: Determine the number of seconds in twenty-seven minutes as follows:

$$27\,\text{min} = \frac{27\,\text{min}}{1} \times \frac{60\,\text{sec}}{1\,\text{min}}$$

$$= \frac{27 \times 60\,\text{sec}}{1 \times 1}$$

$$= \frac{1,620\,\text{sec}}{1}$$

$$= 1,620\,\text{sec}$$

Therefore, there are 1,620 seconds in twenty-seven minutes.

Step 3: Add the 18,000 sec in 5 hr, the 1,620 sec in 27 min, and the 17 sec:

18,000 sec + 1,620 sec + 17 sec
= (18,000 + 1,620 + 17) sec
= 19,637 sec

Therefore, there are 19,637 seconds in five hours, twenty-seven minutes, seventeen seconds.

Exercise 7.26
MATH 100 meets three days a week for seventy-five minutes per class period. The term is fifteen weeks long, with no breaks, exclusive of the final examination period. How many hours per term are spent in class for this course?

Solution 7.26

The class meets three times per week for fifteen weeks. Hence, $3 \times 15 = 45$ is the number of class periods for the course. Because each class period meets for seventy-five minutes, there are $45 \times 75 = 3{,}375$ minutes per term for the course. So, because there are sixty minutes in an hour, we have $3{,}375 \div 60 = 56.25$ hours of class time for the course.

COMBINING DIFFERENT MEASURES

In the following exercises, we combine different measures such as time and length.

Exercise 7.27

A car travels 55 miles per hour (mph). What is its rate of speed in feet per second?

Solution 7.27

$1 \text{ mi} = 5{,}280 \text{ ft}$, $1 \text{ hr} = 60 \text{ min}$, and $1 \text{ min} = 60 \text{ sec}$. So, changing mi to ft, changing hours to minutes, and changing minutes to seconds:

$$55 \text{ mph} = \frac{\overset{11}{\cancel{55}} \text{ mi}}{1 \text{ hr}} \times \frac{\overset{88}{\cancel{5{,}280}} \text{ ft}}{1 \text{ mi}} \times \frac{1 \text{ hr}}{\underset{12}{\cancel{60}} \text{ min}} \times \frac{1 \text{ min}}{\underset{1}{\cancel{60}} \text{ sec}}$$

$$= \frac{11 \times \overset{22}{\cancel{88}} \times 1 \times 1 \text{ ft}}{1 \times 1 \times \underset{3}{\cancel{12}} \times 1 \text{ sec}}$$

$$= \frac{11 \times 22 \times 1 \times 1 \text{ ft}}{1 \times 1 \times 3 \times 1 \text{ sec}}$$

$$= \frac{242 \text{ ft}}{3 \text{ sec}}$$

$$= 80\frac{2}{3} \text{ feet per second}$$

Therefore, 55 miles per hour $= 80\frac{2}{3}$ feet per second.

Exercise 7.28

If the speed limit is posted as 70 km per hour, what is the speed limit in miles per hour?

Solution 7.28

We have $1 \text{ km} = 0.6 \text{ mi}$. Hence, $70 \text{ km} = 70(0.6 \text{ mi}) = 42 \text{ mi}$.
Therefore, 70 km per hour is equal to 42 mph.

In this chapter, we discussed units of measure. Included were units of length, weight, liquid capacity, and time. Both the English and metric systems of measurement were discussed. Conversions within each system and between the two systems were also examined.

TEST YOURSELF
Convert each of the following measurements as indicated.
1) 34 ft to inches
2) 42 cm to decimeters
3) 82 mi to yards
4) 36 dkm to meters
5) 22 L to quarts
6) 14 qt to liters
7) 282 pt to liters
8) 3.5 gallons to pints
9) 80 lb to grams
10) 100 oz to pounds
11) 0.8 kg to ounces
12) 84 g to pounds
13) 9.5 hours to minutes
14) 3,760 seconds to hours
15) 86 hours to seconds
16) 17 days to minutes
17) 23 weeks to minutes
18) 91,000 sec to hours
19) 16 km/hr to ft/min
20) Convert 55 mph to kilometers per minute.
21) Kim Yung is 5 ft, 7 in. tall. How tall is Kim in centimeters?
22) Peaches are selling for 79 cents per pound. Approximately, how many cents per kilogram would that be?
23) Pamela commutes 19.8 miles each way to work. Approximately, how many kilometers is this?
24) A bag of groceries weighs 11.6 pounds. About how many kilograms is that?
25) A container holds 12 liters. How many 96 ml glasses can be filled from the contents of the container?

TEST YOURSELF WITH THE HANDHELD CALCULATOR
Convert each of the following measurements as indicated.
1) 160.3 pounds to kilograms
2) 3,468.9 pounds to kilograms
3) 124.3 kilograms to pounds
4) 1,069.9 kilograms to pounds

5) 468.4 kilograms to ounces
6) 952.5 kilograms to ounces
7) 23.96 miles to kilometers
8) 138.75 miles to kilometers
9) 160.4 kilometers to miles
10) 2,146.6 kilometers to miles
11) 986.4 kilometers to feet
12) 1,027.1 kilometers to feet
13) 762 kilometers to inches
14) 862.3 kilometers to inches
15) 17.8 quarts to liters
16) 238.3 gallons to liters
17) 29.8 liters to pints
18) 163.86 liters to quarts
19) 203.4 liters to gallons
20) 86 yards to meters
21) 132 yards to centimeters
22) 2,134 centimeters to yards
23) Jose can run the 100-yard dash in 9.8 seconds. Maria can run the 100-meter dash in 10.7 seconds. Who runs faster, Jose or Maria?
24) Manuel used 200 liters of gasoline to drive 1,400 kilometers. To the nearest mile, how many miles per gallon did the car average?
25) A truck has a gross weight of 7,400 pounds. What is the gross weight in kilograms?

TEST YOURSELF ANSWERS

1) $34 \text{ ft} = \dfrac{34 \text{ ft}}{1} \times \dfrac{12 \text{ in}}{1 \text{ ft}} = 408 \text{ in.}$

2) $42 \text{ cm} = \dfrac{42 \text{ cm}}{1} \times \dfrac{1 \text{ dm}}{10 \text{ cm}} = \dfrac{21}{1} \times \dfrac{1 \text{ dm}}{5} = \dfrac{21 \text{ dm}}{5} = 4.2 \text{ dm}$

3) $82 \text{ mi} = \dfrac{82 \text{ mi}}{1} \times \dfrac{5280 \text{ ft}}{1 \text{ mi}} \times \dfrac{1 \text{ yd}}{3 \text{ ft}} = \dfrac{82}{1} \times \dfrac{1760}{1} \times \dfrac{1 \text{ yd}}{1} = \dfrac{144{,}320 \text{ yd}}{1} = 144{,}320 \text{ yd}$

4) $36 \text{ dkm} = \dfrac{36 \text{ dkm}}{1} \times \dfrac{10 \text{ m}}{1 \text{ dkm}} = \dfrac{36}{1} \times \dfrac{10 \text{ m}}{1} = 360 \text{ m}$

5) $22 \text{ L} = \dfrac{22 \text{ L}}{1} \times \dfrac{1.06 \text{ qt}}{1 \text{ L}} = \dfrac{22}{1} \times \dfrac{1.06 \text{ qt}}{1} = 23.32 \text{ qt}$

6) $14 \text{ qt} = \dfrac{14 \text{ qt}}{1} \times \dfrac{1 \text{ L}}{1.06 \text{ qt}} = \dfrac{7}{1} \times \dfrac{1 \text{ L}}{0.53} \approx 13.2 \text{ L}$

7) $282 \text{ pt} = \dfrac{282 \text{ pt}}{1} \times \dfrac{1 \text{ qt}}{2 \text{ pt}} \times \dfrac{1 \text{ L}}{1.06 \text{ qt}} = \dfrac{141}{1} \times \dfrac{1}{1} \times \dfrac{1 \text{ L}}{1.06} = \dfrac{141 \text{ L}}{1.06} \approx 133.02 \text{ L}$

8) $3.5 \text{ gal} = \dfrac{3.5 \text{ gal}}{1} \times \dfrac{4 \text{ qt}}{1 \text{ gal}} \times \dfrac{2 \text{ pt}}{1 \text{ qt}} = \dfrac{3.5}{1} \times \dfrac{4}{1} \times \dfrac{2 \text{ pt}}{1} = \dfrac{28 \text{ pt}}{1} = 28 \text{ pt}$

9) $80 \text{ lb} = \dfrac{80 \text{ lb}}{1} \times \dfrac{454 \text{ g}}{1 \text{ lb}} = \dfrac{80}{1} \times \dfrac{454 \text{ g}}{1} = \dfrac{36{,}320 \text{ g}}{1} = 36{,}320 \text{ g}$

10) $100 \text{ oz} = \dfrac{100 \text{ oz}}{1} \times \dfrac{1 \text{ lb}}{16 \text{ oz}} = \dfrac{25}{1} \times \dfrac{1 \text{ lb}}{4} = \dfrac{25 \text{ lb}}{4} = 6.25 \text{ lb}$

11) $0.8 \text{ kg} = \dfrac{0.8 \text{ kg}}{1} \times \dfrac{2.2 \text{ lb}}{1 \text{ kg}} \times \dfrac{16 \text{ oz}}{1 \text{ lb}} = \dfrac{0.8}{1} \times \dfrac{2.2}{1} \times \dfrac{16 \text{ oz}}{1} = \dfrac{28.16 \text{ oz}}{1} = 28.16 \text{ oz}$

12) $84 \text{ g} = \dfrac{84 \text{ g}}{1} \times \dfrac{1 \text{ lb}}{454 \text{ g}} = \dfrac{42}{1} \times \dfrac{1 \text{ lb}}{227} = \dfrac{42 \text{ lb}}{227} \approx 0.185 \text{ lb}$

13) $9.5 \text{ hr} = \dfrac{9.5 \text{ hr}}{1} \times \dfrac{60 \text{ min}}{1 \text{ hr}} = \dfrac{9.5}{1} \times \dfrac{60 \text{ min}}{1} = \dfrac{570 \text{ min}}{1} = 570 \text{ min}$

14) $3{,}760 \text{ sec} = \dfrac{3760 \text{ sec}}{1} \times \dfrac{1 \text{ min}}{60 \text{ sec}} \times \dfrac{1 \text{ hr}}{60 \text{ min}} = \dfrac{3760}{1} \times \dfrac{1}{60} \times \dfrac{1 \text{ hr}}{60} = \dfrac{3760 \text{ hr}}{3600} \approx 1.04 \text{ hr}$

15) $86 \text{ hr} = \dfrac{86 \text{ hr}}{1} \times \dfrac{60 \text{ min}}{1 \text{ hr}} \times \dfrac{60 \text{ sec}}{1 \text{ min}} = \dfrac{86}{1} \times \dfrac{60}{1} \times \dfrac{60 \text{ sec}}{1} = \dfrac{309{,}600 \text{ sec}}{1} = 309{,}600 \text{ sec}$

16) $17 \text{ days} = \dfrac{17 \text{ days}}{1} \times \dfrac{24 \text{ hr}}{1 \text{ day}} \times \dfrac{60 \text{ min}}{1 \text{ hr}} = \dfrac{17}{1} \times \dfrac{24}{1} \times \dfrac{60 \text{ min}}{1} = \dfrac{24{,}480 \text{ min}}{1} = 24{,}480 \text{ min}$

17) $23 \text{ wk} = \dfrac{23 \text{ wk}}{1} \times \dfrac{7 \text{ days}}{1 \text{ wk}} \times \dfrac{24 \text{ hr}}{1 \text{ day}} \times \dfrac{60 \text{ min}}{1 \text{ hr}} = \dfrac{23}{1} \times \dfrac{7}{1} \times \dfrac{24}{1} \times \dfrac{60 \text{ min}}{1} = \dfrac{231{,}840 \text{ min}}{1} = 231{,}840 \text{ min}$

18) $91{,}000 \text{ sec} = \dfrac{91{,}000 \text{ sec}}{1} \times \dfrac{1 \text{ min}}{60 \text{ sec}} \times \dfrac{1 \text{ hr}}{60 \text{ min}} = \dfrac{91{,}000}{1} \times \dfrac{1}{60} \times \dfrac{1 \text{ hr}}{60} = \dfrac{91{,}000 \text{ hr}}{3600} \approx 25.28 \text{ hr}$

19) $\dfrac{16 \text{ km}}{1 \text{ hr}} \times \dfrac{1 \text{ hr}}{60 \text{ min}} \times \dfrac{0.6 \text{ mi}}{1 \text{ km}} \times \dfrac{5280 \text{ ft}}{1 \text{ mi}} = \dfrac{16}{1} \times \dfrac{1}{60 \text{ min}} \times \dfrac{0.6}{1} \times \dfrac{5280 \text{ ft}}{1} = \dfrac{50{,}688 \text{ ft}}{60 \text{ min}} \approx 844.8 \text{ ft/min}$

20) $\dfrac{55 \text{ mi}}{1 \text{ hr}} \times \dfrac{1 \text{ hr}}{60 \text{ min}} \times \dfrac{1.61 \text{ km}}{1 \text{ mi}} = \dfrac{55}{1} \times \dfrac{1}{60 \text{ min}} \times \dfrac{1.61 \text{ km}}{1} = \dfrac{88.55 \text{ km}}{60 \text{ min}} \approx 1.476 \text{ km/min}$

21) $5 \text{ ft}, 7 \text{ in.} = 67 \text{ in} = \dfrac{67 \text{ in.}}{1} \times \dfrac{2.54 \text{ cm}}{1 \text{ in.}} = \dfrac{67}{1} \times \dfrac{2.54 \text{ cm}}{1} = \dfrac{170.18 \text{ cm}}{1} = 170.18 \text{ cm}$

22) $\dfrac{79 \text{ cents}}{1 \text{ lb}} \times \dfrac{1 \text{ lb}}{454 \text{ g}} \times \dfrac{1000 \text{ g}}{1 \text{ kg}} = \dfrac{79 \text{ cents}}{1} \times \dfrac{1}{454} \times \dfrac{1000}{1 \text{ kg}} = \dfrac{79{,}000 \text{ cents}}{454 \text{ kg}} \approx 174.01 \text{ cents/kg}$

23) $19.8 \text{ mi} = \dfrac{19.8 \text{ mi}}{1} \times \dfrac{1.61 \text{ km}}{1 \text{ mi}} = \dfrac{19.8}{1} \times \dfrac{1.61 \text{ km}}{1} = \dfrac{31.878 \text{ km}}{1} = 31.878 \text{ km}$

24) $11.6 \text{ lb} = \dfrac{11.6 \text{ lb}}{1} \times \dfrac{454 \text{ g}}{1 \text{ lb}} \times \dfrac{1 \text{ kg}}{1000 \text{ g}} = \dfrac{11.6}{1} \times \dfrac{454}{1} \times \dfrac{1 \text{ kg}}{1000} = \dfrac{5266.4 \text{ kg}}{1000} = 5.2664 \text{ kg}$

25) $12 \text{ L} = \dfrac{12 \text{ L}}{1} \times \dfrac{1{,}000 \text{ ml}}{1 \text{ L}} \times \dfrac{1 \text{ glass}}{96 \text{ ml}} = \dfrac{12}{1} \times \dfrac{1000}{1} \times \dfrac{1 \text{ glass}}{96} = \dfrac{12{,}000 \text{ glasses}}{96} = 125 \text{ glasses}$

TEST YOURSELF ANSWERS WITH THE HANDHELD CALCULATOR

1) $160.3 \text{ lb} = \dfrac{160.3 \text{ lb}}{1} \times \dfrac{454 \text{ g}}{1 \text{ lb}} \times \dfrac{1 \text{ kg}}{1{,}000 \text{ g}} = 72.7762 \text{ kg}$

2) $3{,}468.9 \text{ lb} = \dfrac{3{,}468.9 \text{ lb}}{1} \times \dfrac{454 \text{ g}}{1 \text{ lb}} \times \dfrac{1 \text{ kg}}{1{,}000 \text{ g}} = 1{,}574.8806 \text{ kg}$

3) $124.3 \text{ kg} = \dfrac{124.3 \text{ kg}}{1} \times \dfrac{1{,}000 \text{ g}}{1 \text{ kg}} \times \dfrac{1 \text{ lb}}{454 \text{ g}} \approx 273.7885 \text{ lb}$

4) $1{,}069.9 \text{ kg} = \dfrac{1{,}069.9 \text{ kg}}{1} \times \dfrac{1{,}000 \text{ g}}{1 \text{ kg}} \times \dfrac{1 \text{ lb}}{454 \text{ g}} \approx 2{,}356.6079 \text{ lb}$

5) $468.4 \text{ kg} = \dfrac{468.4 \text{ kg}}{1} \times \dfrac{2.2 \text{ lb}}{1 \text{ kg}} \times \dfrac{16 \text{ oz}}{1 \text{ lb}} = 16{,}487.68 \text{ oz}$

6) $952.5 \text{ kg} = \dfrac{952.5 \text{ kg}}{1} \times \dfrac{2.2 \text{ lb}}{1 \text{ kg}} \times \dfrac{16 \text{ oz}}{1 \text{ lb}} = 33{,}528 \text{ oz}$

7) $23.96 \text{ mi} = \dfrac{23.96 \text{ mi}}{1} \times \dfrac{1.61 \text{ km}}{1 \text{ mi}} = 38.5756 \text{ km}$

8) $138.75 \text{ mi} = \dfrac{138.75 \text{ mi}}{1} \times \dfrac{1.61 \text{ km}}{1 \text{ mi}} = 223.3875 \text{ km}$

9) $160.4 \text{ km} = \dfrac{160.4 \text{ km}}{1} \times \dfrac{0.6 \text{ mi}}{1 \text{ km}} = 96.24 \text{ mi}$

10) $2{,}146.6 \text{ km} = \dfrac{2{,}146.6 \text{ km}}{1} \times \dfrac{0.6 \text{ mi}}{1 \text{ km}} = 1{,}287.96 \text{ mi}$

11) $986.4 \text{ km} = \dfrac{986.4 \text{ km}}{1} \times \dfrac{0.6 \text{ mi}}{1 \text{ km}} \times \dfrac{5{,}280 \text{ ft}}{1 \text{ mi}} = 3{,}124{,}915.2 \text{ ft}$

12) $1{,}027.1 \text{ km} = \dfrac{1{,}027.1 \text{ km}}{1} \times \dfrac{0.6 \text{ mi}}{1 \text{ km}} \times \dfrac{5{,}280 \text{ ft}}{1 \text{ mi}} = 3{,}253{,}852.8 \text{ ft}$

13) $762 \text{ km} = \dfrac{762 \text{ km}}{1} \times \dfrac{0.6 \text{ mi}}{1 \text{ km}} \times \dfrac{5{,}280 \text{ ft}}{1 \text{ mi}} \times \dfrac{12 \text{ in.}}{1 \text{ ft}} = 28{,}968{,}192 \text{ in.}$

14) $862.3 \text{ km} = \dfrac{862.3 \text{ km}}{1} \times \dfrac{0.6 \text{ mi}}{1 \text{ km}} \times \dfrac{5{,}280 \text{ ft}}{1 \text{ mi}} \times \dfrac{12 \text{ in.}}{1 \text{ ft}} = 32{,}781{,}196.8 \text{ in.}$

15) $17.8 \text{ qt} = \dfrac{17.8 \text{ qt}}{1} \times \dfrac{1 \text{ L}}{1.06 \text{ qt}} \approx 16.79 \text{ L}$

16) $238.3 \text{ gal} = \dfrac{238.3 \text{ gal}}{1} \times \dfrac{4 \text{ qt}}{1 \text{ gal}} \times \dfrac{1 \text{ L}}{1.06 \text{ qt}} \approx 899.245 \text{ L}$

17) $29.8 \text{ L} = \dfrac{29.8 \text{ L}}{1} \times \dfrac{1.06 \text{ qt}}{1 \text{ L}} \times \dfrac{2 \text{ pt}}{1 \text{ qt}} = 63.176 \text{ pt}$

18) $163.86 \text{ L} = \dfrac{163.86 \text{ L}}{1} \times \dfrac{1.06 \text{ qt}}{1 \text{ L}} = 173.6916 \text{ qt}$

19) $203.4 \text{ L} = \dfrac{203.4 \text{ L}}{1} \times \dfrac{1.06 \text{ qt}}{1 \text{ L}} \times \dfrac{1 \text{ gal}}{4 \text{ qt}} = 53.901 \text{ gal}$

20) $86 \text{ yd} = \dfrac{86 \text{ yd}}{1} \times \dfrac{36 \text{ in.}}{1 \text{ yd}} \times \dfrac{1 \text{ m}}{39.4 \text{ in.}} \approx 78.58 \text{ m}$

21) $132 \text{ yd} = \dfrac{132 \text{ yd}}{1} \times \dfrac{36 \text{ in.}}{1 \text{ yd}} \times \dfrac{2.54 \text{ cm}}{1 \text{ in.}} = 12{,}070.08 \text{ cm}$

22) $2{,}134 \text{ cm} = \dfrac{2134 \text{ cm}}{1} \times \dfrac{1 \text{ in.}}{2.54 \text{ cm}} \times \dfrac{1 \text{ yd}}{36 \text{ in.}} \approx 23.34 \text{ yd}$

23) Jose: 100 yd in 9.8 sec

Maria: $\dfrac{100 \text{ m}}{10.7 \text{ sec}} = \dfrac{100 \text{ m}}{10.7 \text{ sec}} \times \dfrac{39.4 \text{ in.}}{1 \text{ m}} \times \dfrac{1 \text{ yd}}{36 \text{ in.}} \approx \dfrac{100 \text{ yd}}{10.23 \text{ sec}} = 100 \text{ yd in } 10.23 \text{ sec}$

Therefore, Jose is the faster runner.

24) $\dfrac{1{,}400 \text{ km}}{200 \text{ L}} = \dfrac{1{,}400 \text{ km}}{200 \text{ L}} \times \dfrac{0.6 \text{ mi}}{1 \text{ km}} \times \dfrac{1 \text{ L}}{1.06 \text{ qt}} \times \dfrac{4 \text{ qt}}{1 \text{ gal}} = \dfrac{3{,}360 \text{ mi}}{212 \text{ gal}} \approx 16 \text{ mpg}$

25) $7{,}400 \text{ lb} = \dfrac{7{,}400 \text{ lb}}{1} \times \dfrac{1 \text{ kg}}{2.2 \text{ lb}} \approx 3{,}363.64 \text{ kg}$

CHAPTER 8

Geometric Measure

In this chapter, we continue our discussion of measurement by considering the measurement of geometric figures. This includes perimeter of polygons, circumference of circles, areas of planar figures, and volumes of solids.

8.1 PERIMETER AND CIRCUMFERENCE

In this section, we consider some basic geometric figures and measurements associated with them. A very basic geometric figure is a line segment. Given two distinct (different) points on a line, a **line segment** is the set of all the points on the line between the two points. It includes the two points, which are called its **endpoints.**

POLYGONS

Definition: A **polygon** is a closed geometric figure that has line segments for its **sides.**

According to the preceding definition, a polygon is a **closed figure,** which means that if you start at any point on the figure, you can trace completely around it and return to the starting point. Polygons are classified as follows:

- A **triangle** is a polygon with **three** sides.
- A **quadrilateral** is a polygon with **four** sides
- A **pentagon** is a polygon with **five** sides.
- A **hexagon** is a polygon with **six** sides, and so on.

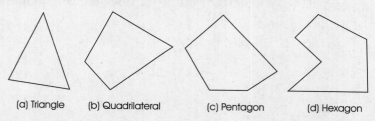

(a) Triangle   (b) Quadrilateral   (c) Pentagon   (d) Hexagon

PERIMETER

Definition: The **perimeter** of a polygon, denoted by the symbol P, is the sum of the lengths of its sides. (All measurements must be in the *same* units.)

Exercise 8.1

Determine the perimeter of a triangle whose sides have lengths 2.5 ft, 4.7 ft, and 5.9 ft.

Solution 8.1

To determine the perimeter P of the triangle, *add* the lengths of its three sides. Hence,
$$P = 2.5 \text{ ft} + 4.7 \text{ ft} + 5.9 \text{ ft}$$
$$= (2.5 + 4.7 + 5.9) \text{ ft}$$
$$= 13.1 \text{ ft}$$
Therefore, the perimeter of the given triangle is 13.1 feet.

Exercise 8.2

A quadrilateral has two sides, each of length 7.2 cm, and two sides, each of length 9.6 cm. Determine its perimeter.

Solution 8.2

A quadrilateral is a four-sided polygon. Because the perimeter is the distance around the quadrilateral, we determine the perimeter by adding the four lengths. Hence,
$$P = 7.2 \text{ cm} + 7.2 \text{ cm} + 9.6 \text{ cm} + 9.6 \text{ cm}$$
$$= (7.2 + 7.2 + 9.6 + 9.6) \text{ cm}$$
$$= 33.6 \text{ cm}$$
Therefore, the perimeter of the given quadrilateral is 33.6 centimeters.

Exercise 8.3

Determine the perimeter of a square whose sides each have length 4.67 yd.

Solution 8.3

A square is a quadrilateral with all four of its sides being of equal length. Hence, instead of adding the four lengths, we can simply multiply the length of its side by 4. Hence,
$$P = 4 \times (4.67 \text{ yd}) = (4 \times 4.67) \text{ yd} = 18.68 \text{ yd}$$
Therefore, the perimeter of the given square is 18.68 yards.

Exercise 8.4

A pentagon has a perimeter of 18.5 cm. If all of its sides are of equal length, what is the length of each side of the pentagon?

Solution 8.4

A pentagon is a polygon having five sides. If all five sides have the same length, we can determine the length of each side by dividing the perimeter by 5. Hence,

$$P = 18.5 \text{ dm} \div 5$$
$$= (18.5 \div 5) \text{ dm}$$
$$= 3.7 \text{ dm}$$

Therefore, the length of each side of the given pentagon is 3.7 dm.

Exercise 8.5

A quadrilateral has sides of lengths 2.1 yd, 4.6 ft, 5.7 ft, and 3.3 yd. Determine its perimeter.

Solution 8.5

A quadrilateral is a polygon having four sides. To determine its perimeter, we add the lengths of the four sides. However, the lengths must be in the *same* units. In this exercise, some lengths are in feet; others are in yards. We must convert one to the other. We will convert yards to feet. Hence,

$$2.1 \text{ yd} = 2.1 \times (3 \text{ ft}) = 6.3 \text{ ft}$$

and

$$3.3 \text{ yd} = 3.3 \times (3 \text{ ft}) = 9.9 \text{ ft}$$

We now add:

$$\mathbf{2.1 \text{ yd}} + 4.6 \text{ ft} + 5.7 \text{ ft} + \mathbf{3.3 \text{ yd}}$$
$$= 6.3 \text{ ft} + 4.6 \text{ ft} + 5.7 \text{ ft} + 9.9 \text{ ft}$$
$$= (6.3 + 4.6 + 5.7 + 9.9) \text{ ft}$$
$$= 26.5 \text{ ft}$$

Therefore, the perimeter of the given quadrilateral is 26.5 feet.

Exercise 8.6

A triangle has sides with measures of 30 in., 2.6 ft, and 1.2 yd. Determine its perimeter.

Solution 8.6

Because the sides of the triangle have measures in different units, we convert all measures to feet. Hence,

$$30 \text{ in} = (30 \div 12) \text{ ft}$$
$$= 2.5 \text{ ft}$$

and

$$1.2 \text{ yd} = (1.2 \times 3) \text{ ft}$$
$$= 3.6 \text{ ft}$$

Then
$$P = 30 \text{ in.} + 2.6 \text{ ft} + 1.2 \text{ yd}$$
$$= 2.5 \text{ ft} + 2.6 \text{ ft} + 3.6 \text{ ft}$$
$$= (2.5 + 2.6 + 3.6) \text{ ft}$$
$$= 8.7 \text{ ft}$$

Therefore, the perimeter of the given triangle is 8.7 feet.

Exercise 8.7

An octagon (an eight-sided polygon) has three sides each of length 10.7 cm, two sides each of length 8.6 cm, two sides each of length 9.9 cm, and one side of length 4.8 cm. Determine the perimeter of the octagon.

Solution 8.7

We have that:
$$P = 3 \times (10.7 \text{ cm}) + 2 \times (8.6 \text{ cm}) + 2 \times (9.9 \text{ cm}) + 4.8 \text{ cm}$$
$$= (32.1 \text{ cm}) + (17.2 \text{ cm}) + (19.8 \text{ cm}) + 4.8 \text{ cm}$$
$$= (32.1 + 17.2 + 19.8 + 4.8) \text{ cm} = 73.9 \text{ cm}$$

Therefore, the perimeter of the given octagon is 73.9 cm.

CIRCLES

Another basic geometric figure is the **circle**. All points on a circle are the same distance from a fixed point in the plane called its **center**. The distance from the center of the circle to any point on it is called the **radius** of the circle. A **diameter** of a circle is any line segment passing through the center and having its endpoints on the circle (see below) A diameter d of a circle is equal to twice the radius of the circle. If r represents the radius of the circle, we have $d = 2r$.

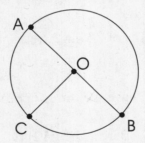

The perimeter of, or distance around, a circle is called its **circumference**. The circumference of a circle is given in terms of a number π (pi). It is equal to 3.14 (to two decimal places) or 3.146 (to three decimal places).

The number π can also be approximated by the fraction $\frac{22}{7}$. It is the ratio of the circumference of a circle to its diameter.

Formula for the circumference of a Circle

To determine the circumference of a circle, multiply its diameter by π. Hence, $C = \pi d$, or, because $d = 2r$, we also have $C = 2\pi r$, where C, r, and d are all measured in the same units.

Exercise 8.8
Determine the circumference of a circle whose radius is 3 yards.

Solution 8.8
Using the formula involving radius, we have
$$C = 2\pi r$$
$$C = 2\pi(3 \text{ yd})$$
$$= 6\pi \text{ yd}$$

Therefore, the circumference of the given circle is 6π yards, which is approximately equal to 6(3.14) yards or 18.84 yards.

Exercise 8.9
Determine the circumference of a circle whose diameter is 14 dm.

Solution 8.9
Using the formula involving diameter, we have
$$C = \pi d$$
$$= \pi(14 \text{ dm})$$
$$= 14\pi \text{ dm}$$

Therefore, the circumference of the given circle is 14π dm, which is approximately equal to 14(3.14) dm or 43.96 dm.

Exercise 8.10
If the circumference of a circle is 37 inches, determine its diameter to the nearest tenth of an inch. (Use the approximation of 3.14 for π).

Solution 8.10
Using the formula involving diameter, we have
$$C = \pi d$$
or
$$37 \text{ in.} = 3.14d$$

To solve for d, divide both sides of the above equation by 3.14, obtaining
$$d = \frac{37 in.}{3.14}$$
$$= 11.8 \text{ in. (to the nearest tenth of an inch)}$$

Therefore, to the nearest tenth of an inch, the diameter of the given circle is 11.8 inches.

Exercise 8.11
Determine the radius of a circle if it circumference is 36 feet. (Use $\frac{22}{7}$ for π.)

Solution 8.11

Using the formula involving radius, we have
$$C = 2\pi r$$
or
$$36 \text{ ft} = 2\left(\frac{22}{7}\right)r$$
$$36 \text{ ft} = \frac{44}{7}r$$

To solve for r, divide both sides of the above equation by $\frac{44}{7}$, obtaining

$$r = 36 \text{ ft} \div \frac{44}{7}$$
$$= 36 \text{ ft} \times \frac{7}{44}$$
$$= (\overset{9}{\cancel{36}} \times \frac{7}{\underset{11}{\cancel{44}}}) \text{ ft}$$
$$= (\frac{9 \times 7}{11}) \text{ ft}$$
$$= \frac{63}{11} \text{ ft}$$
$$= 5\frac{8}{11} \text{ ft}$$

Therefore, the given circle has a radius of $5\frac{8}{11}$ feet.

8.2 ANGLES AND TRIANGLES

In geometry, a **point** is the most basic figure, yet we make no attempt to define it. All geometric figures can be defined in terms of points. In this section, we discuss geometric figures called angles and triangles. First, we give some basic definitions.

DEFINITIONS AND SYMBOLS

A **line** is a set of points that extends indefinitely in opposite directions (see figure below). A line is named by any two points on it. If the points A, B, C, and D are on the same line, we can name the line by the symbols $\overleftrightarrow{AB}, \overleftrightarrow{AC}, \overleftrightarrow{BC}, \overleftrightarrow{BD}, \overleftrightarrow{DA}$, and so on.

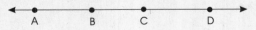

A **line segment** is the set of all points on a line between two distinct points. The points are called its **endpoints.** A line is named by its endpoints (see figure below). If the endpoints of a line segment are the points A and B, the line segment can be named by the symbols $\overline{AB}$ or $\overline{BA}$.

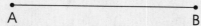

A **ray** is that part of a line that starts at a point on the line and extends indefinitely in one direction. The point is called its **endpoint.** In the figure below, the ray starts from the point A and extends in the direction of B; it is denoted by the symbol $\overrightarrow{AB}$. In Figure 8-6, the ray starts from the point B and extends in the direction of A; it is denoted by the symbol $\overleftarrow{BA}$.

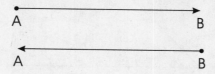

Exercise 8.12

Name each of the geometric figures in the following figure.

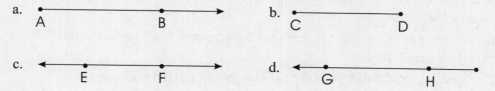

Solution 8.12

a) This figure is a *ray* starting from A and extending in the direction of B; it is denoted by the symbol $\overrightarrow{AB}$.

b) This figure is a *line segment* with endpoints C and D; it is denoted by the symbol $\overline{CD}$ or $\overline{DC}$.

c) This figure is a *line* determined by the two points, E and F; it is denoted by the symbol $\overleftrightarrow{EF}$ or $\overleftrightarrow{FE}$.

d) This figure is a *ray* starting from H and extending in the direction of G; it is denoted by the symbol $\overrightarrow{HG}$.

ANGLES

Two rays with a common endpoint have a special name. Two rays with a common endpoint form a geometric figure called an **angle**. Each of the rays is called a **side** of the angle, and the common endpoint is called the **vertex** of the angle.

We use the symbol ∠ to denote an angle. In the following figure, we denote the angle with vertex at O and having as sides the rays $\overrightarrow{OA}$ and $\overrightarrow{OB}$.

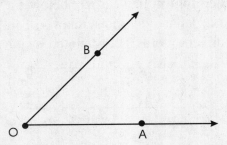

An angle can be measured in either degrees or radians; degrees are most often used in basic math. The symbol ° is used to note degree measure. Hence 47° is read, "47 degrees."

Angles are measured using an instrument called a **protractor**. Angles can be classified using their measures, as follows.

1. If the measure of an angle is greater than or equal to 0° but less than 90°, the angle is called an **acute** angle.
2. If the measure of an angle is 90°, the angle is called a **right** angle.
3. If the measure of an angle is greater than 90° but less than 180°, the angle is called an **obtuse** angle.
4. If the measure of an angle is 180°, the angle is called a **straight** angle.

Consider the following

- An angle whose measure is 37° is an acute angle.
- An angle whose measure is 139° is an obtuse angle.
- An angle whose measure is 179° is an obtuse angle.
- An angle whose measure is 78° is an acute angle.
- An angle whose measure is 90° is a right angle.

Angles can be parts of triangles, which, as noted in Section 8.1, are polygons having three sides.

PERIMETER AND AREA OF TRIANGLES

We now consider the perimeter and area formulas of triangles. **Area** is measured in square units. A **square unit** is a square whose sides each have 1 unit.

Consider a triangle with sides a, b, and c and height h.

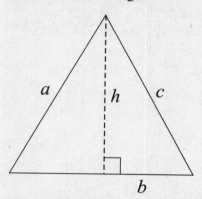

The **perimeter** of a triangle (see above figure) is equal to the sum of the lengths of its sides:
$$P = a + b + c$$
The **area** of a triangle is equal to one-half the product of its base (b) and height (h):
$$A = \frac{1}{2}bh$$

Exercise 8.13

A triangle has a base that measures 13 cm. The other two sides measure 11 cm and 9 cm. The height of the triangle is 8 cm. Determine the perimeter and the area of the triangle.

Solution 8.13

The perimeter, P, is equal to the sum of the lengths of the three sides of the triangle. We have
$$P = 13 \text{ cm} + 11 \text{ cm} + 9 \text{ cm}$$
$$= (13 + 11 + 9) \text{ cm}$$
$$= 33 \text{ cm}$$

The area A is equal to one-half the product of the base and the height. We have
$$A = \frac{1}{2}bh$$
$$= \frac{1}{2}(13 \text{ cm})(8 \text{ cm})$$
$$= 52 \text{ cm}^2 \text{ (where cm}^2 \text{ means squared cm)}$$

Therefore, the perimeter of the given triangle is 33 cm, and its area is 52 cm².

Exercise 8.14
Determine the area of a triangle whose base is 1.8 dm and whose height is 15 cm.

Solution 8.14
Because the measurements are given in different units, we convert one of them to the other. In particular, we convert 1.8 dm to centimeters. We have

1.8 dm = 1.8 × 10 cm = 18 cm

We now have that
$$A = \frac{1}{2}bh$$
$$= \frac{1}{2}(18 \text{ cm})(15 \text{ cm})$$
$$= 135 \text{ cm}^2$$

Therefore, the area of the given triangle is 135 cm².

Exercise 8.15
Determine the base of a triangle whose area is 70 sq in. and whose height is 17.5 in.

Solution 8.15
Because $A = 70$ sq in. and $h = 17.5$ in., we have
$$A = \frac{1}{2}bh$$
$$70 \text{ in.}^2 = \frac{1}{2}(b)(17.5 \text{ in.})$$
$$140 \text{ in.}^2 = 17.5b \text{ in.}$$
$$b = \frac{140 \text{ sq in}}{17.5 \text{ in}}$$
$$b = 8 \text{ in.}$$

Therefore, the base of the given triangle is 8 inches.

Exercise 8.16
Determine the height of a triangle with area 84 sq cm if its base is 12 cm.

Solution 8.16
Because $A = 84$ sq cm and $b = 12$ cm, we have
$$A = \frac{1}{2}bh$$
$$84 \text{ cm}^2 = \frac{1}{2}(12 \text{ cm})(h)$$
$$168 \text{ cm}^2 = 12h \text{ cm}$$

$$h = \frac{168 \text{ sq cm}}{12 \text{ cm}}$$
$$h = 14 \text{ cm}$$

Therefore, the height of the given triangle is 14 cm.

8.3 QUADRILATERALS

In this section, we consider the perimeter and area of quadrilaterals. Recall that in Section 8.1 we defined a quadrilateral to be a four-sided polygon.

1. If a quadrilateral has **both** pairs of its opposite sides parallel, the quadrilateral is called a **parallelogram.**
2. If a parallelogram has four **right** angles, it is called a **rectangle.**
3. If a rectangle has all four of its sides of the same length, it is called a **square.**

It should be noted that opposite sides of parallelograms, rectangles, and squares have the same length.

RECTANGLES

Consider a rectangle with length L and width W, as shown in the following figure.

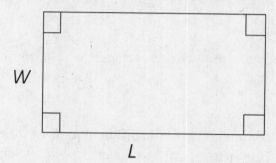

The perimeter P is equal to the sum of twice the length and twice the width. We have
$$P = 2L + 2W$$
The area is equal to the product of the length and width. We have
$$A = LW$$

Exercise 8.17

A rectangle has a length (L) of 13 ft and a width (W) of 11 ft. Determine its perimeter and area.

Solution 8.17

Because $L = 13$ ft and $W = 11$ ft, we have
$$P = 2L + 2W$$
$$P = (2 \times 13 \text{ ft}) + (2 \times 11 \text{ ft})$$

$$= 26 \text{ ft} + 22 \text{ ft}$$
$$= 48 \text{ ft}$$

and

$$A = LW$$
$$A = (13 \text{ ft})(11 \text{ ft})$$
$$= 143 \text{ sq ft}$$

Therefore, the perimeter of the given rectangle is 48 feet, and its area is 143 square feet.

Exercise 8.18
Determine the length (L) of a rectangle if its width (W) is 14 cm and its area (A) is 266 cm^2.

Solution 8.18
Because $W = 14$ cm and $A = 266$ cm^2, we have
$$A = LW$$
$$266 \text{ cm}^2 = (L)(14 \text{ cm})$$

and

$$L = \frac{266 \text{ sq cm}}{14 \text{ cm}}$$
$$= 19 \text{ cm}$$

Therefore, the length of the given rectangle is 19 cm.

Exercise 8.19
Determine the perimeter and the area of a rectangle whose length (L) is 2 meters and whose width (W) is 171 cm.

Solution 8.19
Because the length and width are given in different units of measure, we must convert one of them to the other. In particular, we convert meters to centimeters as follows:
$$2 \text{ m} = 2 \times 100 \text{ cm}$$
$$= 200 \text{ cm}$$

We now have
$$P = 2L + 2W$$
$$P = (2 \times 200 \text{ cm}) + (2 \times 171 \text{ cm})$$
$$= 400 \text{ cm} + 342 \text{ cm}$$
$$= 742 \text{ cm}$$

and

$$A = LW$$
$$A = (200 \text{ cm})(171 \text{ cm})$$
$$= 34,200 \text{ cm}^2$$

Therefore, the perimeter of the given rectangle is 742 cm, and its area is 34,200 cm^2.

Exercise 8.20

A rectangle has length equal to 4.5 cm and width equal to 2.3 cm. Determine its perimeter and area.

Solution 8.20

Because $L = 4.5$ cm and $W = 2.3$ cm, we have
$$P = 2L + 2W$$
$$P = (2 \times 4.5 \text{ cm}) + (2 \times 2.3 \text{ cm})$$
$$= 9 \text{ cm} + 4.6 \text{ cm}$$
$$= 13.6 \text{ cm}$$

and
$$A = LW$$
$$A = (4.5 \text{ cm})(2.3 \text{ cm})$$
$$= 10.35 \text{ cm}^2$$

Therefore, the perimeter of the given rectangle is 13.6 cm, and its area is 10.35 cm^2.

Exercise 8.21

A rectangle has length equal to 8.19 mm and width equal to 6.04 mm. Determine its perimeter and area.

Solution 8.21

Because $L = 8.19$ mm and $W = 6.04$ mm, we have
$$P = 2L + 2W$$
$$P = (2 \times 8.19 \text{ mm}) + (2 \times 6.04 \text{ mm})$$
$$= 16.38 \text{ mm} + 12.08 \text{ mm}$$
$$= 28.46 \text{ mm}$$

and
$$A = LW$$
$$A = (8.19 \text{ mm})(6.04 \text{ mm})$$
$$= 49.4676 \text{ mm}^2$$

Therefore, the perimeter of the given rectangle is 28.46 mm, and its area is 49.4676 mm^2.

SQUARES

A square is a special case of a rectangle having all of its sides the same length. That is, $L = W = s$, where s is the side of the square.

Consider a square with side s, as shown in the following figure.

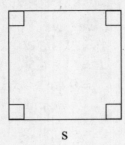

s

The perimeter is equal to 4 times the length of its side:
$$P = 4s$$
The area is equal to the square of the length of its side:
$$A = s^2$$

Exercise 8.22
A side of a square is 4.3 yd long. Determine the perimeter and the area of the square.

Solution 8.22
Because $s = 4.3$ yd, we have:
$$P = 4s$$
$$P = 4 \times (4.3 \text{ yd})$$
$$= 17.2 \text{ yd}$$

and

$$A = s^2$$
$$= (4.3 \text{ yd})^2$$
$$= 18.49 \text{ yd}^2$$

Therefore, the perimeter of the given square is 17.2 yd, and its area is 18.49 yd².

Exercise 8.23
A square has sides of length $5\frac{3}{8}$ inches. Determine the perimeter and area of the square.

Solution 8.23
Because $s = 5\frac{3}{8}$ in., we have:
$$P = 4s$$
$$P = 4 \times 5\frac{3}{8} \text{ in.}$$
$$= 21.5 \text{ in.}$$

and
$$A = s^2$$
$$A = \left(5\frac{3}{8} \text{ in.}\right)^2$$
$$= 28\frac{57}{64} \text{ sq in.}$$

Therefore, the perimeter of the given square is 21.5 in., and its area is $28\frac{57}{64}$ in².

PARALLELOGRAMS

Consider the following parallelogram with base b, width a, and height h.

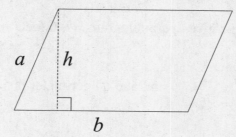

The perimeter is equal to the sum of twice the base and twice the width:
$$P = 2a + 2b$$
The area is equal to the product of the base and the height:
$$A = bh$$

Exercise 8.24

A parallelogram has width equal to 11.2 cm, base equal to 15.1 cm, and height equal to 8.9 cm. Determine the perimeter and the area of the parallelogram.

Solution 8.24

Because $a = 11.2$ cm, $b = 15.1$ cm, and $h = 8.9$ cm, we have
$$P = 2a + 2b$$
$$P = (2 \times 11.2 \text{ cm}) + (2 \times 15.1 \text{ cm})$$
$$= 22.4 \text{ cm} + 30.2 \text{ cm}$$
$$= 52.6 \text{ cm}$$
and
$$A = bh$$
$$A = (15.1 \text{ cm})(8.9 \text{ cm})$$
$$= 134.39 \text{ cm}^2.$$

Therefore, the perimeter of the given parallelogram is 52.6 cm, and the area is 134.39 cm².

Exercise 8.25

Determine the measure of the height of a parallelogram with base 10.9 ft, if its area is 104.64 ft².

Solution 8.25

Because $b = 10.9$ ft and $A = 104.64$ ft², we have
$$A = bh$$
$$104.64 \text{ ft}^2 = (10.9 \text{ ft})h$$
$$h = \frac{104.64 \text{ ft}^2}{10.9 \text{ ft}}$$
$$h = 9.6 \text{ ft}$$

Therefore, the height of the given parallelogram is 9.6 feet.

Exercise 8.26

A parallelogram has a base of 17.4 cm, and its perimeter is 58.2 cm. Determine the width of the parallelogram.

Solution 8.26

Because $b = 17.4$ cm and $P = 58.2$ cm, we have
$$P = 2a + 2b$$
$$58.2 \text{ cm} = 2a + (2 \times 17.4 \text{ cm})$$
$$58.2 \text{ cm} = 2a + 34.8 \text{ cm}$$
$$2a = 58.2 \text{ cm} - 34.8 \text{ cm}$$
$$2a = 23.4 \text{ cm}$$
$$a = 11.7 \text{ cm}$$

Therefore, the width of the given parallelogram is 11.7 cm.

TRAPEZOIDS

A parallelogram is a quadrilateral with two pairs of parallel sides. There are also quadrilaterals with only one pair of parallel sides. If a quadrilateral has exactly one pair of parallel sides, it is called a **trapezoid.** The parallel sides of the trapezoid are called its **bases.**

Consider a trapezoid with bases b and B and height h.

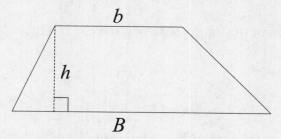

The area of the trapezoid is found by taking one-half the product of the height with the sum of the bases:

$$A = \frac{1}{2}h(b + B)$$

Exercise 8.27

Determine the area of the trapezoid whose bases measure 7 dm and 9 dm and whose height is 6.5 dm.

Solution 8.27

Because $b = 7$ dm, $B = 9$ dm, and $h = 6.5$ dm, we have

$$A = \frac{1}{2}(6.5 \text{ dm})(7 \text{ dm} + 9 \text{ dm})$$
$$= \frac{1}{2}(6.5 \text{ dm})(16 \text{ dm})$$
$$= 52 \text{ dm}^2$$

Therefore, the area of the given trapezoid is 52 square decimeters.

Exercise 8.28

A trapezoid with bases 18 inches and 32 inches has an area of 225 square inches. Determine the height of the trapezoid.

Solution 8.28

Because $b = 18$ in., $B = 32$ in., and $A = 225$ in.^{2}, we have

$$A = \frac{1}{2}h(b + B)$$
$$225 \text{ in.}^2 = \frac{1}{2}h(18 \text{ in.} + 32 \text{ in.})$$
$$450 \text{ in.}^2 = h(50 \text{ in.})$$

$$h = \frac{450 \text{ in}^2}{50 \text{ in}}$$
$$h = 9 \text{ in.}$$

Therefore, the height of the given trapezoid is 9 inches.

Exercise 8.29

Determine the measure of the longer base of a trapezoid with area 366 ft² if its height is 12 ft and its shorter base is 11 ft.

Solution 8.29

Because $A = 366$ ft², $h = 12$ ft, and $b = 11$ ft, we have

$$A = \frac{1}{2}h(b+B)$$

$$366 \text{ ft}^2 = \frac{1}{2}(12 \text{ ft})(11 \text{ ft} + B)$$

$$732 \text{ ft}^2 = (12 \text{ ft})(11 \text{ ft} + B)$$

$$732 \text{ ft}^2 = 132 \text{ ft}^2 + 12B \text{ ft}$$

$$600 \text{ ft}^2 = 12B \text{ ft}$$

$$B = \frac{600 \text{ ft}^2}{12 \text{ ft}}$$

$$B = 50 \text{ ft}$$

Therefore, the longer base of the given trapezoid measures 50 feet.

8.4 CIRCLES

We have already discussed the circumference of a circle. In this section, we introduce the area of a circle. Consider a circle with radius r and diameter d.

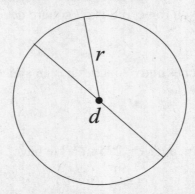

The **circumference** is equal to the product of twice its radius and π:
$$C = 2\pi r \text{ (or, } C = \pi d\text{)}$$

The **area** is equal to the product of π and the square of the radius:
$$A = \pi r^2$$

Exercise 8.30

Determine the circumference and area of a circle that has a diameter of 16 ft. (Leave your answers in terms of π.)

Solution 8.30

$$C = \pi d$$
$$C = \pi(16 \text{ ft})$$
$$= 16\pi \text{ ft}^2$$

Because $d = 16$ ft, $r = 8$ ft

$$A = \pi r^2$$
$$A = \pi(8 \text{ ft})^2$$
$$= \pi(64 \text{ ft}^2)$$
$$= 64\pi \text{ ft}^2$$

Therefore, the circumference of the given circle is 16π feet, and the area is 64π square feet.

Exercise 8.31

Determine the circumference and area of a circle that has a diameter of 15 cm. (Use $\frac{22}{7}$ for π.)

Solution 8.31

$$C = \pi d$$
$$C = \left(\frac{22}{7}\right)(15 \text{ cm})$$
$$= \frac{330}{7} \text{ cm}$$
$$= 47\frac{1}{7} \text{ cm}$$

Because $d = 15$ cm, $r = 7.5$ cm.

$$A = \pi r^2$$
$$A = \left(\frac{22}{7}\right)(7.5 \text{ cm})^2$$
$$= \left(\frac{22}{7}\right)(56.25 \text{ cm})$$
$$= 176\frac{11}{14} \text{ cm}^2$$

Therefore, the circumference of the given circle is $47\frac{1}{7}$ cm and its area is $176\frac{11}{14}$ cm².

Exercise 8.32

Determine the circumference and area of a circle that has a radius of 4.3 yd. (Use 3.14 for π.)

Solution 8.32

$$C = 2\pi r$$
$$C = 2(3.14)(4.3 \text{ yd})$$
$$= (6.28)(4.3 \text{ yd})$$
$$= 27.004 \text{ yd}$$

and

$$A = \pi r^2$$
$$A = (3.14)(4.3 \text{ yd})^2$$
$$= (3.14)(18.49 \text{ yd})$$
$$= 58.0586 \text{ yd}^2$$

Therefore, the circumference of the given circle is 27.004 yards, and the area is 58.0586 square yards.

Exercise 8.33

Determine the area of a circle whose diameter is 10.4 mm. (Use 3.14 for π.)

Solution 8.33

Because $d = 10.4$ mm, $r = 5.2$ mm.

$$A = \pi r^2$$
$$A = (3.14)(5.2 \text{ mm})^2$$
$$= (3.14)(27.04 \text{ mm}^2)$$
$$= 84.9056 \text{ mm}^2$$

Therefore, the area of the given circle is 84.9056 square millimeters.

Exercise 8.34

What is the area of a circular pond that is 24.5 m in diameter? (Use 3.14 for π.)

Solution 8.34

Because $d = 24.5$ m, $r = 12.25$ m.

$$A = \pi r^2$$
$$A = (3.14)(12.25 \text{ m})^2$$
$$= (3.14)(150.0625 \text{ m}^2)$$
$$= 471.19625 \text{ m}^2$$

Therefore, the area of the given circle is 471.19625 square meters.

Exercise 8.35

If the radius of a circle is doubled, how many times larger is the circumference of the new circle than the circumference of the original circle?

Solution 8.35

Let r be the radius of the original circle. Then its circumference is $C = 2\pi r$.
If the radius is doubled, the radius of the new circle is $2r$, and its circumference is
$$C(\text{new circle}) = 2\pi(2r)$$
$$= 4\pi r$$
$$= 2(2\pi r)$$
Therefore, if the radius of a circle is doubled, the circumference is also doubled.

Exercise 8.36

If the radius of a circle is doubled, how many times larger is the area of the new circle than the area of the original circle?

Solution 8.36

Let r be the radius of the original circle. Its area is
$$A = \pi r^2$$
If the radius is doubled, the radius of the new circle is $2r$, and its area is
$$A(\text{new circle}) = \pi(2r)^2$$
$$= \pi(4r^2)$$
$$= 4\pi r^2$$
$$= 4(\pi r^2)$$
Therefore, if the radius of a circle is doubled, the area of the new circle is four times as large as the area of the original circle.

8.5 VOLUMES OF SOLIDS

In this section, we introduce a measure called volume. We look at the volumes of some basic solids.

1. **Volume** is the measure of the space enclosed by a three-dimensional solid. Volume is given in cubic units.
2. A **cubic unit** is the measure of a cube whose sides each have a measure of 1 unit.

RECTANGULAR SOLIDS

The first solid we consider is a **rectangular solid.** A rectangular solid has rectangles for its faces (sides).

Consider a rectangular solid with length L, width W, and height h (see the figure below). The volume V of the rectangular solid is given by the formula

$$V = LWh$$

where L, W, and h are all measured in the same units, and V is measured in cubic units.

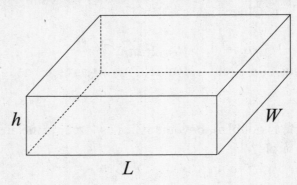

Exercise 8.37

Determine the volume of each of the rectangular solids with indicated dimensions:

a) length 7 yd, width 5.5 yd, and height 2.6 yd
b) length 3 yd, width 4 ft, and height 18 inches
c) length 8 cm, width 7 cm, and height 4.2 cm

Solution 8.37

a) Because $L = 7$ yd, $W = 5.5$ yd, and $h = 2.6$ yd, we have
$V = LWh$
$V = (7 \text{ yd})(5.5 \text{ yd})(2.6 \text{ yd})$
$= (38.5 \text{ sq yd})(2.6 \text{ yd})$
$= 100.1 \text{ cu yd}$
Therefore, the volume of the given rectangular solid is 100.1 cubic yards.

b) We first note that the dimensions are given in three different units. We convert all units to feet:
$3 \text{ yd} = 3 \times 3 \text{ ft} = 9 \text{ ft}$
$18 \text{ inches} = (18 \div 12) \text{ ft} = 1.5 \text{ ft}$
We now have $L = 9$ ft, $W = 4$ ft, and $h = 1.5$ ft.
$V = LWh$
$V = (9 \text{ ft})(4 \text{ ft})(1.5 \text{ ft})$
$= (36 \text{ sq ft})(1.5 \text{ ft})$
$= 54 \text{ cu ft}$
Therefore, the volume of the given solid is 54 cubic feet.

c) Because $L = 8$ cm, $W = 7$ cm, and $h = 4.2$ cm, we have
$$V = LWh$$
$$V = (8 \text{ cm})(7 \text{ cm})(4.2 \text{ cm})$$
$$= (56 \text{ sq cm})(4.2 \text{ cm})$$
$$= 235.2 \text{ cu cm}$$
Therefore, the volume of the given rectangular solid is 235.2 cubic centimeters.

Exercise 8.38

Determine the height of a rectangular solid with length 3.9 m, width 2.8 m, and volume 18.564 cu meters.

Solution 8.38

Because $V = 18.564$ cu m, $L = 3.9$ m, and $W = 2.8$ m, we have
$$V = LWh$$
$$18.564 \text{ cu m} = (3.9 \text{ m})(2.8 \text{ m})h$$
$$18.564 \text{ cu m} = (10.92 \text{ sq m})h$$
$$h = \frac{18.564 \text{ cu m}}{10.92 \text{ sq m}}$$
$$= 1.7 \text{ m}$$
Therefore, the height of the given rectangular solid is 1.7 meters.

CUBES

We now consider a special case of a rectangular solid. A **cube** is a rectangular solid with all of its dimensions being equal. That is, $L = W = h$. Consider a cube with edge of length e (see the figure below). The volume (V) of the cube is given by the formula
$$V = e^3,$$
where e is measured in units and V is measured in cubic units.

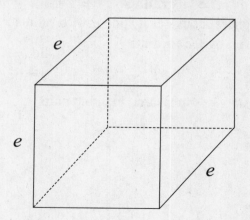

Exercise 8.39

Determine the volume of each of the following cubes:

a) $e = 3$ yd
b) $e = 5.6$ m
c) $e = 26$ cm

Solution 8.39

a) Because $e = 3$ yards,
$$V = e^3$$
$$V = (3 \text{ yd})^3$$
$$= 27 \text{ cu yd}$$
The volume of the given cube is 27 cubic yards.

b) Because e = 5.6 m,
$$V = e^3$$
$$V = (5.6 \text{ m})^3$$
$$= 175.616 \text{ cu m}$$
The volume of the given cube is 175.616 cubic meters.

c) Because $e = 26$ cm,
$$V = e^3$$
$$V = (26 \text{ cm})^3$$
$$= 17{,}576 \text{ cu cm}$$
The volume of the given cube is 17,576 cubic centimeters.

SPHERE

The last solid we consider is a sphere. A **sphere** is a set of all the points in a three-dimensional space that are the same distance away from a fixed point in the space. The fixed point is called its **center;** the same distance is called its **radius.**

To find the volume of a sphere, consider a sphere with radius r (see the figure below). The volume V of the sphere is given by the formula

$$V = \frac{4}{3}\pi r^3$$

where r is measured in units and V is measured in cubic units.

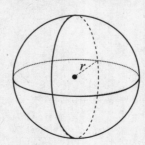

Geometric Measure 241

Exercise 8.40

Determine the volume of each of the following spheres with indicated radius:

a) 8 inches (Leave your answer in terms of π.)

b) 3.6 feet (Use $\dfrac{22}{7}$ for π.)

c) 23 cm (Use 3.14 for π.)

Solution 8.40

a) Because $r = 8$ in.,

$$V = \frac{4}{3}\pi r^3$$

$$V = \frac{4}{3}\pi (8 \text{ in.})^3$$

$$V = \frac{4}{3}\pi (512 \text{ cu in.})$$

$$= \frac{2{,}048\pi}{3}$$

The volume of the given sphere is $\dfrac{2{,}048\pi}{3}$ cubic inches.

b) Because $r = 3.6$ ft,

$$V = \frac{4}{3}\pi r^3$$

$$V = \frac{4}{3}\left(\frac{22}{7}\right)(3.6 \text{ ft})^3$$

$$= \left(\frac{88}{21}\right)(46.656 \text{ cu ft})$$

$$= \left(\frac{88}{7}\right)(15.552 \text{ cu ft})$$

$$= \frac{1368.576}{7} \text{ cu ft}$$

$$= 195.5 \text{ cu ft (to the nearest tenth)}$$

To the nearest tenth, the volume of the given sphere is 195.5 cubic feet.

c) Because $r = 23$ cm,

$$V = \frac{4}{3}\pi r^3$$

$$V = \frac{4}{3}(3.14)(23 \text{ cm})^3$$

$$= \frac{4}{3}(3.14)(12{,}167 \text{ cu cm})$$

242 Basic Math

$$= \frac{12.56}{3}(12{,}167 \text{ cu cm})$$

$$= \frac{152{,}817.52}{3} \text{ cu cm}$$

$$= 50{,}939.2 \text{ cu cm (to the nearest tenth)}$$

To the nearest tenth, the volume of the given sphere is 50,939.2 cubic centimeters.

8.6 THE PYTHAGOREAN FORMULA

If a triangle has a right angle, it is called a **right triangle.** The longest side of a right triangle (that is, the side that is opposite the right angle) is called the **hypotenuse.** The other two sides are called its **legs.**

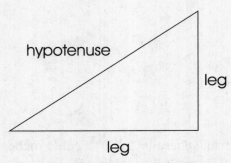

In this section, we introduce the **Pythagorean formula** (also called the **Pythagorean theorem**), which applies to right triangles. Given a right triangle with legs of lengths a and b and hypotenuse of length c,

$$c^2 = a^2 + b^2$$

Exercise 8.41
Determine whether the following are measures of the sides of a right triangle:
a) 3 ft, 4 ft, and 5 ft
b) 10 cm, 24 cm, and 26 cm
c) 8 yd, 10 yd, and 15 yd

Solution 8.41
a) If the given measures are to be measures of the sides of a right triangle, and if $a = 3$ ft, $b = 4$ ft, and $c = 5$ ft (c must be the largest measure), we need $c^2 = a^2 + b^2$ to be true.
$$c^2 \; ? \; a^2 + b^2$$
$$(5 \text{ ft})^2 \; ? \; (3 \text{ ft})^2 + (4 \text{ ft})^2$$
$$25 \text{ sq ft} \; ? \; 9 \text{ sq ft} + 16 \text{ sq ft}$$
$$25 \text{ sq ft} = 25 \text{ sq ft},$$
which checks. Therefore, 3 ft, 4 ft, and 5 ft are the measures of the sides of a right triangle.

b) Let $a = 10$ cm, $b = 24$ cm, and $c = 26$ cm. Then
$$c^2 \; ? \; a^2 + b^2$$
$$(26 \text{ cm})^2 \; ? \; (10 \text{ cm})^2 + (24 \text{ cm})^2$$
$$676 \text{ cm}^2 \; ? \; 100 \text{ cm}^2 + 576 \text{ cm}^2$$
$$676 \text{ cm}^2 = 676 \text{ cm}^2,$$
which checks. Therefore, 10 cm, 24 cm, and 26 cm are measures for the sides of a right triangle.

c) Let $a = 8$ yd, $b = 10$ yd, and $c = 15$ yd.
$$c^2 \; ? \; a^2 + b^2$$
$$(15 \text{ yd})^2 \; ? \; (8 \text{ yd})^2 + (10 \text{ yd})^2$$
$$225 \text{ yd}^2 \; ? \; 64 \text{ yd}^2 + 100 \text{ yd}^2$$
$$225 \text{ yd}^2 \neq 164 \text{ yd}^2$$
Therefore, 8 yd, 10 yd, and 15 yd are not measures of the sides of a right triangle.

SQUARE ROOTS

If you are given the measures for two sides of a right triangle, by using the Pythagorean Formula, you can find the measure of the third side. However, you need to use what is called the square root of a number. The **square root** of a **positive** number n, denoted by $\sqrt{n}$, is the *positive* number b so that $b^2 = n$.

Using the above definition, we have:
- $\sqrt{4} = 2$, because $2^2 = 4$.
- $\sqrt{49} = 7$, because $7^2 = 49$.
- $\sqrt{144} = 12$, because $12^2 = 144$.
- $\sqrt{625} = 25$, because $25^2 = 625$.

The square root of a positive number is not always a whole number. For example, $\sqrt{15}$ is not a whole number; $3^2 = 9$ and $4^2 = 16$, so $\sqrt{15}$ is a number between 3 and 4, because 15 is between 9 and 16. To determine the square root of a positive number, you may use either a table of square roots or a calculator with a square root key.

Exercise 8.42

For each of the following, determine the measure of the hypotenuse of a right triangle whose legs have the indicated measures:

a) 9 in. and 12 in.

b) 4 ft and 9 ft

c) 10 yd and 10 yd

Solution 8.42

a) Using the Pythagorean Formula with $a = 9$ in. and $b = 12$ in., we determine the value of c as follows:
$$c^2 = a^2 + b^2$$
$$c^2 = (9 \text{ in.})^2 + (12 \text{ in.})^2$$
$$= (81 \text{ sq in.}) + (144 \text{ sq in.})$$
$$= 225 \text{ sq in.}$$

Because $c^2 = 225$ sq in., $c = \sqrt{225}$ in. $= 15$ in.

The measure of the hypotenuse of the given right triangle is 15 inches.

b) Let $a = 4$ ft and $b = 9$ ft. Then
$$c^2 = a^2 + b^2$$
$$c^2 = (4 \text{ ft})^2 + (9 \text{ ft})^2$$
$$= 16 \text{ sq ft} + 81 \text{ sq ft}$$
$$= 52 \text{ sq ft}$$

Hence, $c = \sqrt{97}$ ft

The hypotenuse of the given triangle will have the measure of $\sqrt{97}$ feet. This is equal to 9.8 feet to the nearest tenth of a foot.

c) Let $a = b = 10$ yd. Then
$$c^2 = a^2 + b^2$$
$$c^2 = (10 \text{ yd})^2 + (10 \text{ yd})^2$$
$$= 100 \text{ sq yd} + 100 \text{ sq yd}$$
$$= 200 \text{ sq yd}$$

Hence, $c = \sqrt{200}$ yd

The hypotenuse of the given right triangle has the measure of $\sqrt{200}$ yards. This is equal to 14.14 yards to the nearest hundredth of a yard.

Exercise 8.43

For each of the following, determine the length of one leg of the right triangle with the measures of the other leg and hypotenuse as indicated:

a) leg of length 12 in. and hypotenuse of 13 in.
b) leg of length 7 cm and hypotenuse of 11 cm
c) leg of length 26 dm and hypotenuse of 34 dm

Solution 8.43

a) Using the Pythagorean Formula with $a = 12$ in. and $c = 13$ in., we determine the value of b as follows:
$$c^2 = a^2 + b^2$$
$$(13 \text{ in.})^2 = (12 \text{ in.})^2 + b^2$$
$$169 \text{ sq in.} = 144 \text{ sq in.} + b^2 \text{ sq in.}$$
$$b^2 = 169 \text{ sq in.} - 144 \text{ sq in.}$$
$$b^2 = 25 \text{ sq in.}$$
$$b = \sqrt{25} \text{ in}$$
$$b = 5 \text{ in.}$$

The other leg of the given right triangle is 5 inches.

b) Let $a = 7$ cm and $c = 11$ cm. Then
$$c^2 = a^2 + b^2$$
$$(11 \text{ cm})^2 = (7 \text{ cm})^2 + b^2$$
$$121 \text{ sq cm} = 49 \text{ sq cm} + b^2 \text{ sq cm}$$
$$b^2 = 121 \text{ sq cm} - 49 \text{ sq cm}$$
$$b^2 = 72 \text{ sq cm}$$

Hence, $b = \sqrt{72}$ cm

The measure of the other leg of the given right triangle is $\sqrt{72}$ cm. This is equal to 8.5 cm to the nearest tenth of a centimeter.

c) Let $a = 26$ dm and $c = 34$ dm. Then
$$c^2 = a^2 + b^2$$
$$(34 \text{ dm})^2 = (26 \text{ dm})^2 + b^2$$
$$1{,}156 \text{ sq dm} = 676 \text{ sq dm} + b^2$$
$$b^2 = 1{,}156 \text{ sq dm} - 676 \text{ sq dm}$$
$$= 480 \text{ sq dm}$$

Hence, $b = \sqrt{480}$ dm

The other leg of the given right triangle has measure $\sqrt{480}$ dm. This is equal to 21.9 dm to the nearest tenth decimeter.

In this chapter, we discussed measurement of geometric figures, including perimeter and area of some polygons, circumference and area of circles, and volumes of some basic solids. We concluded the chapter with a brief introduction to the Pythagorean Formula.

TEST YOURSELF

1) Determine the perimeter of a triangle with sides of lengths 4.9 cm, 5.4 cm, and 7.1 cm.
2) Determine the perimeter of a square whose sides are each 19.7 in. long.
3) A rectangle has length 16 ft and width 4.6 yd. Determine its perimeter.
4) A parallelogram has a base of 29 in. and a width of 1.6 ft. Determine its perimeter.
5) A **hexagon** (a six-sided figure) has three sides each measuring 5.1 dm, two sides each measuring 6.2 dm, and the other side measuring 5.9 dm. Determine its perimeter.
6) Determine the perimeter of an isosceles triangle with two sides of 21.3 cm each and a third side of 17.9 cm. (*Definition:* An **isosceles triangle** is a triangle having *exactly* two sides of equal length.)
7) Determine the perimeter of an isosceles triangle with two sides of 13.7 ft each and a third side of 2.7 yd.
8) Determine the perimeter of an equilateral triangle all of whose sides measure 14.68 inches each. (*Definition:* An **equilateral triangle** is a triangle having *all* three sides of equal length.)
9) Determine the perimeter of an equilateral triangle all of whose sides measure 23.71 millimeters each.
10) Determine the circumference of a circle whose diameter is 49 yards. (Use $\frac{22}{7}$ for π.)
11) Determine the radius of a circle whose circumference is 137.67 meters. (Use 3.14 for π.)
12) Determine the diameter of a circle whose circumference is 256.78 ft. (Use 3.14 for π and give your answer to the nearest tenth.)
13) Determine the area of a square whose sides each measure 43.5 cm.
14) Determine the area of a rectangle whose length is 23 in. and whose width is 1.2 feet.
15) Determine the height of a parallelogram whose area is 580 sq yd and whose base is 40 yd.
16) The bases of a trapezoid are 2.4 yd and 5.3 ft. Determine the area of the trapezoid if its height is 6.3 ft.
17) Determine the area of a circle whose radius is 1.2 cm. (Use $\frac{22}{7}$ for π and give your result to the nearest tenth.)
18) Determine the area of a circle whose diameter is 14 dm. (Use 3.14 for π and give your result to the nearest tenth.)
19) Determine the volume of the rectangular solid that measures 4.3 ft by 6.1 ft by 7.4 ft.
20) Determine the volume of a cube all of whose edges measure 8.3 yd.
21) Determine the volume of a sphere whose radius is 7.7 cm. (Use $\frac{22}{7}$ for π and give your answer to the nearest whole number.)
22) Determine the volume of a sphere whose diameter is 12 inches. (Use 3.14 for π and give your answer to the nearest whole number.)
23) Determine the measure of the hypotenuse of a right triangle whose legs have measures 11 cm and 14 cm.

24) Determine the measure of a leg of a right triangle if its hypotenuse has measure of 24 yd and the other leg has measure of 17 yd.

25) Two legs of a right triangle have measures of 5 ft and 12 ft. Determine the perimeter of the triangle.

TEST YOURSELF WITH THE HANDHELD CALCULATOR

1) Determine the perimeter of a triangle with sides of lengths 251.63 dm, 197.23 dm, and 201.19 dm.

2) Determine the perimeter of a square whose sides each measure 1,209.68 mm. (Give your result in cm.)

3) A rectangle has length 126.8 ft and width 35.13 yd. Determine its perimeter.

4) Determine the area of the rectangle in Exercise 3 above.

5) An **octagon** (an eight-sided figure) has three sides each measuring 117.54 in., two sides each measuring 97.9 in., two sides each measuring 89.2 in., and one side measuring 103.47 inches. Determine its perimeter.

6) Determine the perimeter of an isosceles triangle with two sides each of 56.37 cm and a third side of 49.56 cm.

7) Determine the circumference of a circle whose radius is 113.4 ft. (Use 3.14 for π.)

8) Determine the area of the circle in Exercise 7.

9) Determine the base of a parallelogram whose area is 4,904.88 sq m and whose height is 64.2 m.

10) A trapezoid has a base of 7.62 ft, a height of 2.31 yd, and an area of 65.5578 sq ft. Determine the length of the other base.

11) Determine the volume of rectangular solid that measures 8.63 yd by 27.12 ft by 319.4 in. (Give your answer to the nearest tenth.)

12) Determine the volume of a sphere whose diameter is 212.6 cm. (Use 3.14 for π and give your answer to the nearest whole number.)

13) Determine the measure of the hypotenuse of a right triangle whose legs have measures 13.6 dm and 17 dm. (Give your answer to the nearest tenth.)

14) A right triangle has a leg of 25 ft and hypotenuse of 65 ft. Determine the perimeter of the triangle.

In Exercises 15–20, use 3.14 for π.
In Exercises 15–16, use the following diagram. (Units are in feet.)

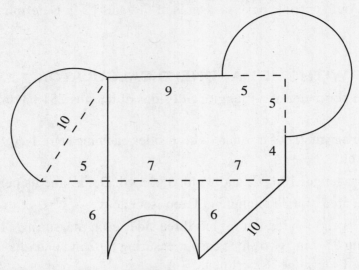

15) Determine the perimeter of the total figure.
16) Determine the area of the enclosed region.

In Exercises 17–18, use the diagram in the figure below, of a semi-circular region with two smaller semi-circular regions cut out of it.

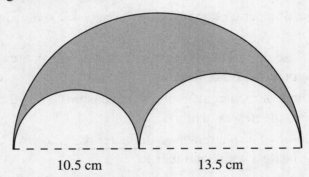

17) Determine the perimeter of the shaded region.
18) Determine the area of the shaded region.
19) Determine the area of the shaded region in the following figure. The overall circular region has a diameter of 18.8 yd. (All units are in yards.)

Geometric Measure **249**

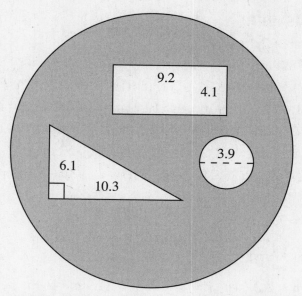

20) Determine the area of the shaded region in the following figure. (All units are in millimeters.)

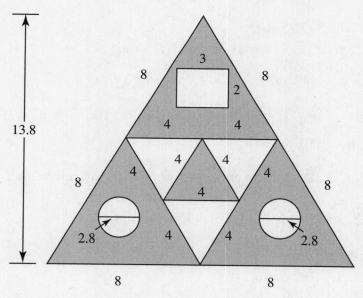

TEST YOURSELF ANSWERS

1) $P = 4.9 \text{ cm} + 5.4 \text{ cm} + 7.1 \text{ cm} = 17.4 \text{ cm}$
2) $P = 4(19.7 \text{ in.}) = 78.8 \text{ in.}$
3) $P = 2(16 \text{ ft}) + 2(4.6 \text{ yd}) = 2(16 \text{ ft}) + 2(13.8 \text{ ft}) = 32 \text{ ft} + 27.6 \text{ ft} = 59.6 \text{ ft}$
4) $P = 2(29 \text{ in.}) + 2(1.6 \text{ ft}) = 2(29 \text{ in.}) + 2(19.2 \text{ in.}) = 58 \text{ in.} + 38.4 \text{ in.} = 96.4 \text{ in.}$
5) $P = 3(5.1 \text{ dm}) + 2(6.2 \text{ dm}) + 5.9 \text{ dm} = 15.3 \text{ dm} + 12.4 \text{ dm} + 5.9 \text{ dm} = 33.6 \text{ dm}$
6) $P = 2(21.3 \text{ cm}) + 17.9 \text{ cm} = 42.6 \text{ cm} + 17.9 \text{ cm} = 60.5 \text{ cm}$
7) $P = 2(13.7 \text{ ft}) + 2.7 \text{ yd} = 2(13.7 \text{ ft}) + 3(2.7 \text{ ft}) = 27.4 \text{ ft} + 8.1 \text{ ft} = 35.5 \text{ ft}$
8) $P = 3(14.68 \text{ in.}) = 44.04 \text{ in.}$
9) $P = 3(23.71 \text{ mm}) = 71.13 \text{ mm}$
10) $C = \pi d = \dfrac{22}{7}(49 \text{ yd}) = \dfrac{1,078}{7} \text{ yd} = 154 \text{ yd}$
11) $C = 2\pi r$; $137.67 \text{ m} = 2(3.14)r$; $r = \dfrac{137.67}{6.28} \text{m} = 21.9 \text{ m}$ (to the nearest tenth meter)
12) $C = \pi d$; $256.78 \text{ ft} = 3.14\, d$; $d = \dfrac{256.78}{3.14} \text{ ft} = 81.8 \text{ ft}$ (to the nearest tenth of a foot)
13) $A = (43.5 \text{ cm})^2 = 1,892.25 \text{ sq cm}$
14) $A = (23 \text{ in.})(1.2 \text{ ft}) = (23 \text{ in.})(14.4 \text{ in.}) = 331.2 \text{ sq in.}$
15) $A = bh$; $580 \text{ sq yd} = (40 \text{ yd})h$; $h = \dfrac{580}{40} \text{ yd} = 14.5 \text{ yd}$
16) $A = \dfrac{1}{2} h(b + B) = \dfrac{1}{2}(6.3 \text{ ft})(5.3 \text{ ft} + 2.4 \text{ yd}) = \dfrac{1}{2}(6.3 \text{ ft})(5.3 \text{ ft} + 7.2 \text{ ft}) =$
    $(3.15 \text{ ft})(12.5 \text{ ft}) = 39.375 \text{ sq ft}$
17) $A = \dfrac{22}{7}(1.2 \text{ cm})^2 = \dfrac{22}{7}(1.44 \text{ sq cm}) = 4.5 \text{ sq cm}$ (to the nearest tenth cm)
18) $A = 3.14\left(\dfrac{14 \text{ dm}}{2}\right)^2 = 3.14(49 \text{ sq dm}) = 153.9 \text{ sq dm}$
19) $V = (4.3 \text{ ft})(6.1 \text{ ft})(7.4 \text{ ft}) = 194.102 \text{ cu ft}$
20) $V = (8.3 \text{ yd})^3 = 571.787 \text{ cu yd}$
21) $V = \dfrac{4}{3}\left(\dfrac{22}{7}\right)(7.7 \text{ cm})^3 = \dfrac{88}{21}(456.533 \text{ cu cm}) = 1,913 \text{ cm}$ (to the nearest cu cm)
22) $V = \dfrac{4}{3}(3.14)(6 \text{ in.})^3 = 904 \text{ cu in.}$ (to the nearest whole number)
23) $(c)^2 = (11 \text{cm})^2 + (14 \text{ cm})^2 = (121 + 196 \text{ sq cm}) = 317 \text{ sq cm}$; $c \approx \sqrt{317} \text{ cm} \approx 17.8 \text{ cm}$
24) $(24 \text{ yd})^2 = (17 \text{ yd})^2 + (b^2)^2$; $576 \text{ sq yd} = 289 \text{ sq yd} + (b^2)^2$; $(b) = 287 \text{ sq yd}$;
    $b = \sqrt{287} \text{ yd} \approx 16.9 \text{ yd}$
25) $(c)^2 = (5 \text{ ft})^2 + (12 \text{ ft})^2 = 25 \text{ sq ft} + 144 \text{ sq ft} = 169 \text{ sq ft}$; $c = \sqrt{169} \text{ ft} = 13 \text{ ft}$. $P = 5 \text{ ft} +$
    $12 \text{ ft} + 13 \text{ ft} = 30 \text{ ft}$

TEST YOURSELF ANSWERS FOR THE HANDHELD CALCULATOR

1) $P = 251.63 \text{ dm} + 197.23 \text{ dm} + 201.19 \text{ dm} = 650.05 \text{ dm}$
2) $P = 4(1{,}209.68 \text{ mm}) = 4{,}838.72 \text{ mm} = 483.872 \text{ cm}$
3) $P = 2(126.8 \text{ ft}) + 2(35.13 \text{ yd}) = 2(126.8 \text{ ft}) + 2(105.39 \text{ ft}) = 463.86 \text{ ft}$
4) $A = (126.8 \text{ ft})(35.13 \text{ yd}) = (126.8 \text{ ft})(105.39 \text{ ft}) = 13{,}363.452 \text{ sq ft}$
5) $P = 3(117.54 \text{ in.}) + 2(97.9 \text{ in.}) + 2(89.2 \text{ in.}) + 103.47 \text{ in.} = 830.29 \text{ in.}$
6) $P = 2(56.37 \text{ cm}) + 49.56 \text{ cm} = 162.3 \text{ cm}$
7) $C = 2(3.14)(113.4 \text{ ft}) = 712.152 \text{ ft}$
8) $A = (3.14)(113.4 \text{ ft})^2 = 40{,}379.0184 \text{ sq ft}$
9) $A = bh;\ 4{,}904.88 \text{ sq m} = b(64.2 \text{ m});\ b = 76.4 \text{ m}$
10) $A = \frac{1}{2} h(b + B);\ 65.5578 \text{ sq ft} = \frac{1}{2}(2.31 \text{ yd})(7.62 \text{ ft} + B) = \frac{1}{2}(6.93 \text{ ft})(7.62 \text{ ft} + B);$
    $65.5578 \text{ sq ft} = 26.4033 \text{ sq ft} + 3.465B \text{ ft};\ 3.465B \text{ ft} = 65.5578 \text{ sq ft} - 26.4033 \text{ sq ft};$
    $B = 11.3 \text{ ft}$
11) $V = (8.63 \text{ yd})(27.12 \text{ ft})(319.4 \text{ in.}) \approx (25.89 \text{ ft})(27.12 \text{ ft})(26.616 \text{ ft}) = 18{,}688.1 \text{ cu ft}$ (to the nearest tenth)
12) $V = \frac{4}{3}(3.14)(106.3 \text{ cm})^3 = 5{,}028{,}844 \text{ cu cm}$ (to the nearest whole number)
13) $h^2 = (13.6 \text{ dm})^2 + (17 \text{ dm})^2 = 184.96 \text{ sq dm} + 289 \text{ sq dm} = 473.96 \text{ sq dm};\ h = \sqrt{473.96} \text{ dm} = 21.8 \text{ dm}$ (to the nearest tenth)
14) $h^2 = (\text{leg})^2 + (\text{leg})^2;\ 4{,}225 \text{ sq ft} = (25 \text{ ft})^2 + (\text{leg})^2;\ (\text{leg})^2 = 3{,}600 \text{ sq ft};\ \text{leg} = \sqrt{3600};\ \text{leg} = 60 \text{ ft}$
15) $P = \frac{1}{2}(2)(3.14)(5 \text{ ft}) + 9 \text{ ft} + \frac{3}{4}(2)(3.14)(5 \text{ ft}) + 4 \text{ ft} + 10 \text{ ft} + \frac{1}{2}(2)(3.14)(3.5 \text{ ft}) + 6 \text{ ft} + 5 \text{ ft} = 84.24 \text{ ft}$
16) $A = \frac{1}{2}(3.14)(5 \text{ ft})^2 + \frac{1}{2}(5 \text{ ft})(9 \text{ ft}) + (14 \text{ ft})(9 \text{ ft}) + \frac{3}{4}(3.14)(5 \text{ ft})^2 + \frac{1}{2}(6 \text{ ft})(10 \text{ ft}) + [(6 \text{ ft})(7 \text{ ft}) - \frac{1}{2}(3.14)(3.5 \text{ ft})^2] = 299.3925 \text{ sq ft}$
17) Circumference of semi-circular region = $\frac{1}{2}(2)(3.14)(12 \text{ cm}) = 37.68 \text{ cm}$.
    Circumference of the two smaller semi-circular regions = $\frac{1}{2}(2)(3.14)(5.25 \text{ cm}) + \frac{1}{2}(2)(3.14)(6.75 \text{ cm}) = 37.68 \text{ cm}$. Perimeter of shaded region = $37.68 \text{ cm} + 37.68 \text{ cm} = 73.36 \text{ cm}$.
18) Area of semi-circular region = $\frac{1}{2}(3.14)(12 \text{ cm})^2 = 226.08 \text{ sq cm}$. Area of two smaller semi-circular regions = $\frac{1}{2}(3.14)(5.25 \text{ cm})^2 + \frac{1}{2}(3.14)(6.75 \text{ cm})^2 = 114.80625 \text{ sq cm}$. Area of shaded region = $(226.08 - 114.80625) \text{ sq cm} = 111.27375 \text{ sq cm}$.

19) Area of large circular region = $(3.14)(9.4 \text{ yd})^2 = 277.4504$ sq yd.

Area of triangular region = $\frac{1}{2}$ (10.3 yd)(6.1 yd) = 31.415 sq yd.

Area of rectangular region = (4.1 yd)(9.2 yd) = 37.72 sq yd.

Area of smaller circular region = $(3.14)(1.95 \text{ yd})^2 = 11.93985$ sq yd.

Area of shaded region = (277.4504 − 31.415 − 37.72 − 11.93985) sq yd = 196.37 sq yd.

20) Area of large triangular region = $\frac{1}{2}$ (13.8 mm)(16 mm) = 110.4 sq mm.

Area of unshaded rectangular region = (2 mm)(3 mm) = 6 sq mm.

Area of three unshaded triangular regions = $3\left(\frac{1}{2}\right)$(4mm)(3.45 mm) = 20.7 sq mm.

Area of two unshaded circular regions = $2(3.14)(1.4 \text{ mm})^2 = 12.3088$ sq mm.

Area of shaded region = (110.4 − 6 − 20.7 − 12.3088) sq mm = 71.3912 sq mm.

CHAPTER 9

Signed Numbers

In this chapter, we introduce integers and rational numbers, which together are called the **signed numbers**. **Integers** are treated as extensions of the whole numbers. **Rational numbers** are treated as extensions of the fractions.

The arithmetic of the signed is also discussed. Applications involving them are considered as well.

9.1 INTEGERS AND SIGNED NUMBERS

In this section, we introduce integers and signed numbers, which are the extensions of the whole numbers and fractions. The absolute value of a number is also discussed.

In earlier chapters, we discussed whole numbers, decimals, and fractions. Whole numbers are located on a line called a **number line,** as indicated in the figure below. The arrow at the right end of the number line indicates that there is no largest whole number and that the number line continues to the right indefinitely.

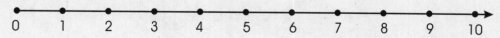

INTEGERS

Whole numbers are used to answer the question, "How many?" We now introduce a number called an **integer** to answer the question, "How many and in which direction?"

We can extend the number line and continue marking off equally spaced points to the left of 0. These new points are named with the numbers –1, –2, –3, –4, –5, and so on, as illustrated below.

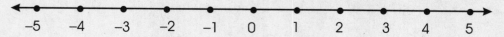

The whole numbers along with these new numbers are called **integers. Integers** are the numbers (–6, –5, –4, –3, –2, –1, 0, 1, 2, 3, 4, 5, 6, 7, and so on). The arrow at the left end of the line indicates that there is no smallest integer and that the number line continues to the left indefinitely.

The integers located to the right of 0 on the number line are called the **positive integers.** Earlier in this book, we called the positive integers the counting numbers or natural numbers. Those located to the left of 0 on the number line are called the **negative integers.** The integer 0 is neither positive nor negative. The + sign before a positive integer is generally not written. However, the negative integers are *always* written with a – sign in front of them.

For every positive integer, there is a negative integer, both of which are the same distance from 0 on the number line but in **opposite** directions. Each of these integers is called the opposite of the other. For example:

- Negative nine (–9) is the opposite of positive nine (+9).
- Eight (8) is the opposite of negative eight (–8).
- Negative seventeen (–17) is the opposite of seventeen (17).
- Positive fifty-four (+54) is the opposite of negative fifty-four (–54).
- Zero (0) is the opposite of zero (0).
- Negative eighty-nine (–89) is the opposite of positive eighty-nine (+89).
- Negative two hundred nineteen (–219) is the opposite of two hundred nineteen (219).
- Five hundred sixty-three (563) is the opposite of negative five hundred sixty-three (–563).

ORDERING OF INTEGERS

In Chapter 1, we ordered the whole numbers by examining their locations on the number line. We order integers in the same manner.

1. Let a and b be two integers. Then a is **greater than** b, denoted by $a > b$, if the integer a is to the right of the integer b on the number line.
2. Let c and d be two integers. Then c is less than d, denoted by $c < d$, if the integer c is to the left of the integer d on the number line.

Exercise 9.1
Order each of the following pairs of integers.
  a) −3 and 2
  b) 1 and −3
  c) −4 and 2
  d) 0 and −5
  e) 7 and −2
  f) −27 and 39
  g) −19 and 0
  h) 0 and −22

Solution 9.1

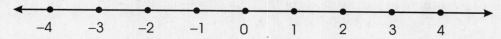

a) As shown above, −3 is to the left of 2 on the number line; hence, −3 is *less than* 2, denoted by −3 < 2.

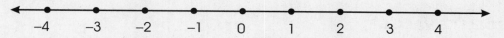

b) As shown above, 1 is to the *right* of −3 on the number line; hence, 1 is *greater than* −3, denoted by 1 > −3.

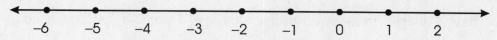

c) As shown above, −4 is to the *left* of 2 on the number line; hence, −4 is *less than* 2, denoted by −4 < 2.

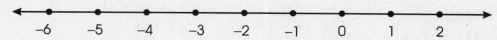

d) As shown above, 0 is to the *right* of −5 on the number line; hence, 0 is *greater than* −5, denoted by 0 > −5.

e) 7 is to the *right* of −2 on the number line; hence, 7 is *greater than* −2, denoted by 7 > −2.

f) −27 is to the *left* of 39 on the number line; hence, −27 is *less than* 39, denoted by −27 < 39.

g) −19 is to the *left* of 0 on the number line; hence, −19 is *less than* 0, denoted by −19 < 0.

h) 0 is to the *right* of −22 on the number line; hence, 0 is *greater than* −22, denoted by 0 > −22.

Exercise 9.2

Order each of the following sets of integers.

a) 4, 0, –3, +11, –23, –37
b) –13, 0, +5, –43, –57, 19, +34
c) 49, –63, 3, +72, 0, –87, 96
d) –194, 0, +672, –423, –745, –6, –1,010, 999, –17, 39

Solution 9.2

a) According to the positions of the integers on the number line, we have:
   –37 < –23 < –3 < 0 < 4 < +11

b) According to the positions of the integers on the number line, we have:
   –57 < –43 < –13 < 0 < +5 < 19 < +34

c) According to the positions of the integers on the number line, we have:
   96 > +72 > 49 > 3 > 0 > –63 > –87

d) According to the positions of the integers on the number line, we have:
   999 > +672 > 39 > 0 > –6 > –17 > –194 > –423 > –745 > –1,010

ABSOLUTE VALUE

In our discussion of integers, we looked at the sign of the integer. However, when performing arithmetic operations with integers, we also use the numerical value of an integer, which is the integer without its sign.

Let n be any integer. Then the **absolute value** of n, denoted by $|n|$, is defined as follows:

1. $|n|$ is equal to n if n is positive or 0.
2. $|n|$ is equal to the opposite of n if n is negative.

For example:
- $|+17| = 17$
- $|-59| = 59$
- $|-68| = 68$
- $|0| = 0$
- $|123| = 123$
- $|-317| = 317$

The absolute value of an integer is also the undirected distance that the integer is from 0 on the number line.

Exercise 9.3

Determine the absolute value of each of the following integers.
a) 6
b) –6
c) +8
d) –21
e) –17
f) 92
g) 0

Solution 9.3

a) $|6| = 6$, because 6 is 6 units from 0 on the number line, as shown below.

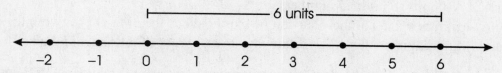

b) $|-6| = 6$, because –6 is 6 units from 0 on the number line, as shown below.

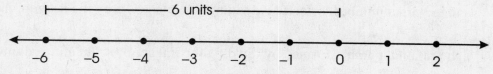

c) Because +8 is 8 units from 0 on the number line, $|+8| = 8$.
d) Because –21 is 21 units from 0 on the number line, $|-21| = 21$.
e) Because –17 is 17 units from 0 on the number line, $|-17| = 17$.
f) Because 92 is 92 units from 0 on the number line, $|92| = 92$.
g) Because 0 is 0 units from 0 on the number line, $|0| = 0$.

Exercise 9.4

For each of the following, insert the symbol >, =, or < between the two given number symbols to make a true statement:
a) $|-4|$ $|+3|$
b) $|-2| \times |-3|$ $|-(2 \times 3)|$
c) $|6| - |2|$ $|-(6 - 2)|$
d) -5 $|-(7 - 4)|$
e) $|3| - |6|$ $|3 - 6|$

Solution 9.4

a) $|-4| = 4$ and $|+3| = 3$. Because $4 > 3$, $|-4| > |+3|$.

b) $|-2| = 2$ and $|-3| = 3$. Therefore, $|-2| \times |-3| = 2 \times 3 = 6$. Also, $|-(2 \times 3)| = |-6| = 6$. Because $6 = 6$, $|-2| \times |-3| = |-(2 \times 3)|$.

c) $|6| = 6$ and $|2| = 2$. Therefore, $|6| - |2| = 6 - 2 = 4$. Also, $|-(6 - 2)| = |-4| = 4$. Because $4 = 4$, $|6| - |2| = |-(6 - 2)|$.

d) $|-(7 - 4)| = |-3| = 3$. Because $-5 < 3$, $-5 < |-(7 - 4)|$.

e) $|3| = 3$ and $|6| = 6$. Therefore, $|3| - |6| = 3 - 6 = -3$. Also, $|3 - 6| = |-3| = 3$. Because $-3 < 3$, $|3| - |6| < |3 - 6|$.

RATIONAL NUMBERS

A fraction is a quotient of whole numbers with a nonzero denominator. We now extend our discussion of fractions to rational numbers.

A **rational number** is a number that can be expressed as the quotient of two integers with a denominator different from zero. The following are examples of rational numbers:

- $\dfrac{-3}{+5}$ is the rational number formed by dividing the integer -3 by the nonzero integer $+5$.

- $\dfrac{-6}{-7}$ is the rational number formed by dividing the integer -6 by the nonzero integer -7.

- $\dfrac{0}{-8}$ is the rational number formed by dividing the integer 0 by the nonzero integer -8.

- $\dfrac{13}{-9}$ is the rational number formed by dividing the integer 13 by the nonzero integer -9.

CLASSIFYING RATIONAL NUMBERS

Rational numbers can be classified as positive, negative, or zero.

- A rational number is **positive** if its numerator and denominator are **both** positive or **both** negative.
- A rational number is **negative** if its numerator and denominator have **different** signs.
- A rational number is **0** if its numerator is 0 and its denominator is not zero.

Exercise 9.5

Classify each of the following rational numbers as being positive, negative, or zero.

a) $\dfrac{-21}{56}$

b) $\dfrac{-63}{-87}$

c) $\dfrac{+41}{27}$

d) $\dfrac{91}{-9}$

e) $\dfrac{0}{31}$

Solution 9.5

a) $\dfrac{-21}{56}$ is *negative*, because the numerator is negative and the denominator is positive.

b) $\dfrac{-63}{-87}$ is *positive*, because *both* the numerator and the denominator are negative.

c) $\dfrac{+41}{27}$ is *positive*, because *both* the numerator and denominator are positive.

d) $\dfrac{91}{-9}$ is *negative*, because the numerator is positive and the denominator is negative.

e) $\dfrac{0}{31}$ is the rational number 0, because the numerator is 0 and the denominator is nonzero.

OPPOSITES OF RATIONAL NUMBERS

If an integer is positive, its opposite is negative. If an integer is negative, its opposite is positive. Every rational number has an opposite. The opposite of a rational number is formed by taking the opposite of its numerator and leaving its denominator alone. It can also be formed by taking the opposite of its denominator and leaving its numerator alone.

Exercise 9.6

Write the opposite for each of the following rational numbers.

a) $\dfrac{49}{52}$

b) $\dfrac{73}{-6}$

c) $\dfrac{0}{-19}$

Solution 9.6

a) The rational number $\dfrac{49}{52}$ is positive. Its opposite can be written either as the negative rational number $\dfrac{-49}{52}$ or as the negative rational number $\dfrac{49}{-52}$.

b) The rational number $\dfrac{73}{-6}$ is negative. Its opposite can be written either as the positive rational number $\dfrac{-73}{-6}$ or as the positive rational number $\dfrac{73}{6}$.

c) The rational number $\dfrac{0}{-19}$ is 0. Its opposite is the rational number $\dfrac{0}{19}$, which is also 0.

ORDERING OF RATIONAL NUMBERS

The integers are ordered according to their locations on the number line. Similarly, the rational numbers are also ordered. On the number line below, the rational number $\dfrac{-5}{2}$ is to the left of the rational number $\dfrac{3}{2}$. Therefore, $\dfrac{-5}{2} < \dfrac{3}{2}$, or $\dfrac{3}{2} > \dfrac{-5}{2}$.

$$\longleftarrow \underset{-3}{\bullet} \; \underset{-5/2}{\bullet} \; \underset{-2}{\bullet} \; \underset{-3/2}{\bullet} \; \underset{-1}{\bullet} \; \underset{-1/2}{\bullet} \; \underset{0}{\bullet} \; \underset{1/2}{\bullet} \; \underset{1}{\bullet} \; \underset{3/2}{\bullet} \; \underset{2}{\bullet} \; \underset{5/2}{\bullet} \; \underset{3}{\bullet} \longrightarrow$$

If we multiply or divide *both* the numerator and the denominator of a rational number by the *same nonzero* number, the resulting fraction is **equivalent** to the original rational number. Note that this is an extension of the Fundamental Law of Fractions. In particular, multiplying both the numerator and the denominator of a rational number by –1 will produce an equivalent rational number.

Instead of using the number line to order rational numbers, we could (re)write the rational numbers as equivalent rational numbers with the *same positive denominators,* and then compare the numerators. For example:

- The rational numbers $\dfrac{-1}{2}$ and $\dfrac{-3}{5}$ can be rewritten as the equivalent rational numbers $\dfrac{-5}{10}$ and $\dfrac{-6}{10}$, respectively, with the same positive denominators, 10. Comparing the numerators of the last two rational numbers, we determine that $-6 < -5$. Therefore, $\dfrac{-6}{10} < \dfrac{-5}{10}$. Hence, looking at the original rational numbers, we conclude that $\dfrac{-3}{5} < \dfrac{-1}{2}$.

- The rational numbers $\dfrac{17}{-3}$ and $\dfrac{11}{-4}$ can be rewritten as the equivalent rational numbers $\dfrac{-68}{12}$ and $\dfrac{-33}{12}$, respectively, with the same positive denominators, 12. Comparing the numerators of the last two rational numbers, we determine that $-68 < -33$. Therefore, $\dfrac{-68}{12} < \dfrac{-33}{12}$. Hence, looking at the original rational numbers, we conclude that $\dfrac{17}{-3} < \dfrac{11}{-4}$.

ABSOLUTE VALUE OF A RATIONAL NUMBER

Earlier in this section, we indicated that the absolute value of an integer is the integer without its algebraic sign. So, too, can we consider the absolute value of a rational number.

Exercise 9.7

Determine the absolute value for each of the following rational numbers.

a) $\dfrac{-34}{29}$

b) $\dfrac{-73}{-47}$

c) $\dfrac{93}{-7}$

d) $\dfrac{0}{-11}$

Solution 9.7

a) $\left|\dfrac{-34}{29}\right| = \dfrac{34}{29}$

b) $\left|\dfrac{-73}{-47}\right| = \dfrac{73}{47}$

c) $\left|\dfrac{93}{-7}\right| = \dfrac{93}{7}$

d) $\left|\dfrac{0}{-11}\right| = 0$

Rational numbers can also be rewritten as equivalent decimal numbers. Some examples are:

- $\dfrac{7}{10} = 0.7$

- $\dfrac{-9}{10} = -\left(\dfrac{9}{10}\right) = -(0.9) = -0.9$

- $\dfrac{4}{-20} = -\left(\dfrac{4}{20}\right) = -(0.2) = -0.2$

- $\dfrac{-60}{75} = -\left(\dfrac{60}{75}\right) = -\left(\dfrac{4}{5}\right) = -(0.8) = -0.8$

The integers and rational numbers together are called **signed numbers,** which also include the decimals.

9.2 ADDITION AND SUBTRACTION OF SIGNED NUMBERS

In this section, we discuss the addition and subtraction of signed numbers. Recall from Chapter 1 that if 0 is added to a whole number or if the whole number is added to 0, the sum is that whole number. This property of addition extends to signed numbers.

If a is *any* signed number, then $a + 0 = a$ and $0 + a = a$. For example:

- $+7 + 0 = +7$
- $-21 + 0 = -21$
- $0 + (-19) = -19$
- $0 + 0 = 0$
- $0 + 3\frac{2}{7} = 3\frac{2}{7}$

To add two signed numbers with the same signs, use the following rule:

1. Disregard the signs and add the numbers as you would add whole numbers.
2. For the sum, use the sign of the numbers.

Exercise 9.8

Add the following:

a) $(+3) + (+9)$
b) $(+6) + (+11)$
c) $(-4) + (-7)$
d) $(-8) + (-9)$

Solution 9.8

a) $(+3) + (+9) = +(3 + 9) = +12$
b) $(+6) + (+11) = +(6 + 11) = +17$
c) $(-4) + (-7) = -(4 + 7) = -11$
d) $(-8) + (-9) = -(8 + 9) = -17$

Exercise 9.9

Add the following:

a) $(+3.2) + (+4.57)$
b) $(+0.981) + (+6.32)$
c) $(-16.4) + (-7.88)$
d) $(-0.015) + (-7.29)$
e) $(-4.1) + (-0.71)$

Solution 9.9

a) $(+3.2) + (+4.57) = +(3.2 + 4.57) = +7.77$
b) $(+0.981) + (+6.32) = +(0.981 + 6.32) = +7.301$
c) $(-16.4) + (-7.88) = -(16.4 + 7.88) = -24.28$
d) $(-0.015) + (-7.29) = -(0.015 + 7.29) = -7.305$
e) $(-4.1) + (-0.71) = -(4.1 + 0.71) = -4.81$

Exercise 9.10

Add the following:

a) $\left(\dfrac{+1}{3}\right) + \left(\dfrac{+3}{4}\right)$

b) $\left(\dfrac{-1}{5}\right) + \left(\dfrac{-1}{7}\right)$

c) $\left(\dfrac{+1}{2}\right) + (8.03)$

Solution 9.10

a) $\left(\dfrac{+1}{3}\right) + \left(\dfrac{+3}{4}\right) = +\left(\dfrac{1}{3} + \dfrac{3}{4}\right) = +\dfrac{13}{12}$ or $+1\dfrac{1}{12}$

b) $\left(\dfrac{-1}{5}\right) + \left(\dfrac{-1}{7}\right) = -\left(\dfrac{1}{5} + \dfrac{1}{7}\right) = \dfrac{-12}{35}$

c) $\left(\dfrac{+1}{2}\right) + (8.03) = +\left(\dfrac{1}{2} + 8.03\right) = 8.53$

To add two signed numbers with different signs, use the following rule:

1. Disregard the signs and subtract the smaller absolute value from the larger as you would subtract whole numbers.
2. For the answer, take the sign of the number with the *larger* absolute value.

Exercise 9.11

Add the following:

a) $(+5) + (-8)$
b) $(-17) + (+26)$

Solution 9.11

a) $(+5) + (-8) = -(8-5) = -3$ (because $|-8| > |+5|$, take $(-)$ sign)

b) $(-17) + (+26) = +(26-17) = +9$ (because $|+26| > |-17|$, take $(+)$ sign)

Exercise 9.12
Add the following:

a) $(-5.1) + (+9.2)$

b) $(-9.8) + (+3.04)$

c) $(+2.87) + (-8.4)$

d) $(-23.09) + (+11.5)$

e) $(-2.73) + (+11.1)$

Solution 9.12

a) $(-5.1) + (+9.2) = +(9.2 - 5.1) = +4.1$ (because $|+9.2| > |-5.1|$, take $(+)$ sign)

b) $(-9.8) + (+3.04) = -(9.8 - 3.04) = -6.76$ (because $|-9.8| > |+3.04|$, take $(-)$ sign)

c) $(+2.87) + (-8.4) = -(8.4 - 2.87) = -5.53$ (because $|-8.4| > |+2.87|$, take $(-)$ sign)

d) $(-23.09) + (+11.5) = -(23.09 - 11.5) = -11.59$ (because $|-23.09| > |+11.5|$, take $(-)$ sign)

e) $(-2.73) + (11.1) = +(11.1 - 2.73) = +8.37$ (because $|11.1| > |-2.73|$, take $(+)$ sign)

Exercise 9.13
Add the following:

a) $\left(\dfrac{+4}{5}\right) + \left(\dfrac{-1}{3}\right)$

b) $\left(\dfrac{-7}{8}\right) + \left(\dfrac{+3}{5}\right)$

c) $\left(\dfrac{-11}{12}\right) + \left(\dfrac{+4}{7}\right)$

d) $\left(\dfrac{-4}{5}\right) + \left(\dfrac{1}{3}\right)$

e) $\left(\dfrac{-2}{9}\right) + \left(\dfrac{+4}{7}\right)$

Solution 9.13

a) $\left(\dfrac{+4}{5}\right)+\left(\dfrac{-1}{3}\right)=+\left(\dfrac{4}{5}-\dfrac{1}{3}\right)=\dfrac{+7}{15}$ (because $\left|\dfrac{+4}{5}\right|>\left|\dfrac{-1}{3}\right|$, take (+) sign)

b) $\left(\dfrac{-7}{8}\right)+\left(\dfrac{+3}{5}\right)=\left(\dfrac{7}{8}-\dfrac{3}{5}\right)=\dfrac{-11}{40}$ (because $\left|\dfrac{-7}{8}\right|>\left|\dfrac{+3}{5}\right|$, take (−) sign)

c) $\left(\dfrac{-11}{12}\right)+\left(\dfrac{+4}{7}\right)=-\left(\dfrac{11}{12}-\dfrac{4}{7}\right)=\dfrac{-29}{84}$ (because $\left|\dfrac{-11}{12}\right|>\left|\dfrac{4}{7}\right|$, take (−) sign)

d) $\left(\dfrac{-4}{5}\right)+\left(\dfrac{1}{3}\right)=-\left(\dfrac{4}{5}-\dfrac{1}{3}\right)=\dfrac{-7}{15}$

e) $\left(\dfrac{-2}{9}\right)+\left(\dfrac{+4}{7}\right)=+\left(\dfrac{4}{7}-\dfrac{2}{9}\right)=\dfrac{22}{63}$

To add two signed numbers, you must first determine whether the numbers have the same signs or different signs, and then apply the appropriate rule.

Exercise 9.14

Add the following:

a) (+15) + (+32)
b) (+24.9) + (−37.8)
c) (−18.2) + (+11.57)
d) (+671) + (−908)

Solution 9.14

a) The signs are the same, (positive) so *add* the absolute values of the numbers and place the common sign in front of the sum. Therefore, (+15) + (+32) = +(15 + 32) = +47.

b) The signs are different, so *subtract* the absolute values and place the sign of the number that has the larger absolute value in front of this difference. Therefore, (+24.9) + (−37.8) = −(37.8 − 24.9) = −12.9.

c) The signs are different, so *subtract* the absolute values and place the sign of the number that has the larger absolute value in front of this difference. Therefore, (−18.2) + (+11.57) = −(18.2 − 11.57) = −6.63.

d) The signs are different, so *subtract* the absolute values and place the sign of the number that has the larger absolute value in front of this difference. Therefore, (+671) + (−908) = −(908 − 671) = −237.

Exercise 9.15

Add the following:

a) $\left(\dfrac{-4}{5}\right) + \left(\dfrac{-2}{7}\right)$

b) $\left(-2\dfrac{1}{2}\right) + \left(+8\dfrac{3}{7}\right)$

c) $\left(\dfrac{+4}{11}\right) + \left(\dfrac{-5}{7}\right)$

d) $\left(7\dfrac{1}{6}\right) + \left(-2\dfrac{4}{5}\right)$

e) $\left(-1\dfrac{3}{4}\right) + \left(-4\dfrac{1}{2}\right)$

Solution 9.15

a) The signs are the same (negative), so *add* the absolute values and place the common sign in front of the sum. Therefore,

$$\left(\dfrac{-4}{5}\right) + \left(\dfrac{-2}{7}\right) = -\left(\dfrac{4}{5} + \dfrac{2}{7}\right) = \dfrac{-38}{35} \text{ or } -1\dfrac{3}{35}$$

b) The signs are different so, *subtract* the absolute values and place the sign of the number that has the larger absolute value in front of this difference. Therefore,

$$\left(-2\dfrac{1}{2}\right) + \left(8\dfrac{3}{7}\right) = +\left(8\dfrac{3}{7} - 2\dfrac{1}{2}\right) = 5\dfrac{13}{14}$$

c) The signs are different, so *subtract* the absolute values and place the sign of the number that has the larger absolute value in front of this difference. Therefore,

$$\left(\dfrac{+4}{11}\right) + \left(\dfrac{-5}{7}\right) = -\left(\dfrac{5}{7} - \dfrac{4}{11}\right) = \dfrac{-27}{77}$$

d) The signs are different, so *subtract* the absolute values and place the sign of the number that has the larger absolute value in front of this difference. Therefore,

$$\left(7\dfrac{1}{6}\right) + \left(-2\dfrac{4}{5}\right) = +\left(7\dfrac{1}{6} - 2\dfrac{4}{5}\right) = 4\dfrac{11}{30}$$

e) The signs are the same, so *add* the absolute values of the numbers and place the common sign in front of the sum. Therefore,

$$\left(-1\dfrac{3}{4}\right) + \left(-4\dfrac{1}{2}\right) = -\left(1\dfrac{3}{4} + 4\dfrac{1}{2}\right) = -6\dfrac{1}{4}$$

Exercise 9.16
Add:
$$(+4) + (-7) + (+13)$$

Solution 9.16
We shall solve this problem by adding the numbers, two at a time, from left to right.
$$(+4) + (-7) + (+13)$$
$$= -(7-4) + (+13)$$
$$= (-3) + (+13)$$
$$= +(13 - 3)$$
$$= +10$$

Exercise 9.17
Add:
$$(-10.2) + (-11.01) + (+9.87) + (-0.89)$$

Solution 9.17

$$(-10.2) + (-11.01) + (+9.87) + (-0.89)$$
$$= -(10.2 + 11.01) + (+9.87) + (-0.89)$$
$$= (-21.21) + (+9.87) + (-0.89)$$
$$= -(21.21 - 9.87) + (-0.89)$$
$$= -(-11.34) + (-0.89)$$
$$= (11.34 + 0.89)$$
$$= -12.23$$

Exercise 9.18
Add:
$$(-7) + (+9) + (-2) + (+8)$$

Solution 9.18
We shall solve this problem by adding two pairs of integers together.
$$(-7) + (+9) + (-2) + (+8)$$
$$= [(-7) + (+9)] + [(-2) + (+8)]$$
$$= +(9 - 7) + [+(8 - 2)]$$
$$= (+2) + (+6)$$
$$= +(2 + 6)$$
$$= 8$$

Exercise 9.19

Add: $$\left(-4\frac{1}{2}\right)+\left(+5\frac{2}{3}\right)+\left(+2\frac{1}{4}\right)+\left(-6\frac{3}{5}\right)$$

Solution 9.19

$$\left(-4\frac{1}{2}\right)+\left(+5\frac{2}{3}\right)+\left(+2\frac{1}{4}\right)+\left(-6\frac{3}{5}\right)$$

$$=\left[\left(-4\frac{1}{2}\right)+\left(+5\frac{2}{3}\right)\right]+\left[\left(+2\frac{1}{4}\right)+\left(-6\frac{3}{5}\right)\right]$$

$$=+\left(5\frac{2}{3}-4\frac{1}{2}\right)+\left[-\left(6\frac{3}{5}-2\frac{1}{4}\right)\right]$$

$$=+\left(1\frac{1}{6}\right)+\left(-4\frac{7}{20}\right)$$

$$=\left(4\frac{7}{20}-1\frac{1}{6}\right)$$

$$=\left(4\frac{21}{60}-1\frac{10}{60}\right)$$

$$=-3\frac{11}{60}$$

Exercise 9.20

Add: $(23.1) + (-19.7) + (-5.78) + (+34.09) + (-15.94)$

Solution 9.20

Solve this problem by grouping the numbers according to their signs.

$(23.1) + (-19.7) + (-5.78) + (+34.09) + (-15.94)$
$= [(23.1) + (+34.09)] + [(-19.7) + (-5.78) + (-15.94)]$
$= +(23.1 + 34.09) + [-(19.7 + 5.78 + 15.94)]$
$= +(57.19) + [-(41.42)]$
$= +(57.19 - 41.42)$
$= 15.77$

Exercise 9.21

Add: $$\left(-1\frac{1}{3}\right)+\left(+2\frac{3}{4}\right)+\left(-4\frac{2}{5}\right)+\left(5\frac{1}{2}\right)+\left(-3\frac{3}{5}\right)+\left(+6\frac{2}{7}\right)$$

Solution 9.21

$$\left(-1\frac{1}{3}\right)+\left(+2\frac{3}{4}\right)+\left(-4\frac{2}{5}\right)+\left(5\frac{1}{2}\right)+\left(-3\frac{3}{5}\right)+\left(+6\frac{2}{7}\right)$$

$$=\left[\left(-1\frac{1}{3}\right)+\left(-4\frac{2}{5}\right)+\left(-3\frac{3}{5}\right)\right]+\left[\left(+2\frac{3}{4}\right)+\left(5\frac{1}{2}\right)+\left(+6\frac{2}{7}\right)\right]$$

$$=\left(-9\frac{1}{3}\right)+\left(+14\frac{15}{28}\right)=\left(-9\frac{28}{84}\right)+\left(+14\frac{45}{84}\right)=+5\frac{17}{84}$$

Subtracting a signed number is the same as adding its opposite: To subtract one signed number from another signed number, replace the number to be subtracted with its opposite and add.

Exercise 9.22
Subtract
  a) $(-19) - (-7)$
  b) $(+13) - (+29)$

Solution 9.22
  a) $(-19) - (-7) = (-19) \mathbf{+ (+7)}$
  Instead of subtracting –7, add its opposite, +7.
  $= -(\mathbf{19 - 7})$
  $= -12$
  Check: $(-7) + (-12) = -(7 + 12) = -19$ ✓
  b) $(+13) - (+29) = (+13) \mathbf{+ (-29)}$
  Instead of subtracting +29, add its opposite, –29.
  $= -(29 - 13)$
  $= -16$
  Check: $(+29) + (-16) = +(29 - 16) = +13$ ✓

Exercise 9.23
Subtract:
  a) $(-8.2) - (4.19)$
  b) $(+6.23) - (-12.5)$
  c) $0 - (-34.69)$
  d) $(-11.45) - (+9.506)$

Solution 9.23

a) $(-8.2) - (+4.19) = (-8.2) + (-4.19)$
Instead of subtracting +4.19, add its opposite, −4.19.
$= -(8.2 + 4.19)$
$= -12.39$
Check: $(+4.19) + (-12.39) = -(12.39 - 4.19) = -8.2$ ✓

b) $(+6.23) - (-12.5) = (+6.23) + (+12.5)$
Instead of subtracting −12.5, add its opposite, +12.5.
$= +(6.23 + 12.5)$
$= 18.73$
Check: $(-12.5) + (18.73) = +(18.73 - 12.5) = 6.23$ ✓

c) $0 - (-34.69) = 0 + (+34.69)$
Instead of subtracting −34.69, add its opposite, +34.69.
$= 34.69$
Check: $(-34.69) + (34.69) = 0$ ✓

d) $(-11.45) - (+9.506) = (-11.45) + (-9.506)$
Instead of subtracting +9.506, add its opposite, −9.506.
$= -(11.45 + 9.506)$
$= -20.956$
Check: $(+9.506) + (-20.956) = -(20.956 - 9.506) = -11.45$ ✓

Exercise 9.24

Subtract:

a) $\left(+7\frac{1}{3}\right) - \left(+3\frac{1}{2}\right)$

b) $\left(\frac{+4}{7}\right) - \left(\frac{+6}{11}\right)$

c) $\left(-4\frac{2}{5}\right) - \left(+5\frac{3}{7}\right)$

Solution 9.24

a) $\left(+7\frac{1}{3}\right) - \left(+3\frac{1}{2}\right) = \left(+7\frac{1}{3}\right) + \left(-3\frac{1}{2}\right)$

Instead of subtracting $+3\frac{1}{2}$, add its opposite, $-3\frac{1}{2}$.

$= +\left(7\frac{1}{3} - 3\frac{1}{2}\right) = +\left(7\frac{2}{6} - 3\frac{3}{6}\right) = +\left(6\frac{8}{6} - 3\frac{3}{6}\right)$

$$= +3\frac{5}{6}$$

Check: $\left(+3\frac{1}{2}\right) + \left(+3\frac{5}{6}\right) = +\left(3\frac{1}{2} + 3\frac{5}{6}\right) = +\left(3\frac{3}{6} + 3\frac{5}{6}\right) = \left(6\frac{8}{6}\right) = \left(6 + 1\frac{2}{6}\right) = \left(6 + 1\frac{1}{3}\right) = 7\frac{1}{3}$ ✓

b) $\left(\frac{+4}{7}\right) - \left(\frac{+6}{11}\right) = \left(\frac{+4}{7}\right) + \left(\frac{-6}{11}\right)$

Instead of subtracting $\frac{+6}{11}$, add its opposite, $\frac{-6}{11}$.

$$= +\left(\frac{4}{7} - \frac{+6}{11}\right) = +\left(\frac{44}{77} - \frac{+42}{77}\right)$$

$$= \frac{2}{77}$$

Check: $\left(\frac{+6}{11}\right) + \left(\frac{2}{77}\right) = +\left(\frac{6}{11} + \frac{2}{77}\right) = +\left(\frac{42}{77} + \frac{2}{77}\right) = \frac{+44}{77} = \frac{+4}{7}$ ✓

c) $\left(-4\frac{2}{5}\right) - \left(+5\frac{3}{7}\right) = \left(-4\frac{2}{5}\right) + \left(-5\frac{3}{7}\right)$

Instead of subtracting $+5\frac{3}{7}$, add its opposite, $-5\frac{3}{7}$.

$$= -\left(4\frac{2}{5} + 5\frac{3}{7}\right) = -\left(4\frac{14}{35} + 5\frac{15}{35}\right)$$

$$= -9\frac{29}{35}$$

Check: $\left(+5\frac{3}{7}\right) + \left(-9\frac{29}{35}\right) = -\left(9\frac{29}{35} - 5\frac{3}{7}\right) = -\left(9\frac{29}{35} - 5\frac{15}{35}\right) = -4\frac{14}{35} = -4\frac{2}{5}$ ✓

The following examples involve both addition and subtraction.

Exercise 9.25

Perform the indicated operations: $(+16) + (-13) - (-12) + (-15) - (+13)$

Solution 9.25

$$(+16) + (-13) - (-12) + (-15) - (+13)$$

Rewriting the subtractions as equivalent additions

$$= (+16) + (-13) + (+12) + (-15) + (-13). \text{ Regrouping:}$$
$$= [(+16) + (+12)] + [(-13) + (-15) + (-13)]$$
$$= +(16 + 12) + [-(13 + 15 + 13)]$$
$$= (+28) + (-41)$$

$$= -(41 - 28)$$
$$= -13$$

Exercise 9.26
Perform the indicated operations: $(-15) - (-24) - (+2) - (-27) + (-16)$

Solution 9.26
$$(-15) - (-24) - (+2) - (-27) + (-16)$$
$$= (-15) + (+24) + (-2) + (+27) + (-16)$$
$$= [(-15) + (-2) + (-16)] + [(+24) + (+27)]$$
$$= -(15 + 2 + 16) + [+(24 + 27)]$$
$$= (-33) + (+51)$$
$$= +(51 - 33)$$
$$= 18$$

Exercise 9.27
Perform the indicated operations: $(-6.02) - (-9.4) - (+5.13)$

Solution 9.27
$$(-6.02) - (-9.4) - (+5.13)$$
$$= (-6.02) + (+9.4) + (-5.13)$$
$$= -(6.02 + 5.13) + (+9.4)$$
$$= (-11.15) + (+9.4)$$
$$= -(11.15 - 9.4)$$
$$= -1.75$$

Exercise 9.28
Perform the indicated operations: $(2.91) + (-3.4) - (-0.59)$

Solution 9.28
$$(2.91) + (-3.4) - (-0.59)$$
$$= (2.91) + (-3.4) + (+0.59)$$
$$= +(2.91 + 0.59) + (-3.4)$$
$$= (+3.5) + (-3.4)$$
$$= +(3.5 - 3.4)$$
$$= 0.1$$

Exercise 9.29
Perform the indicated operations: $\left(-2\dfrac{1}{2}\right) + \left(+3\dfrac{1}{4}\right) - \left(4\dfrac{1}{5}\right)$

Solution 9.29

$$\left(-2\frac{1}{2}\right)+\left(+3\frac{1}{4}\right)-\left(4\frac{1}{5}\right)$$

$$=\left(-2\frac{1}{2}\right)+\left(+3\frac{1}{4}\right)+\left(-4\frac{1}{5}\right)$$

$$=-\left(2\frac{1}{2}+4\frac{1}{5}\right)+\left(+3\frac{1}{4}\right)$$

$$=\left(-6\frac{7}{10}\right)+\left(+3\frac{1}{4}\right)$$

$$=-\left(6\frac{7}{10}-3\frac{1}{4}\right)$$

$$=-\left(6\frac{14}{20}-3\frac{5}{20}\right)$$

$$=-3\frac{9}{20}$$

9.3 MULTIPLICATION AND DIVISION OF SIGNED NUMBERS

In this section, we discuss multiplication and division of signed numbers. Recall from Chapter 1 that if a whole number is multiplied by 0 or if 0 is multiplied by a whole number, the product is 0. This property extends to signed numbers as well. If a is any signed number, then

$$a \times 0 = 0 \quad \text{and} \quad 0 \times a = 0$$

For example:
- $(+7) \times 0 = 0$
- $(-8.31) \times 0 = 0$
- $0 \times \left(\dfrac{-4}{9}\right) = 0$
- $0 \times \left(-5\dfrac{1}{6}\right) = 0$
- $0 \times 0 = 0$

We multiply signed numbers according to the following rules:
1. Disregard the signs and multiply the numbers as you would multiply whole numbers.
2. The product will be:
    a. 0, if either number is 0.
    b. Positive, if the two numbers have the *same* sign.
    c. Negative, if the two numbers have *different* signs.

Exercise 9.30
Multiply:
a) $(+2) \times (+3)$
b) $(-13) \times (+50)$

Solution 9.30
a) $(+2) \times (+3) = +(2 \times 3) = 6$ (*same* signs)
b) $(-13) \times (+50) = -(13 \times 50) = -650$ (*different* signs)

Exercise 9.31
Multiply:
a) $(+2.3) \times (-3.1)$
b) $(-1.2) \times (+1.3)$
c) $(6) \times (-0.7)$
d) $(+9.27) \times 0$
e) $(-1.4) \times (-8)$
f) $(+2.1) \times (-48)$
g) $(-3.5) \times (-2.7)$

Solution 9.31
a) $(+2.3) \times (-3.1) = -(2.3 \times 3.1) = -7.13$ (*different* signs)
b) $(-1.2) \times (+1.3) = -(1.2 \times 1.3) = -1.56$ (*different* signs)
c) $(6) \times (-0.7) = -(6 \times 0.7) = -4.2$ (*different* signs)
d) $(+9.27) \times 0 = 0$ (one factor is *0*)
e) $(-1.4) \times (-8) = +(1.4 \times 8) = +11.2$ (*same* signs)
f) $(+2.1) \times (-48) = -(2.1 \times 48) = -100.8$ (*different* signs)
g) $(-3.5) \times (-2.7) = +(3.5 \times 2.7) = 9.45$ (*same* signs)

Exercise 9.32
Multiply:
a) $\left(\dfrac{+1}{3}\right) \times \left(\dfrac{+5}{7}\right)$

b) $\left(-3\dfrac{1}{2}\right) \times \left(-4\dfrac{1}{3}\right)$

c) $\left(\dfrac{-3}{7}\right) \times \left(\dfrac{+4}{9}\right)$

d) $\left(+5\dfrac{1}{3}\right) \times \left(+6\dfrac{3}{4}\right)$

e) $\left(\dfrac{-3}{8}\right) \times \left(+4\dfrac{1}{7}\right)$

f) $\left(-7\dfrac{1}{2}\right) \times \left(3\dfrac{2}{3}\right)$

g) $0 \times \left(-7\dfrac{1}{9}\right)$

Solution 9.32

a) $\left(\dfrac{+1}{3}\right) \times \left(\dfrac{+5}{7}\right) = +\left(\dfrac{1}{3} \times \dfrac{5}{7}\right) = \dfrac{5}{21}$ (*same* signs)

b) $\left(-3\dfrac{1}{2}\right) \times \left(-4\dfrac{1}{3}\right) = +\left(3\dfrac{1}{2} \times 4\dfrac{1}{3}\right) = +\left(\dfrac{7}{2} \times \dfrac{13}{3}\right) = \dfrac{91}{6} = 15\dfrac{1}{6}$ (*same* signs)

c) $\left(\dfrac{-3}{7}\right) \times \left(\dfrac{+4}{9}\right) = -\left(\dfrac{3}{7} \times \dfrac{4}{9}\right) = \dfrac{-12}{63} = \dfrac{-4}{21}$ (*different* signs)

d) $\left(+5\dfrac{1}{3}\right) \times \left(+6\dfrac{3}{4}\right) = +\left(5\dfrac{1}{3} \times 6\dfrac{3}{4}\right) = +\left(\dfrac{16}{3} \times \dfrac{27}{4}\right) = +\left(\dfrac{4}{1} \times \dfrac{9}{1}\right) = \dfrac{36}{1} = 36$ (*same* signs)

e) $\left(\dfrac{-3}{8}\right) \times \left(+4\dfrac{1}{7}\right) = -\left(\dfrac{3}{8} \times 4\dfrac{1}{7}\right) = -\left(\dfrac{3}{8} \times \dfrac{29}{7}\right) = -\dfrac{87}{56} = -1\dfrac{31}{56}$ (*different* signs)

f) $\left(-7\dfrac{1}{2}\right) \times \left(3\dfrac{2}{3}\right) = -\left(7\dfrac{1}{2} \times 3\dfrac{2}{3}\right) = -\left(\dfrac{15}{2} \times \dfrac{11}{3}\right) = -\left(\dfrac{5}{2} \times \dfrac{11}{1}\right) = -\dfrac{55}{2} = -27\dfrac{1}{2}$
(*different* signs)

g) $0 \times \left(-7\dfrac{1}{9}\right) = 0$ (one factor is *0*)

In the following examples, we consider multiplying three or more signed numbers.

Exercise 9.33

Multiply: $(+0.7) \times (-9) \times (+0.4)$

Solution 9.33

$(+0.7) \times (-9) \times (+0.4)$
$= -(0.7 \times 9) \times (+0.4)$ (*different* signs)
$= (-6.3) \times (+0.4)$
$= -(6.3 \times 0.4)$ (*different* signs)
$= -2.52$

Exercise 9.34

Multiply: $(-28) \times (+0.4) \times (-35) \times (-7) \times (-0.3)$

Solution 9.34

$(-28) \times (+0.4) \times (-35) \times (-7) \times (-0.3)$
$= -(28 \times 0.4) \times (-35) \times (-7) \times (-0.3)$ (*different* signs)
$= (-11.2) \times (-35) \times (-7) \times (-0.3)$
$= +(11.2 \times 35) \times (-7) \times (-0.3)$ (*same* signs)
$= (+392) \times (-7) \times (-0.3)$
$= -(392 \times 7) \times (-0.3)$ (*different* signs)
$= (-2,744) \times (-0.3)$
$= +(2,744 \times 0.3)$ (*same* signs)
$= 823.2$

Exercise 9.35

Multiply: $\left(\dfrac{-2}{5}\right) \times \left(\dfrac{+6}{11}\right) \times \left(\dfrac{+7}{9}\right)$

Solution 9.35

$\left(\dfrac{-2}{5}\right) \times \left(\dfrac{+6}{11}\right) \times \left(\dfrac{+7}{9}\right)$

$= -\left(\dfrac{2}{5} \times \dfrac{6}{11}\right) \times \left(\dfrac{+7}{9}\right)$ (*different* signs)

$= \left(\dfrac{-12}{55}\right) \times \left(\dfrac{+7}{9}\right)$

$= -\left(\dfrac{\overset{4}{\cancel{12}}}{55} \cdot \dfrac{7}{\underset{3}{\cancel{9}}}\right) = -\left(\dfrac{4}{55} \cdot \dfrac{7}{3}\right)$ (*different* signs)

$= \dfrac{-28}{165}$

When multiplying three or more signed numbers, disregard the signs and multiply as you would with whole numbers. The product will be

- 0, if at least one of the factors is 0.
- Positive, if there is an even number of negative factors.
- Negative, if there is an odd number of negative factors.

Exercise 9.36
Multiply: $(-7) \times (-8) \times (+3) \times (-9) \times (+5)$

Solution 9.36
There are three (which is an *odd* number) negative factors in the product and no 0 factor; the product will be negative.
$$(-7) \times (-8) \times (+3) \times (-9) \times (+5)$$
$$= -(7 \times 8 \times 3 \times 9 \times 5)$$
$$= -7{,}560$$

Exercise 9.37
Multiply: $(-0.4) \times (+1.3) \times (-2.1) \times (-3.4) \times (-0.3)$

Solution 9.37
There are four (which is an *even* number) negative factors in the product and no 0 factor; the product will be *positive*.
$$(-0.4) \times (+1.3) \times (-2.1) \times (-3.4) \times (-0.3)$$
$$= +(0.4 \times 1.3 \times 2.1 \times 3.4 \times 0.3)$$
$$= 1.11384$$

Division and multiplication are inverse operations. Therefore, we can convert a division problem into an equivalent multiplication problem, as we have done in earlier chapters. Hence, the rules for dividing signed numbers are basically the same as those for multiplying signed numbers: Disregard the signs and divide the numbers as you would divide whole numbers. The quotient will be:

- 0, if the numerator is 0 and the denominator is nonzero.
- Positive, if the numerator and denominator have the *same* sign.
- Negative, if the numerator and denominator have *different* signs.

Exercise 9.38
Divide:

a) $(+27) \div (+3)$
b) $(-50) \div (-5)$
c) $0 \div (-57)$
d) $(-54) \div (+3)$
e) $(+26) \div (-2)$
f) $(+47) \div (-10)$
g) $(+29) \div (+8)$
h) $0 \div (-0.23)$

Solution 9.38

a) $(+27) \div (+3) = +(27 \div 3) = 9$ (*same* signs)
b) $(-50) \div (-5) = +(50 \div 5) = 10$ (*same* signs)
c) $0 \div (-57) = 0$ (numerator = 0, denominator ≠ 0)
d) $(-54) \div (+3) = -(54 \div 3) = -18$ (*different* signs)
e) $(+26) \div (-2) = -(26 \div 2) = -13$ (*different* signs)
f) $(+47) \div (-10) = -(47 \div 10) = -4.7$ (*different* signs)
g) $(+29) \div (+8) = +(29 \div 8) = 3.265$ (*same* signs)
h) $0 \div (-0.23) = 0$ (numerator = 0, denominator ≠ 0)

Exercise 9.39

Divide:

a) $(-3.6) \div (+0.04)$
b) $(+0.26) \div (+1.6)$

Solution 9.39

a) $(-3.6) \div (+0.04) = -(3.6 \div 0.04) = -90$ (*different* signs)
b) $(+0.26) \div (+1.6) = +(0.26 \div 1.6) = 0.1625$ (*same* signs)

We divide rational numbers as we do fractions but include the appropriate sign for the quotient.

Exercise 9.40

Divide:

a) $\left(\dfrac{+2}{5}\right) \div \left(\dfrac{+3}{8}\right)$

b) $\left(\dfrac{-4}{9}\right) \div \left(\dfrac{-5}{7}\right)$

c) $\left(\dfrac{0}{-8}\right) \div \left(\dfrac{-3}{7}\right)$

d) $\left(\dfrac{-11}{15}\right) \div \left(\dfrac{+2}{3}\right)$

e) $\left(\dfrac{-3}{5}\right) \div \left(\dfrac{-2}{9}\right)$

f) $\left(\dfrac{-5}{7}\right) \div \left(\dfrac{+4}{9}\right)$

Solution 9.40

a) $\left(\dfrac{+2}{5}\right) \div \left(\dfrac{+3}{8}\right) = \left(\dfrac{+2}{5}\right) \times \left(\dfrac{+8}{3}\right)$

$$= \dfrac{(+2) \times (+8)}{5 \times 3}$$

$$= \dfrac{+(2 \times 8)}{15}$$

$$= \dfrac{+16}{15} \left(\text{or } +1\dfrac{1}{15}\right)$$

b) $\left(\dfrac{-4}{9}\right) \div \left(\dfrac{-5}{7}\right) = \left(\dfrac{-4}{9}\right) \times \left(\dfrac{-7}{5}\right)$

$$= \dfrac{(-4) \times (-7)}{9 \times 5}$$

$$= \dfrac{+(4 \times 7)}{45}$$

$$= \dfrac{+28}{45}$$

c) $\left(\dfrac{0}{-8}\right) \div \left(\dfrac{-3}{7}\right) = \left(\dfrac{0}{-8}\right) \times \left(\dfrac{-7}{3}\right)$

$$= \dfrac{0 \times (-7)}{(-8) \times 3}$$

$$= \dfrac{0}{-(8 \times 3)}$$

$$= \dfrac{0}{-24}$$

$$= 0$$

d) $\left(\dfrac{-11}{15}\right) \div \left(\dfrac{+2}{3}\right) = \left(\dfrac{-11}{15}\right) \times \left(\dfrac{+3}{2}\right)$

$$= \dfrac{(-11) \times (+3)}{15 \times 2}$$

$$= \dfrac{-(11 \times 3)}{30}$$

$$= \frac{-33}{30} = \frac{-11}{10} \left(\text{or } -1\frac{1}{10}\right)$$

e) $\left(\frac{-3}{5}\right) \div \left(\frac{-2}{9}\right) = \left(\frac{-3}{5}\right) \times \left(\frac{-9}{2}\right)$

$$= \frac{(-3) \times (-9)}{5 \times 2}$$

$$= \frac{+(3 \times 9)}{5 \times 2}$$

$$= \frac{27}{10}$$

f) $\left(\frac{-5}{7}\right) \div \left(\frac{+4}{9}\right) = \left(\frac{-5}{7}\right) \times \left(\frac{+9}{4}\right)$

$$= \frac{(-5) \times (+9)}{7 \times 4}$$

$$= \frac{-(5 \times 9)}{7 \times 4}$$

$$= \frac{-45}{28} \left(\text{or } -1\frac{17}{28}\right)$$

To divide signed mixed numbers, first rewrite them as rational numbers.

Exercise 9.41
Divide:

a) $\left(-9\frac{1}{2}\right) \div \left(+4\frac{2}{3}\right)$

b) $\left(-4\frac{1}{2}\right) \div \left(2\frac{2}{3}\right)$

c) $\left(2\frac{1}{5}\right) \div \left(-3\frac{1}{4}\right)$

Solution 9.41

a) $\left(-9\dfrac{1}{2}\right) \div \left(+4\dfrac{2}{3}\right) = \left(\dfrac{-19}{2}\right) \div \left(\dfrac{+14}{3}\right)$

$\qquad\qquad\qquad = \left(\dfrac{-19}{2}\right) \times \left(\dfrac{+3}{14}\right)$

$\qquad\qquad\qquad = \dfrac{(-19) \times (+3)}{2 \times 14}$

$\qquad\qquad\qquad = \dfrac{-(19 \times 3)}{28}$

$\qquad\qquad\qquad = \dfrac{-57}{28}$

$\qquad\qquad\qquad = -2\dfrac{1}{28}$

b) $\left(-4\dfrac{1}{2}\right) \div \left(2\dfrac{2}{3}\right) = \left(\dfrac{-9}{2}\right) \div \left(\dfrac{+8}{3}\right)$

$\qquad\qquad\qquad = \left(\dfrac{-9}{2}\right) \times \left(\dfrac{+3}{8}\right)$

$\qquad\qquad\qquad = \dfrac{(-9) \times (+3)}{2 \times 8}$

$\qquad\qquad\qquad = \dfrac{-(9 \times 3)}{2 \times 8}$

$\qquad\qquad\qquad = \dfrac{-27}{16}$

$\qquad\qquad\qquad = -1\dfrac{11}{16}$

c) $\left(-2\dfrac{1}{5}\right) \div \left(-3\dfrac{1}{4}\right) = \left(\dfrac{-11}{5}\right) \div \left(\dfrac{-13}{4}\right)$

$\qquad\qquad\qquad = \left(\dfrac{-11}{5}\right) \times \left(\dfrac{-4}{13}\right)$

$\qquad\qquad\qquad = \dfrac{(-11) \times (-4)}{5 \times 13}$

$$=\frac{+(11\times 4)}{5\times 13}$$

$$=\frac{44}{65}$$

9.4 SIMPLIFYING SIGNED NUMBERS

In this section, we discuss simplifying signed numbers. If we multiply or divide *both* the numerator and the denominator of a rational number by the same *nonzero* integer, the resulting rational number is equivalent to the original rational number. For example:

- The rational numbers $\frac{2}{-7}$ and $\frac{4}{-14}$ are equivalent, because

$$\frac{2}{-7}=\frac{2\times 2}{-7\times 2}=\frac{4}{-14}.$$

- The rational numbers $\frac{3}{8}$ and $\frac{-18}{-48}$ are equivalent, because

$$\frac{3}{8}=\frac{3\times(-6)}{8\times(-6)}=\frac{-18}{-48}.$$

- The rational numbers $\frac{0}{9}$ and $\frac{0}{-63}$ are equivalent, because

$$\frac{0}{9}=\frac{0\times(-7)}{9\times(-7)}=\frac{0}{-63}.$$

- The rational numbers $\frac{-5}{+7}$ and $\frac{+5}{-7}$ are equivalent, because

$$\frac{-5}{+7}=\frac{(-5)\times(-1)}{(+7)\times(-1)}=\frac{+5}{-7}.$$

- The rational numbers $\frac{-11}{7}$ and $\frac{+88}{-56}$ are equivalent, because

$$\frac{-11}{7}=\frac{(-11)\times(-8)}{(7)\times(-8)}=\frac{+88}{-56}.$$

- The rational numbers $\frac{8}{-32}$ and $\frac{2}{-8}$ are equivalent, because

$$\frac{8}{-32}=\frac{8\div 4}{-32\div 4}=\frac{2}{-8}.$$

- The rational numbers $\frac{-18}{-27}$ and $\frac{6}{9}$ are equivalent, because

$$\frac{-18}{-27}=\frac{-18\div(-3)}{-27\div(-3)}=\frac{6}{9}.$$

- The rational numbers $\dfrac{38}{-57}$ and $\dfrac{-2}{3}$ are equivalent, because
$$\dfrac{38}{-57}=\dfrac{38\div(-19)}{-57\div(-19)}=\dfrac{-2}{3}.$$
- The rational numbers $\dfrac{-5}{11}$ and $\dfrac{5}{-11}$ are equivalent, because
$$\dfrac{-5}{11}=\dfrac{-5\div(-1)}{11\div(-1)}=\dfrac{5}{-11}.$$
- The rational numbers $\dfrac{42}{-14}$ and $\dfrac{-6}{2}$ are equivalent, because
$$\dfrac{42}{-14}=\dfrac{42\div(-7)}{-14\div(-7)}=\dfrac{-6}{2}.$$

A rational number is said to be in its **simplest form,** or **simplified,** if the denominator is **positive** and the only divisors of **both** the numerator and denominator are 1 and –1.

Exercise 9.42

Determine the simplest form for each of the following:

a) $\dfrac{5}{-9}$

b) $\dfrac{-9}{19}$

c) $\dfrac{28}{-6}$

d) $\dfrac{-16}{24}$

e) $\dfrac{0}{-3}$

f) $\dfrac{105}{-20}$

g) $\dfrac{-31}{11}$

h) $\dfrac{122}{-80}$

i) $\dfrac{-70}{-16}$

j) $\dfrac{256}{-200}$

Solution 9.42

a) The simplest form of the rational number $\dfrac{5}{-9}$ is $\dfrac{-5}{9}$.

b) The rational number $\dfrac{-9}{19}$ *is* in its simplest form.

c) The simplest form of the rational number $\dfrac{28}{-6}$ is $\dfrac{-14}{3}$.

d) The simplest form of the rational number $\dfrac{-16}{24}$ is $\dfrac{-2}{3}$.

e) The simplest form of the rational number $\dfrac{0}{-3}$ is 0.

f) The simplest form of the rational number $\dfrac{105}{-20}$ is $\dfrac{-21}{4}$.

g) The rational number $\dfrac{-31}{11}$ *is* in its simplest form.

h) The simplest form of the rational number $\dfrac{122}{-80}$ is $\dfrac{-61}{40}$.

i) The simplest form of the rational number $\dfrac{-70}{-16}$ is $\dfrac{35}{8}$.

j) The simplest form of the rational number $\dfrac{256}{-200}$ is $\dfrac{-32}{25}$.

9.5 MIXED OPERATIONS WITH SIGNED NUMBERS; APPLICATIONS

In this section, we discuss mixed operations with signed numbers. We also examine some applications involving signed numbers. To perform mixed operations on signed numbers, we use the same basic procedure as with whole numbers.

Exercise 9.43

Perform the indicated operations: $(+3) \times (-2)^2 - [(-6) + (-3)] \times [(+4) - (-7)]$

Solution 9.43

$$(+3) \times (-2)^2 - [(-6) + (-3)] \times [(+4) - (-7)]$$
$$= (+3) \times (-2)^2 - [-(6 + 3)] \times [(+4) + (+7)]$$
$$= (+3) \times (-2)^2 - (-9) \times [+(4 + 7)]$$
$$= (+3) \times (-2)^2 - (-9) \times (+11)$$
$$= \mathbf{(+3) \times (+4)} - (-9) \times (+11)$$
$$= +(3 \times 4) - (-9) \times (+11)$$
$$= (+12) - \mathbf{(-9) \times (+11)}$$
$$= (+12) - [-(9 \times 11)]$$

$$= (+12) - (-99)$$
$$= (+12) + (+99)$$
$$= +(12 + 99)$$
$$= +111$$

Exercise 9.44
Evaluate:
a) $(-3)^2$
b) -3^2

Solution 9.44
a) $(-3)^2 = (-3) \times (-3) = +9$
b) $-3^2 = -(3 \times 3) = -9$

Note that $-3^2 \neq (-3)^2$; $(-3)^2$ means the square of -3; -3^2 means the opposite of the square of 3.

Exercise 9.45
Perform the indicated operations:
$$-2^3 \div (-2)^2 + (2.7)^2 - \left[(-6.0) \times (-0.7)\right]^2$$

Solution 9.45
$$-2^3 \div (-2)^2 + (2.7)^2 - [(-6.0) \times (-0.7)]^2$$
$$= -2^3 \div (-2)^2 + (2.7)^2 - [+(6.0 \times 0.7)]^2$$
$$= -2^3 \div (-2)^2 + (2.7)^2 - [4.2)^2$$
$$= -8 \div (+4) + (7.29) - (17.64)$$
$$= -(8 \div 4) + (7.29) - (17.64)$$
$$= (-2) + (7.29) - (17.64)$$
$$= +(7.29 - 2) - (17.64)$$
$$= (+5.29) - (17.64)$$
$$= (+5.29) + (-17.64)$$
$$= -(17.64 - 5.29)$$
$$= -12.35$$

Exercise 9.46
Perform the indicated operations:
$$\left[-7^3 - (-7)\right] \div \left[(-21) \div (+3)\right]$$

Solution 9.46

$$[-7^3 - (-7)] \div [(-21) \div (+3)]$$
$$= [-343 - (-7)] \div [-(21 \div 3)]$$
$$= [-343 + (+7)] \div (-7)$$
$$= -(343 - 7) \div (-7)$$
$$= (-336) \div (-7)$$
$$= 48$$

Exercise 9.47

Perform the indicated operations:

$$\left(\frac{-1}{3}\right)^3 \times \left(\frac{-1}{2}\right)^2 - \left(\frac{1}{5}\right)^2$$

Solution 9.47

$$\left(\frac{-1}{3}\right)^3 \times \left(\frac{-1}{2}\right)^2 - \left(\frac{1}{5}\right)^2$$
$$= \left(\frac{-1}{27}\right) \times \left(\frac{1}{4}\right) - \left(\frac{1}{25}\right)$$
$$= -\left(\frac{1}{27} \times \frac{1}{4}\right) - \left(\frac{1}{25}\right)$$
$$= -\left(\frac{1 \times 1}{27 \times 4}\right) - \left(\frac{1}{25}\right)$$
$$= \left(\frac{-1}{108}\right) - \left(\frac{1}{25}\right)$$
$$= \left(\frac{-1}{108}\right) + \left(\frac{-1}{25}\right)$$
$$= -\left(\frac{1}{108} + \frac{1}{25}\right)$$
$$= -\left(\frac{25}{2,700} + \frac{108}{2,700}\right)$$
$$= \frac{-133}{2,700}$$

Exercise 9.48

Perform the indicated operations:

$$\left[\left(\frac{-3}{8}\right) \times \left(\frac{-1}{3}\right)\right]^2$$

Solution 9.48

$$\left[\left(\frac{-3}{8}\right)\times\left(\frac{-1}{3}\right)\right]^2$$

$$=\left[+\left(\frac{3}{8}\times\frac{1}{3}\right)\right]^2$$

$$=\left[+\left(\frac{3\times 1}{8\times 3}\right)\right]^2$$

$$=\left(\frac{1\times 1}{8\times 1}\right)^2$$

$$=\left(\frac{1}{8}\right)^2$$

$$=\frac{1}{64}$$

Exercise 9.49

Perform the indicated operations:

$$\left(-2\frac{1}{2}\right)^2\times\left(3\frac{1}{3}\right)-\left(-4\frac{1}{4}\right)$$

Solution 9.49

$$\left(-2\frac{1}{2}\right)^2\times\left(3\frac{1}{3}\right)-\left(-4\frac{1}{4}\right)$$

$$=\left(\frac{-5}{2}\right)^2\times\left(\frac{10}{3}\right)-\left(\frac{-17}{4}\right)$$

$$=\left(\frac{25}{4}\right)\times\left(\frac{10}{3}\right)-\left(\frac{-17}{4}\right)$$

$$=\left(\frac{25\times 5}{2\times 3}\right)-\left(\frac{-17}{4}\right)$$

$$=\left(\frac{125}{6}\right)-\left(\frac{-17}{4}\right)$$

$$=\left(\frac{250}{12}\right)+\left(\frac{17}{4}\right)$$

$$=\left(\frac{250}{12}+\frac{51}{12}\right)$$

$$= \frac{301}{12}$$

$$= 25\frac{1}{12}$$

Exercise 9.50

Perform the indicated operations:

$$\left(-5\frac{1}{2}\right)+\left(-2\frac{1}{4}\right)^2 \times \left(4\frac{1}{3}\right)^2 \div \left(\frac{-3}{5}\right)$$

Solution 9.50

$$\left(-5\frac{1}{2}\right)+\left(-2\frac{1}{4}\right)^2 \times \left(4\frac{1}{3}\right)^2 \div \left(\frac{-3}{5}\right)$$

$$=\left(\frac{-11}{2}\right)+\left(\frac{-9}{4}\right)^2 \times \left(\frac{13}{3}\right)^2 \div \left(\frac{-3}{5}\right)$$

$$=\left(\frac{-11}{2}\right)+\left(\frac{81}{16}\right) \times \left(\frac{169}{9}\right) \div \left(\frac{-3}{5}\right)$$

$$=\left(\frac{-11}{2}\right)+\left(\frac{81}{16} \times \frac{169}{9}\right) \div \left(\frac{-3}{5}\right)$$

$$=\left(\frac{-11}{2}\right)+\left(\frac{9 \times 169}{16 \times 1}\right) \div \left(\frac{-3}{5}\right)$$

$$=\left(\frac{-11}{2}\right)+\left(\frac{1,521}{16}\right) \div \left(\frac{-3}{5}\right)$$

$$=\left(\frac{-11}{2}\right)+\left(\frac{1,521}{16}\right) \times \left(\frac{-5}{3}\right)$$

$$=\left(\frac{-11}{2}\right)+\left(\frac{507}{16} \times \frac{-5}{1}\right)$$

$$=\left(\frac{-11}{2}\right)+\left(\frac{-2,535}{16}\right)$$

$$=-\left(\frac{11}{2}+\frac{2,535}{16}\right)$$

$$=-\left(\frac{88}{16}+\frac{2,535}{16}\right)$$

$$= -\frac{-2,623}{16}$$

$$= -163\frac{15}{16}$$

Exercise 9.51

Michelle has $372 in her savings account. She makes two deposits, one for $92 and the other for $137. Later, she withdraws $265. How much money is left in her savings account? (Disregard interest earned.)

Solution 9.51

Michelle starts with money in her savings account. The two deposits are *added* to the amount already in her account. The withdrawal is *subtracted* from the new amount. Hence,

$372	(the original amount)
92	(the first deposit)
+137	(the second deposit)
$601	(the new balance after deposits)
−265	(the withdrawal)
$336	(the new balance after withdrawal)

Hence, there is $336 left in her savings account.

Exercise 9.52

A new boat lists for $8,967. The dealer preparation charges are $465, taxes amount to $437, and registration fees are $55. The manufacturer of the boat offers a $1,200 rebate. Brian buys the boat and receives a trade-in allowance of $2,400 for his old boat. What is Brian's actual cost for his new boat? (Disregard finance charges.)

Solution 9.52

To determine Brian's actual cost for his new boat, add the list price, the dealer preparation charges, the taxes, and the registration fees. Then the manufacturer's rebate and the trade-in allowance for his old boat must be subtracted. Hence,

$8,967	(list price)
465	(dealer preparation charges)
437	(taxes)
+55	(registration fees)
$9,924	(total costs)
−1,200	(manufacturer's rebate)
$8,724	
−2,400	(trade-in allowance)
$6,324	(Brian's actual cost)

Hence, Brian actually spent $6,324 for his new boat.

Exercise 9.53

A gravel pit contains 1,000,000 cubic yards of gravel. From this pit each week are shipped 3,500 cubic yards of gravel to Location A, 4,000 cubic yards of gravel to Location B, and 5,500 cubic yards of gravel to Location C. For how many weeks can the three locations be supplied with gravel from the pit until no additional gravel is available?

Solution 9.53

First, we determine how many cubic yards of gravel are shipped to the three locations per week, as follows:

$$\begin{array}{rl} 3{,}500 & \text{cubic yards to Location A} \\ 4{,}000 & \text{cubic yards to Location B} \\ +5{,}500 & \text{cubic yards to Location C} \\ \hline 13{,}000 & \text{cubic yards shipped to all three locations per week} \end{array}$$

We next determine how many 13,000 cubic yards are contained in 1,000,000 cubic yards as follows:

$$\begin{array}{r} 76 \\ 13{,}000 \overline{)1{,}000{,}000} \\ \underline{910{,}000} \\ 90{,}000 \\ \underline{78{,}000} \\ 12{,}000 \end{array}$$

$$\text{or } (1{,}000{,}000) \div (13{,}000) = 76\frac{12}{13}.$$

Therefore, the three locations can be supplied with gravel from the pit for 76 weeks.

Exercise 9.54

The product of −64 and +34 is divided by a third integer. If the quotient is −26, what is the third integer?

Solution 9.54

What is being divided by a third integer is a product. Therefore, we start with finding the *product* of −65 and +34:

$$(-65) \times (+34) = -(65 \times 34)$$
$$= -2{,}210$$

Next, we *divide* −2,210 by −26:

$$-2{,}210 \div (-26) = +(2{,}210 \div 26)$$
$$= +85$$

Hence, the third integer is +85.

Exercise 9.55

The Stewart family owed their credit card company $3,125.60. They charged an additional $1,325.50, repaid $1,735.75, and charged another $800. What is the amount owed the credit card company after these transactions? (Disregard interest charges.)

Solution 9.55

We can solve this problem as follows:

$3,125.60   initial amount owed
+1,325.50   additional amount charged
$4,451.10   new amount owed
 1,735.75   amount repaid
$2,715.35   new amount owed
+800.00   additional amount charged
$3,515.35   new balance owed

Therefore, after the three transactions, the Stewart family owes the credit card company $3,515.35.

Exercise 9.56

The consumer price index had the following changes during six months: a loss of 0.42%, a loss of 0.23%, a gain of 0.5%, a gain of 0.37%, a loss of 0.34%, and a loss of 0.26%. What was the net change in the index over these six months?

Solution 9.56

Using (−) for loss and (+) for gain, we add the six changes:
$(-0.42\%) + (-0.23\%) + (+0.5\%) + (+0.37\%) + (-0.34\%) + (-0.26\%)$
$= [(-0.42) + (-0.23) + (+0.5) + (+0.37) + (-0.34) + (-0.26)]\%$
$= [-(0.42 + 0.23 + 0.34 + 0.26) + (0.5 + 0.37)]\%$
$= [(-1.25) + (+0.87)]\%$
$= [-(1.25 - 0.87)]\%$
$= -0.38\%$

Therefore, the net change in the index over these six months is −0.38%.

Exercise 9.57

Joyce, Jan, and Paula each invested $6,000 in a gift shop. During the first year, their total income from the business was $192,600, and their net profits, after taxes, amounted to $58,800. If the initial investment was subtracted from the net profits and the balance divided equally among the three, how much would each receive?

Solution 9.57

There are three parts to this problem.

1. Determine the *total initial investment*:

   ($6,000 per person) × (3 persons)

   = $18,000

2. *Subtract* the total initial investment *from the net profits*:

   $58,800 − $18,000

   = $40,800

3. Divide $40,800 by 3:

   $40,800 ÷ 3

   = $13,600

Therefore, each person would receive $13,600. (Did you notice that the total income of $192,600 was not used to solve this problem?)

In this chapter, we introduced and discussed signed numbers, together with arithmetic operations on them. Applications involving signed numbers were also considered.

TEST YOURSELF

In Exercise 1–3, determine the value for each expression.

1) $|+9| - |-5|$
2) $|-4| \times |-6|$
3) $|-27| \div |-9|$

In Exercises 4–6, write the opposite of each of the given rational numbers.

4) $\dfrac{-5}{+9}$
5) $\dfrac{+18}{-23}$
6) $\dfrac{-37}{-54}$

In Exercises 7–9, classify the indicated rational numbers as being positive, negative, or zero.

7) $\dfrac{+89}{-76}$
8) $\dfrac{-57}{-45}$
9) $\dfrac{0}{-32}$

In Exercises 10–13, insert the symbol "<", "=", or ">" between the two number symbols to make a true statement.

10) $-13 \quad |-27|$
11) $|-57| \quad 0$
12) $|-3| + |-9| \quad |-12|$
13) $|-36| \div |+4| \quad |-56| \div |-7|$

In Exercises 14–25, perform the indicated operations and simplify your results.

14) $(-7.2) + (+9.8)$
15) $(+12.6) - (-34.2)$
16) $(-23.1) \times (+14.7)$
17) $(-11) - (-25) + (-31)$
18) $\left(-5\dfrac{1}{2}\right) - \left(+7\dfrac{2}{5}\right)$
19) $\left(-6\dfrac{2}{3}\right) \times \left(+4\dfrac{1}{2}\right)$
20) $\left(6\dfrac{2}{5}\right) \div \left(-7\dfrac{1}{2}\right)$
21) $(+78) + (-39) \times (-4.1)$
22) $[(-3.4) - (-7.9)] \times (-23.6)$
23) $(-3)^2 - (-2)^3 \div (-4) - (-5)$
24) $-3\dfrac{1}{3} \times \left(-4\dfrac{1}{2} + 2\dfrac{1}{4}\right)$
25) $\left(\dfrac{-3}{8} - \dfrac{-4}{7}\right) \times \left(\dfrac{-1}{9} + \dfrac{-2}{5}\right)$

TEST YOURSELF WITH THE HANDHELD CALCULATOR

In Exercises 1–5, add.

1) (+17) + (–36) + (–49) + (+86) + (–93) + (–67)
2) (–68) + (–19) + (+692) + (–867) + (+56) + (–396) + (–37)
3) (–18,297) + (–37,693) + (+43,269) + (–101,203) + (+269,123) + (–301,609)
4)   –91,296
     –72,769
      87,205
      69,873
  + –53,196

5)   312,674
   –451,693
   –329,674
     56,980
   –876,992
  + 976,054

In Exercises 6–10, subtract.

6) (+96) – (+137)
7) (+207) – (–189)
8) (+906) – (–809) – (+516)
9)   –23,462
   –14,984

10) –3,672,819
    –2,904,607

In Exercises 11–15, multiply.

11) (+116) × (–97) × (+63)
12) (–237) × (–119) × (–86)
13) +1,296
  × –236

14) –32,609
  × –107

15) $\begin{array}{r}-642{,}651 \\ \times\ +89 \\ \hline \end{array}$

In Exercises 16–20, divide.

16) $(-109{,}968) \div (-87)$
17) $(-204{,}048) \div (-234)$
18) $(-1{,}575{,}840) \div (+672)$
19) $+576 \overline{\smash{)}-4{,}004{,}352}$
20) $-913 \overline{\smash{)}-11{,}270{,}985}$

In Exercises 21–25, perform the indicated operations.

21) $(-213) + (-2{,}691) \div (+23) - (+37) \times (-41)$
22) $(-219{,}936) \div (-87) \times (-19) + (-316) - (-8{,}064) \div (+18)$
23) $(-23{,}692) - (-13{,}462) + (+303) \times (-1{,}209) \div (-101)$
24) $(+899) \times (-123) - (-631) \times (+234) + (-456) \times (-862) \times (-37)$
25) $(-393{,}960) \div (+168) \div (-7) \times (-119) - (-23{,}699) - (-37{,}684)$

TEST YOURSELF ANSWERS

1) $|+9| - |-5| = 9 - 5 = 4$
2) $|-4| \times |-6| = 4 \times 6 = 24$
3) $|-27| \div |-9| = 27 \div 9 = 3$
4) The opposite of $\dfrac{-5}{+9}$ is $\dfrac{5}{9}$.
5) The opposite of $\dfrac{+18}{-23}$ is $\dfrac{18}{23}$.
6) The opposite of $\dfrac{-37}{-54}$ is $\dfrac{-37}{54}$.
7) $\dfrac{+89}{-76}$ is negative.
8) $\dfrac{-57}{-45}$ is positive.
9) $\dfrac{0}{-32}$ is zero.
10) $-13 < |-27|$, because $-13 < 27$.
11) $|-57| > 0$, because $57 > 0$.
12) $|-3| + |-9| = |-12|$, because $3 + 9 = 12$.
13) $|-36| \div |+4| > |-56| \div |-7|$, because $(36 \div 4 = 9) > (56 \div 7 = 8)$.

14) $(-7.2) + (+9.8) = +(9.8 - 7.2) = +2.6$

15) $(+12.6) - (-34.2) = (+12.6) + (+34.2) = +(12.6 + 34.2) = +46.8$

16) $(-23.1) \times (+14.7) = -(23.1 \times 14.7) = -339.57$

17) $(-11) - (-25) + (-31) = (-11) + (+25) + (-31) = +(25 - 11) + (-31) = (+14) + (-31) = -(31 - 14) = -17$

18) $\left(-5\dfrac{1}{2}\right) - \left(+7\dfrac{2}{5}\right) = \left(-5\dfrac{1}{2}\right) + \left(-7\dfrac{2}{5}\right) = -\left(5\dfrac{1}{2} + 7\dfrac{2}{5}\right) = -12\dfrac{9}{10}$

19) $\left(-6\dfrac{2}{3}\right) \times \left(+4\dfrac{1}{2}\right) = -\left(6\dfrac{2}{3} \times 4\dfrac{1}{2}\right) = -29\dfrac{7}{10}$

20) $\left(6\dfrac{2}{5}\right) \div \left(-7\dfrac{1}{2}\right) = -\left(6\dfrac{2}{5} \div 7\dfrac{1}{2}\right) = \dfrac{-64}{75}$

21) $(+78) + (-39) \times (-4.1) = (+78) + [+(39 \times 4.1)] = (+78) + (+159.9) = +237.9$

22) $[(-3.4) - (-7.9)] \times (-23.6) = [(-3.4) + (+7.9)] \times (-23.6) = +(7.9 - 3.4) \times (-23.6) = (+4.5) \times (-23.6) = -(4.5 \times 23.6) = -106.2$

23) $(-3)^2 - (-2)^3 \div (-4) - (-5) = (+9) - (-8) \div (-4) - (-5) = (+9) - (8 \div 4) - (-5) = (+9) - (+2) + (+5) = (+7) + (+5) = +12$

24) $-3\dfrac{1}{3} \times \left(-4\dfrac{1}{2} + 2\dfrac{1}{4}\right) = -3\dfrac{1}{3} \times \left(-\left(4\dfrac{1}{2} - 2\dfrac{1}{4}\right)\right) = -3\dfrac{1}{3} \times \left(-2\dfrac{1}{4}\right) = +\left(3\dfrac{1}{3} \times 2\dfrac{1}{4}\right) = 7\dfrac{1}{2}$

25) $\left(\dfrac{-3}{8} - \dfrac{-4}{7}\right) \times \left(\dfrac{-1}{9} + \dfrac{-2}{5}\right) = \left(\dfrac{-3}{8} + \dfrac{4}{7}\right) \times \left(-\left(\dfrac{1}{9} + \dfrac{2}{5}\right)\right) = +\left(\dfrac{4}{7} - \dfrac{3}{8}\right) \times \left(-\dfrac{23}{45}\right)$

$= \left(+\dfrac{11}{56}\right) \times \left(-\dfrac{23}{45}\right) = -\left(\dfrac{11}{56} \times \dfrac{23}{45}\right) = -\dfrac{253}{2520}$

TEST YOURSELF ANSWERS WITH THE HANDHELD CALCULATOR

1) $(+17) + (-36) + (-49) + (+86) + (-93) + (-67) = -142$
2) $(+68) + (+19) + (-692) + (+867) + (-56) + (+396) + (+37) = +639$
3) $(-18,297) + (-37,693) + (+43,269) + (-101,203) + (+269,123) + (-301,609) = -146,410$
4)
```
    -91,296
    -72,769
     87,205
     69,873
 + -53,196
   ────────
    -60,183
```
5)
```
    -312,674
     451,693
     329,674
     -56,980
     876,992
 + -976,054
   ─────────
    +312,651
```
6) $(+96) - (+137) = -41$
7) $(+207) - (-189) = 396$
8) $(+906) - (-809) - (+516) = 1,199$
9)
```
   -23,462
   -14,984
   ───────
   -38,446
```
10)
```
    -3,672,819
   + 2,904,607
     ─────────
      -768,212
```
11) $(+116) \times (-97) \times (+63) = -708,576$
12) $(-237) \times (-119) \times (-86) = -2,425,458$
13)
```
       1,296
     × -236
     ──────
        7776
        3888
        2592
     ────────
     -305,856
```

14) $\begin{array}{r} -32{,}609 \\ \times\ -107 \\ \hline 228263 \\ 32609 \\ \hline 3{,}489{,}163 \end{array}$

15) $\begin{array}{r} -642{,}651 \\ \times 89 \\ \hline 5783859 \\ 5141208 \\ \hline -57{,}195{,}939 \end{array}$

16) $(-109{,}968) \div (-87) = 1{,}264$
17) $(-204{,}048) \div (-234) = 872$
18) $(-1{,}575{,}840) \div (+672) = -2{,}345$
19) $(-4{,}004{,}352) \div (+576) = -6{,}952$
20) $(-11{,}270{,}985) \div (-913) = 12{,}345$
21) $(-213) + (-2{,}691) \div (+23) - (+37) \times (-41) = 1{,}187$
22) $(-219{,}936) \div (-87) \times (-19) + (-316) - (8{,}064) \div (+18) = -47{,}900$
23) $(-23{,}692) - (-13{,}462) + (+303) \times (-1{,}209) \div (-101) = -6{,}603$
24) $(+899) \times (-123) - (-631) \times (+234) + (-456) \times (-862) \times (-37) = -14{,}506{,}587$
25) $(-393{,}960) \div (+168) \div (-7) \times (-119) - (-23{,}699) - (-37{,}684) = 21{,}518$

CHAPTER 10

Exponents and Scientific Notation

In this chapter, we extend our discussion of exponents to include negative and zero integer exponents. With exponential notation, we can write algebraic expressions more compactly. Basic rules for using exponents are discussed. We also introduce scientific notation as an application of exponents.

10.1 POSITIVE INTEGER EXPONENTS CONTINUED

In this section, we extend our discussion of positive integer exponents by considering some basic rules. Let x be any number and let n be a **positive integer** greater than 1. Then

$$x^n = \underbrace{x \cdot x \cdot x \cdot \ldots \cdot x,}_{n \text{ times}}$$

If $n = 1$, $x^n = x^1 = x$.

For example:
- $a^4 = (a)(a)(a)(a)$
- $6^3 = (6)(6)(6)$
- $p^5 = (p)(p)(p)(p)(p)$
- $(-4)^7 = (-4)(-4)(-4)(-4)(-4)(-4)(-4)$
- $b^6 = (b)(b)(b)(b)(b)(b)$
- $u^8 = (u)(u)(u)(u)(u)(u)(u)(u)$
- $\left(\dfrac{1}{3}\right)^4 = \left(\dfrac{1}{3}\right)\left(\dfrac{1}{3}\right)\left(\dfrac{1}{3}\right)\left(\dfrac{1}{3}\right)$

PRODUCT RULE

To multiply two powers of the same base, we use the following rule. Let x be any number and let m and n be any **positive integers.** Then,

$x^m x^n = x^{m+n}$.

Exercise 10.1

Simplify the following:

a) $a^4 \cdot a^3$

b) $u^7 \cdot u^4$

c) $8^5 \cdot 8^6$

d) $p^3 \cdot p^9$

e) $b^{13} \cdot b^6$

f) $(-4)^6 \cdot (-4)^7$

g) $w^6 \cdot w^{11}$

h) $q^9 \cdot q$

Solution 10.1

a) $a^4 \cdot a^3 = a^{4+3} = a^7$

b) $u^7 \cdot u^4 = u^{7+4} = u^{11}$

c) $8^5 \cdot 8^6 = 8^{5+6} = 8^{11}$

d) $p^3 \cdot p^9 = p^{3+9} = p^{12}$

e) $b^{13} \cdot b^6 = b^{13+6} = b^{19}$

f) $(-4)^6 \cdot (-4)^7 = (-4)^{6+7} = (-4)^{13}$

g) $w^6 \cdot w^{11} = w^{6+11} = w^{17}$

h) $q^9 \cdot q = q^{9+1} = q^{10}$

Caution: The product rule involves *multiplying* powers of the *same base*. Hence, $y^3 + y^5$ is a *sum* of two powers of the same base. It *cannot* be simplified further using the product rule. Similarly, $u^4 \cdot v^7$ is a product of powers of two *different bases*. It generally cannot be simplified further.

The *Product Rule for Exponents* can be extended to a product of three or more factors that are powers of the same base.

Exercise 10.2

Simplify the following:

a) $a^3 \cdot a^2 \cdot a^5$

b) $(b^2)(b^4)(b^7)$

c) $(5^3)(5^4)(5^5)(5^3)(5)$

Solution 10.2

a) $a^3 \cdot a^2 \cdot a^5 = (a^3 \cdot a^2) \cdot a^5$
$= (a^{3+2}) \cdot a^5$
$= a^5 \cdot a^5$
$= a^{5+5}$
$= a^{10}$

b) $(b^2)(b^4)(b^7) = b^{2+4+7}$
$= b^{13}$

c) $(5^3)(5^4)(5^5)(5^3)(5) = 5^{3+4+5+3+1}$
$= 5^{16}$

POWER RULE

The expression $(p^3)^2$ is a power of a power of p. We can simplify it by using the following rule. Let x be any number and let m and n be positive integers. Then
$(x^m)^n = x^{mn}$

Exercise 10.3

Simplify the following:

a) $(y^3)^4$
b) $[(-3)^6]^5$
c) $(w^8)^5$
d) $(9^6)^7$
e) $(p^9)^4$

Solution 10.3

a) $(y^3)^4 = y^{3 \times 4} = y^{12}$
b) $[(-3)^6]^5 = (-3)^{6 \times 5} = (-3)^{30}$
c) $(w^8)^5 = w^{8 \times 5} = w^{40}$
d) $(9^6)^7 = 9^{6 \times 7} = 9^{42}$
e) $(p^9)^4 = p^{9 \times 4} = p^{36}$

Exercise 10.4

Simplify the following:

a) $(x^2)(x^7)$
b) $(x^2)^7$

Solution 10.4

a) $(x^2)(x^7) = x^{2+7} = x^9$ (use the Product Rule)

b) $(x^2)^7 = x^{2\times 7} = x^{14}$ (use the Power Rule)

Exercise 10.5

Simplify the following:

a) $(u^6)(u^7)$

b) $(u^6)^7$

Solution 10.5

a) $(u^6)(u^7) = u^{6+7} = u^{13}$ (use the Product Rule)

b) $(u^6)^7 = u^{6\times 7} = u^{42}$ (use the Power Rule)

Exercise 10.6

Simplify the following:

a) $(3^8)(3^5)$

b) $(3^8)^5$

Solution 10.6

a) $(3^8)(3^5) = 3^{8+5} = 3^{13}$ (use the Product Rule)

b) $(3^8)^5 = 3^{8\times 5} = 3^{40}$ (use the Power Rule)

QUOTIENT RULE

To simplify an expression of the form $x^m \div x^n$, consider the following rule. Let x be any **nonzero** number and let m and n be **positive integers**. Then

a) $x^m \div x^n = x^{m-n}$, if $m > n$

b) $x^m \div x^n = \dfrac{1}{x^{n-m}}$, if $m < n$

c) $x^m \div x^n = 1$, if $m = n$.

Exercise 10.7

Simplify the following:

a) $5^6 \div 5^4$

b) $a^3 \div a^7$

c) $x^7 \div x^7$

d) $(-4)^9 \div (-4)^5$

Solution 10.7

a) $5^6 \div 5^4 = 5^{6-4} = 5^2$ (part a of the Quotient Rule)

b) $a^3 \div a^7 = \dfrac{1}{a^{7-3}} = \dfrac{1}{a^4}$, if $a \neq 0$ (part b of the Quotient Rule)

c) $x^7 \div x^7 = 1$, if $x \neq 0$ (part c of the Quotient Rule)

d) $(-4)^9 \div (-4)^5 = (-4)^{9-5} = (-4)^4$ (part a of the Quotient Rule)

Exercise 10.8

Evaluate each of the following:

a) 3^2

b) $(-4)^3$

c) $\left(\dfrac{1}{2}\right)^3$

d) $\left(\dfrac{-1}{3}\right)^4$

e) $(-2)^3(3)^2$

f) $\left(\dfrac{1}{4}\right)^2\left(\dfrac{-2}{3}\right)^3$

Solution 10.8

a) $3^2 = (3)(3) = 9$

b) $(-4)^3 = (-4)(-4)(-4) = -64$

c) $\left(\dfrac{1}{2}\right)^3 = \left(\dfrac{1}{2}\right)\left(\dfrac{1}{2}\right)\left(\dfrac{1}{2}\right) = \dfrac{1}{8}$

d) $\left(\dfrac{-1}{3}\right)^4 = \left(\dfrac{-1}{3}\right)\left(\dfrac{-1}{3}\right)\left(\dfrac{-1}{3}\right)\left(\dfrac{-1}{3}\right) = \dfrac{1}{81}$

e) $(-2)^3(3)^2 = \left[(-2)(-2)(-2)\right]\left[(3)(3)\right]$

$\qquad = (-8)(9)$

$\qquad = -72$

f) $\left(\dfrac{1}{4}\right)^2\left(\dfrac{-2}{3}\right)^3 = \left[\left(\dfrac{1}{4}\right)\left(\dfrac{1}{4}\right)\right]\left[\left(\dfrac{-2}{3}\right)\left(\dfrac{-2}{3}\right)\left(\dfrac{-2}{3}\right)\right]$

$\qquad = \left(\dfrac{1}{16}\right)\left(\dfrac{-8}{27}\right) = \left(\dfrac{1}{2}\right)\left(\dfrac{-1}{27}\right)$

$\qquad = \dfrac{-1}{54}$

Exercise 10.9
Simplify each of the following:

a) $\dfrac{(u^2)^3 (u^3)^4}{u^5}, u \neq 0$

b) $\dfrac{\left[(-5)^3 (-5)^4\right]^2}{(-5)^{18}}$

Solution 10.9

a) $= \dfrac{(u^2)^3 (u^3)^4}{u^5}, u \neq 0$

$= \dfrac{(u^6)(u^{12})}{u^5}$ (Power Rule)

$= \dfrac{u^{18}}{u^5}$ (Product Rule)

$= u^{13}$ (Quotient Rule, part a)

b) $\dfrac{\left[(-5)^3 (-5)^4\right]^2}{(-5)^{18}}$

$= \dfrac{\left[(-5)^7\right]^2}{(-5)^{18}}$ (Product Rule)

$= \dfrac{(-5)^{14}}{(-5)^{18}}$ (Power Rule)

$= \dfrac{1}{(-5)^4}$ (Quotient Rule, part b)

$= \dfrac{1}{625}$ (Rewrite)

10.2 POWERS OF PRODUCTS AND QUOTIENTS

In this section, we take up some additional rules pertaining to positive integer exponents.

POWER OF A PRODUCT

Let a and b be any numbers and let m, n, and p be **positive integers**. Then
$$(ab)^m = a^m b^m$$
and
$$(a^m b^n)^p = a^{mp} b^{np}.$$

Exercise 10.10

Simplify each of the following:

a) $(pq)^5$
b) $(3a)^4$
c) $(u^3 v^4)^6$
d) $(-4t^4)^3$
e) $(b^3 c^4)^5$
f) $\left[\left(\frac{1}{2}\right)\left(\frac{-2}{3}\right)\right]^3$
g) $\left[\left(\frac{-1}{4}\right)\left(\frac{1}{2}\right)^2\right]^3$

Solution 10.10

a) $(pq)^5 = p^5 q^5$
b) $(3a)^4 = (3)^4 (a)^4 = 81 a^4$
c) $(u^3 v^4)^6 = (u^3)^6 (v^4)^6 = u^{18} v^{24}$
d) $(-4t^4)^3 = (-4)^3 (t^4)^3 = -64 t^{12}$
e) $(b^3 c^4)^5 = (b^3)^5 (c^4)^5 = b^{15} c^{20}$
f) $\left[\left(\frac{1}{2}\right)\left(\frac{-2}{3}\right)\right]^3 = \left(\frac{1}{2}\right)^3 \left(\frac{-2}{3}\right)^3$
$= \left(\frac{1}{8}\right)\left(\frac{-8}{27}\right) = \left(\frac{1}{1}\right)\left(\frac{-1}{27}\right)$
$= \frac{-1}{27}$

g) $\left[\left(\frac{-1}{4}\right)\left(\frac{1}{2}\right)^2\right]^3 = \left(\frac{-1}{4}\right)^3\left[\left(\frac{1}{2}\right)^2\right]^3$

$= \left(\frac{-1}{64}\right)\left(\frac{1}{2}\right)^6$

$= \left(\frac{-1}{64}\right)\left(\frac{1}{64}\right)$

$= \frac{-1}{4,096}$

Exercise 10.11
Simplify:
a) $(p^2p^3)^4(p^3p^2)^3$
b) $(2m^3)^3(3n^4)^4$

Solution 10.11
a) $(p^2p^3)^4(p^3p^2)^3 = (p^{2+3})^4(p^{3+2})^3$ (Product Rule)
$\qquad = (p^5)^4(p^5)^3$ (simplifying)
$\qquad = p^{5\times4}p^{5\times3}$ (Power Rule)
$\qquad = p^{20}p^{15}$ (simplifying)
$\qquad = p^{20+15}$ (Product Rule)
$\qquad = p^{35}$ (simplifying)

b) $(2m^3)^3(3n^4)^4 = (2^3m^9)(3n^4)^4$ (Power of a Product Rule)
$\qquad = (2^3m^9)(3^4n^{16})$ (Power of a Product Rule)
$\qquad = (8m^9)(81n^{16})$ (simplifying)
$\qquad = (8)(81)m^9n^{16}$ (regrouping)
$\qquad = 648m^9n^6$ (simplifying)

POWER OF A QUOTIENT

We simplify a power of a quotient by using the following rule. Let a be any number, b be any **nonzero** number, and n be any **positive integer**. Then $\left(\frac{a}{b}\right)^n = \frac{a^n}{b^n}$.

Exercise 10.12
Simplify:
a) $\left(\frac{x}{y}\right)^6$, if $y \neq 0$
b) $\left(\frac{-3}{7}\right)^4$

c) $\left(\dfrac{b}{c}\right)^{10}$, if $c \neq 0$

d) $\left(\dfrac{-3}{8}\right)^{13}$

e) $\left(\dfrac{-10}{w}\right)^{7}$, if $w \neq 0$

Solution 10.12

a) $\left(\dfrac{x}{y}\right)^6 = \dfrac{x^6}{y^6}$, if $y \neq 0$

b) $\left(\dfrac{-3}{7}\right)^4 = \dfrac{(-3)^4}{7^4}$

c) $\left(\dfrac{b}{c}\right)^{10} = \dfrac{b^{10}}{c^{10}}$, if $c \neq 0$

d) $\left(\dfrac{-3}{8}\right)^{13} = \dfrac{(-3)^{13}}{8^{13}}$

e) $\left(\dfrac{-10}{w}\right)^{7} = \dfrac{(-10)^{7}}{w^{7}}$, if $w \neq 0$

Exercise 10.13

Simplify:

$\left(\dfrac{r^3}{2s}\right)^4 \left(\dfrac{2r}{s^2}\right)^3$, if $s \neq 0$

Solution 10.13

$\left(\dfrac{r^3}{2s}\right)^4 \left(\dfrac{2r}{s^2}\right)^3 = \left(\dfrac{(r^3)^4}{(2s)^4}\right)\left(\dfrac{(2r)^3}{(s^2)^3}\right)$ (Power of a Quotient Rule)

$= \left(\dfrac{r^{12}}{(2s)^4}\right)\left(\dfrac{(2r)^3}{s^6}\right)$ (Power Rule)

$= \left(\dfrac{r^{12}}{2^4 s^4}\right)\left(\dfrac{2^3 r^3}{s^6}\right)$ (Power of a Product Rule)

$= \left(\dfrac{r^{12}}{16 s^4}\right)\left(\dfrac{8 r^3}{s^6}\right)$ (simplifying)

$$= \frac{(r^{12})(8r^3)}{(16s^4)(s^6)} \text{ (multiplying fractions)}$$

$$= \frac{8r^{15}}{16s^{10}} \text{ (Product Rule)}$$

$$= \frac{r^{15}}{2s^{10}} \text{ (simplifying)}$$

Exercise 10.14
Simplify:
$$\left(\frac{-3a^2}{4b}\right)^3 \left(\frac{2c^3}{d^2}\right)^2, \text{ if } b \neq 0 \text{ and } d \neq 0$$

Solution 10.14
$$\left(\frac{-3a^2}{4b}\right)^3 \left(\frac{2c^3}{d^2}\right)^2 = \left(\frac{-27a^6}{64b^3}\right)\left(\frac{4c^6}{d^4}\right)$$

$$= \frac{(-27)(4)a^6 c^6}{64b^3 d^4}$$

$$= \frac{-27a^6 c^6}{16b^3 d^4} (b \neq 0, d \neq 0)$$

10.3 ZERO AND NEGATIVE INTEGER EXPONENTS

In this section, we consider the basic rules for working with zero and negative integer exponents.

ZERO EXPONENT

Let x be any **nonzero** number. Then $x^0 = 1$.
For example,
- $5^0 = 1$ (because $5 \neq 0$)
- $(-7)^0 = 1$ (because $-7 \neq 0$)
- $b^0 = 1$, if $b \neq 0$
- $(3u)^0 = 1$, if $u \neq 0$
- $3u^0 = 3(u^0) = 3(1) = 3$, if $u \neq 0$
- $(xy)^0 = 1$, if $x \neq 0$ and $y \neq 0$
- $(-2uv)^0 = 1$, if $u \neq 0$ and $v \neq 0$
- $-2(uv)^0 = -2(1) = -2$, if $u \neq 0$ and $v \neq 0$

NEGATIVE INTEGER EXPONENTS

Simplifying expressions involving negative integer exponents is done using the following rule. Let a be any **nonzero** number and let n be a **positive integer** exponent. Then $a^{-n} = \dfrac{1}{a^n}$.

Exercise 10.15
Evaluate each of the following:
 a) 3^{-3}
 b) 4^{-1}
 c) 5^{-4}
 d) 6^{-2}
 e) 7^{-3}

Solution 10.15
 a) $3^{-3} = \dfrac{1}{3^3} = \dfrac{1}{27}$
 b) $4^{-1} = \dfrac{1}{4}$
 c) $5^{-4} = \dfrac{1}{5^4} = \dfrac{1}{625}$
 d) $6^{-2} = \dfrac{1}{6^2} = \dfrac{1}{36}$
 e) $7^{-3} = \dfrac{1}{7^3} = \dfrac{1}{343}$

The *Negative Integer Power Rule* can be used if the base is a variable, provided that the base is not equal to zero.

Exercise 10.16
Rewrite each of the following without negative exponents:
 a) y^{-4}
 b) b^{-5}
 c) $(uv)^{-2}$
 d) $3x^{-5}$

Solution 10.16
 a) $y^{-4} = \dfrac{1}{y^4}$, if $y \neq 0$
 b) $b^{-5} = \dfrac{1}{b^5}$, if $b \neq 0$

c) $(uv)^{-2} = \dfrac{1}{(uv)^2}$, if $u \neq 0$ and $v \neq 0$

$3x^{-5} = 3\left(\dfrac{1}{x^5}\right) = \dfrac{3}{x^5}$, if $x \neq 0$

Exercise 10.17
Rewrite $3a^{-2}b^2c^{-3}$ without negative exponents.

Solution 10.17
$3a^{-2}b^2c^{-3} = (3)(a^{-2})(b^2)(c^{-3})$

$= (3)\left(\dfrac{1}{a^2}\right)(b^2)\left(\dfrac{1}{c^3}\right)$ (if $a \neq 0$ and $c \neq 0$)

$= \dfrac{(3)(1)(b^2)(1)}{(a^2)(c^3)}$

$= \dfrac{3b^2}{a^2 c^3}$ (if $a \neq 0$ and $c \neq 0$)

Exercise 10.18
Rewrite $\dfrac{a^4 b^{-5} c^7}{d^{-2} e f^{-5}}$ without negative exponents.

Solution 10.18

$\dfrac{a^4 b^{-5} c^7}{d^{-2} e f^{-3}} = \dfrac{(a^4)\left(\dfrac{1}{b^5}\right)(c^7)}{\left(\dfrac{1}{d^2}\right)(e)\left(\dfrac{1}{f^3}\right)}$, if $b \neq 0, d \neq 0, e \neq 0, f \neq 0$

$= \dfrac{\dfrac{a^4 c^7}{b^5}}{\dfrac{e}{d^2 f^3}}$

$= \left(\dfrac{a^4 c^7}{b^5}\right)\left(\dfrac{d^2 f^3}{e}\right)$

$= \dfrac{a^4 c^7 d^2 f^3}{b^5 e}$ ($b \neq 0, d \neq 0, e \neq 0, f \neq 0$)

10.4 SCIENTIFIC NOTATION

We often find very large numbers or very small numbers in applications, especially in science. Numbers such as 0.0000002358; 5,880,000,000; and 9,461,000,000,000 are encountered but can be written with fewer digits using **scientific notation.**

A **positive** number N is said to be in scientific notation when it is written in the form
$$N = p \times 10^k,$$
where k is an **integer** and p is a **positive** number so that $1 \leq p < 10$.

REWRITING A NUMBER IN SCIENTIFIC NOTATION

To rewrite a **positive** number N in scientific notation:

1. Place a decimal point in the number N after the first **nonzero** digit from the left. (This new number is p.)
2. Count the number of digits that the decimal point must be moved to get back to the original number. (This is the value of k.)
   a. k will be **positive** if the decimal point is moved to the **right.**
   b. k will be **negative** if the decimal point is moved to the **left.**
   c. k will be **0** if the decimal point is not moved at all.

Exercise 10.19

Rewrite each of the following positive numbers in scientific notation:
a) 857
b) 0.000087
c) 0.001992
d) 28,157
e) 467,000,000
f) 41.17×10^{-8}
g) 0.57×10^{17}
h) 632.4×10^6

Solution 10.19

a) $857 = 8.57 \times 100$
   $= 8.57 \times 10^2$

b) $0.000087 = 8.7 \times \dfrac{1}{100,000}$
   $= 8.7 \times 10^{-5}$

c) $0.001992 = 1.992 \times \dfrac{1}{1,000}$
   $= 1.992 \times 10^{-3}$

d) $28{,}157 = 2.8157 \times 10{,}000$
$= 2.8157 \times 10^4$

e) $467{,}000{,}000 = 4.67 \times 100{,}000{,}000$
$= 4.67 \times 10^8$

f) $41.17 \times 10^{-8} = (4.117 \times 10) \times 10^{-8}$
$= 4.117 \times (10 \times 10^{-8})$
$= 4.117 \times 10^{-7}$

g) $0.57 \times 10^{17} = (5.7 \times 10^{-1}) \times 10^{17}$
$= 5.7 \times (10^{-1} \times 10^{17}) = 5.7 \times 10^{16}$

h) $632.4 \times 10^6 = (6.324 \times 10^2) \times 10^6 = 6.324 \times 10^8$

A number given in scientific notation can be rewritten without exponents.

Exercise 10.20

Rewrite each of the following numbers without exponents:

a) 5.63×10^3

b) 2.79×10^{-4}

c) 17×10^5

d) 23.51×10

e) 8.79×10^0

f) 8.43×10^{-6}

Solution 10.20

a) $5.63 \times 10^3 = 5{,}630$

b) $2.79 \times 10^{-4} = 0.000279$

c) $17 \times 10^5 = 1{,}700{,}000$

d) $23.51 \times 10 = 23.51 \times 10^1 = 235.1$

e) $8.79 \times 10^0 = 8.79$

f) $8.43 \times 10^{-6} = 0.00000843$

SIGNIFICANT DIGITS

When a number is written in the scientific notation $p \times 10k$, the digits in the number p are called **significant digits**. For example,

$$0.00047$$
$$47$$
$$4{,}700$$
$$470{,}000$$
$$47{,}000{,}000$$

all have two significant digits, because
$$0.00047 = 4.7 \times 10^{-4}$$
$$47 = 4.7 \times 10^{1}$$
$$4{,}700 = 4.7 \times 10^{3}$$
$$470{,}000 = 4.7 \times 10^{5}$$
$$47{,}000{,}000 = 4.7 \times 10^{7},$$
where, in each case, $p = 4.7$ has two digits.

ACCURACY

If a number has n significant digits, we say that the number is **accurate to n places** or that it has **n-place accuracy.**

Exercise 10.21
Determine the accuracy for each of the following numbers:
- a) 532
- b) 0.037
- c) 156,000
- d) 29,890,000,000
- e) 730,900
- f) 80009.3

Solution 10.21
a) The number 532 has three-place accuracy, because $532 = 5.32 \times 10^{2}$ and $p = 5.32$ has three digits.

b) The number 0.037 has two-place accuracy, because $0.037 = 3.7 \times 10^{-2}$ and $p = 3.7$ has two digits.

c) The number 156,000 has three-place accuracy, because $156{,}000 = 1.56 \times 10^{5}$ and $p = 1.56$ has three digits.

d) The number 29,890,000,000 has four-place accuracy, because $29{,}890{,}000{,}000 = 2.989 \times 10^{10}$ and $p = 2.989$ has four digits.

e) The number 730,900 has four-place accuracy, because $730{,}900 = 7.309 \times 10^{5}$. (The 0 in 7.309 *is* significant.)

f) The number 80,009.3 has six-place accuracy, because $80{,}009.3 = 8.00093 \times 10^{4}$. (*All* digits are significant.)

We do not always keep all the significant digits in a number. The number may be rounded to fewer significant digits, using the rules for rounding given in Section 4.3. For example:
- The number 29.4376 has six significant digits. It can be rounded to three significant digits as 29.4.
- The number 8.7796 has five significant digits. It can be rounded to three significant digits as 8.78.

- The number 798.5312 has seven significant digits. It can be rounded to four significant digits as 798.5.
- The number 29.750 has four significant digits. It can be rounded to two significant digits as 30.
- The number 0.000976500 has four significant digits. It can be rounded to three significant digits as 0.000977.

In this chapter, exponents were extended to include zero and negative integer exponents. We also discussed scientific notation, significant digits, and rounding.

TEST YOURSELF

In Exercises 1–10, simplify each of the given expressions, if possible, using the rules of exponents. If not possible, give an appropriate reason.

1) $x^4 x^5$
2) $u^3 + u^5$
3) $(u^2 v^3)^3$
4) $-3(x^2)^3(y^3)^2$
5) $(-3a)^2(-2b)^3$
6) $(4a^3 b^2) \div (-2ab)$
7) $(p^2 q)^3 (p^{-1} q^3)^{-2} (pq^2)^2$
8) $2u^2 v + 3uv^2$
9) $x^0 y^4 \div 2x^{-3} y^{-1}$
10) $(-2p^{-2} q^3)(-3p^5 q^{-2}) \div 4p^2 q^2$

In Exercises 11–16, rewrite each of the given numbers in scientific notation.

11) 38.6
12) 0.0067
13) 3^4
14) 56,700,000
15) 0.0000589
16) 0.987

In Exercises 17–18, determine the number N for each of the numbers given in scientific notation.

17) 7.23×10^4
18) 5.12×10^{-3}

In Exercises 19–22, evaluate each of the given expressions and write your results without exponents.

19) 0.546×10^4
20) 125.7×10^{-8}

21) $(4^5) \div (-4)^3$
22) $(5^2)^3 \div (5^4)^2$
23) Round 54.74809 to three significant digits.
24) Round 0.00020386 to four significant digits.
25) Round 2.003976 to four significant digits.

TEST YOURSELF WITH THE HANDHELD CALCULATOR

Evaluate each of the following expressions and write your results in scientific notation.
1) $(1.23 \times 10^3) + (2.014 \times 10^{-2}) + (4.1 \times 10^5)$
2) $(3.5 \times 10^{-6}) - (4.71 \times 10^4) - (8.9 \times 10^7)$
3) $(1.983 \times 10^1) + (7.6 \times 10^{-5}) - (9.164 \times 10^4)$
4) $(5.67 \times 10^3) \times (7.02 \times 10^{-5})$
5) $(8.762 \times 10^{-2}) \times (-4.05 \times 10^6)$
6) $(-6.54 \times 10^3) \times (-9.9 \times 10^{-7})$
7) $(16,506.18 \times 10^{-2}) \div (-0.0054 \times 10^3)$
8) $(578.6522 \times 10^4) \div (3.98 \times 10^2) \div (0.67 \times 10^1)$
9) $(12.64 \times 10^{-4}) + (31,567 \times 10^{-5})$
10) $(231.4 \times 10^{-3}) - (79.28 \times 10^2) + (0.0123 \times 10^5)$

TEST YOURSELF ANSWERS

1) $(x^4)(x^5) = x^{4+5} = x^9$
2) $u^3 + u^5$ is not possible; addition of unlike terms
3) $(u^2v^3)^3 = u^{2\times 3}v^{3\times 3} = u^6v^9$
4) $-3(x^2)^3(y^3)^2 = -3x^{2\times 3}y^{3\times 2} = -3x^6y^6$
5) $(-3a)^2(-2b)^3 = (9a^2)(-8b^3) = (9)(-8)(a^2)(b^3) = -72a^2b^3$
6) $(4a^3b^2) \div (-2ab) = [4 \div (-2)](a^3 \div a)(b^2 \div b) = -2a^2b$, if $a \neq 0$ and $b \neq 0$
7) $(p^2q)^3(p^{-1}q^3)^{-2}(pq^2)^2 = (p^{(2)(3)}q^3)(p^{(-1)(-2)}q^{(3)(-2)})(p^2q^{(2)(2)}) = (p^6q^3)(p^2q^{-6})(p^2q^4) =$
$(p^6)(p^2)(p^2)(q^3)(q^{-6})(q^4) = (p^{6+2+2})(q^{3-6+4}) = p^{10}q$, if $p \neq 0$ and $q \neq 0$
8) $2u^2v + 3uv^2$ is not possible; addition of unlike terms
9) $x^0y^4 \div 2x^{-3}y^{-1} = (1 \div 2)(x^0 \div x^{-3})(y^4 \div y^{-1}) = (1/2)(x^{0+3})(y^{4-1}) = (1/2)x^3y^3$, if $x \neq 0$ and $y \neq 0$
10) $(-2p^{-2}q^3)(-3p^5q^{-2}) \div 4p^2q^2 = [(-2)(-3)p^{-2+5})(q^{3-2})] \div 4p^2q^2 = (6p^3q) \div (4p^2q^2) = (6 \div 4)(p^3 \div p^2)(q \div q^2) = \left(\dfrac{3}{2}\right)(p^{3-2})(q^{1-2}) = \left(\dfrac{-3}{2}\right)pq^{-1} = \dfrac{3p}{2q}$, if $p \div 0$ and $q \div 0$
11) $38.6 = 3.86 \times 10^1$
12) $0.0067 = 6.7 \times 10^{-3}$
13) $3^4 = 81 = 8.1 \times 10^1$
14) $56,700,000 = 5.67 \times 10^7$

15) $0.0000589 = 5.89 \times 10^{-5}$

16) $0.987 = 9.87 \times 10^{-1}$

17) $7.23 \times 10^4 = 72,300$

18) $5.12 \times 10^{-3} = 0.00512$

19) $0.546 \times 10^4 = 5,460$

20) $125.7 \times 10^{-8} = 0.0000001257$

21) $(4^5) \div (-4)^3 = 1,024 \div (-64) = -16$

22) $(5^2)^3 \div (5^4)^2 = 5^6 \div 5^8 = 5^{-2} = \dfrac{1}{25}$

23) 54.7

24) 0.0002039

25) 2.004

TEST YOURSELF WITH THE HANDHELD CALCULATOR

1) $(1.23 \times 10^3) + (2.014 \times 10^{-2}) + (4.1 \times 10^5) = 1,230 + 0.02014 + 410,000 = 411230.0201 = 4.11230201 \times 10^5$

2) $(3.5 \times 10^{-6}) - (4.71 \times 10^4) - (8.9 \times 10^7) = 0.0000035 - 47,100 - 89,000,000 = -89,047,100 = -8.90471 \times 10^7$

3) $(1.983 \times 10) + (7.6 \times 10^{-5}) - (9.164 \times 10^4) = 19.83 + 0.000076 - 91,640 = -91,920.16992 = -9.192016992 \times 10^4$

4) $(5.67 \times 10^3) \times (7.02 \times 10^{-5}) = (5.67)(7.02)(10^3)(10^{-5}) = (5.67)(7.02)(10^{-2}) = 39.8034 \times 10^{-2} = 3.98034 \times 10^{-1}$

5) $(8.762 \times 10^{-2}) \times (-4.05 \times 10^6) = (8.762)(-4.05)(10^{-2})(10^6) = (8.762)(-4.05)(10^4) = -35.4861 \times 10^4 = -3.54861 \times 10^5$

6) $(-6.54 \times 10^3) \times (-9.9 \times 10^{-7}) = (-6.54)(-9.9)(10^3)(10^{-7}) = (-6.54)(-9.9)(10^{-4}) = 64.746 \times 10^{-4} = 6.4746 \times 10^{-3}$

7) $(16,506.18 \times 10^{-2}) \div (-0.0054 \times 10^3) = [16,506.18 \div (-0.0054)] \times (10^{-2} \div 10^3) = -3,056,700 \times 10^{-5} = -3.0567 \times 10^1$

8) $(578.6522 \times 10^4) \div (3.98 \times 10^2) \div (0.67 \times 10^1) = [(578.6522 \div 3.98) \times (10^4 \div 10^2)] \div (0.67 \times 10^1) = (145.39 \times 10^2) \div (0.67 \times 10^1) = (145.39 \div 0.67) \times (10^2 \div 10^1) = 217 \times 10^1 = 2.17 \times 10^3$

9) $(12.64 \times 10^{-4}) + (31,567 \times 10^{-5}) = 0.001264 + 0.31567 = 0.316934 = 3.16934 \times 10^{-1}$

10) $(231.4 \times 10^{-3}) - (79.28 \times 10^2) + (0.0123 \times 10^5) = 0.2314 - 7,928 + 1,230 = -6697.7686 = -6.6977686 \times 10^3$

CHAPTER 11

An Introduction to Algebra

In this chapter, we introduce algebra as a generalization of arithmetic. In algebra, we use letters to represent numbers. Algebraic expressions are discussed, including evaluating and combining them. Removing symbols of grouping is also discussed. In addition, the language of algebra is introduced. Linear equations in a single variable is also discussed, together with the applied problems they can help solve.

11.1 EVALUATING ALGEBRAIC EXPRESSIONS

In this section, you learn to classify algebraic expressions as sums or products and to evaluate an algebraic expression.

ALGEBRAIC EXPRESSIONS

In algebra, letters are used to represent numbers. Such letters are called **variables.** The numbers are called **constants.** For example, in the expression $x + 3y$, the constant is 3; the variables are x and y. An **algebraic expression** is an expression that consists of constants and variables. Operation symbols and grouping symbols may also be included in the expression.

The following are algebraic expressions:

- $x + y^2 + z^3$
- $-2a + 3bc$
- $\dfrac{pq}{p-q}$
- $5m^3 - 3q^5$
- $-7xyz$
- $u^3 - 8u + u + \sqrt{u}$

TERMS AND FACTORS

For our purposes, a difference is treated as an algebraic sum. Hence, $x - y$ means $x + (-y)$. The expression $3u - 2v$ is called a sum of two terms, which are $3u$ and $-2v$.

- The algebraic expression $4t^2 - 3t$ can be written as $4t^2 + (-3t)$. The constants are 4, 2, and -3; the variable is t.
- The algebraic expression $3s^4 - 2s - 3 + \sqrt{s}$ can be written as $3s^4 + (-2s) + (-3) + \sqrt{s}$. The constants are 3, -2, and -3; the variable is s.

In an algebraic expression, a **term** is part of a **sum**. Terms are separated by + or − signs. A **factor** is part of a **product.**

Exercise 11.1

Determine whether each of the following expressions is a sum or a product. If a sum, determine the number of terms; if a product, determine the number of factors.

a) $5m - 7n$
b) $x^3y + z^4$
c) $4u^5vw^3$
d) $5x^2(x - 5)$
e) $2(a - b) - 3(a + b)$

Solution 11.1

a) The expression is a *sum* of *two terms,* which are $5m$ and $-7n$.
b) The expression is a *sum* of *two terms,* which are x^3y and z^4.
c) The expression is *a product* of *four factors,* which are 4, u^5, v, and w^3.
d) The expression is a single term that is *a product* of *three factors*. The factors are 5, x^2, and $x - 5$.
e) The expression is a *sum* of *two terms,* which are $2(a - b)$ and $-3(a + b)$

COEFFICIENTS

In an algebraic expression, the **numerical coefficient,** or simply **coefficient,** is the number factor.

- The expression $-8x^4y^5$ is a product of the factors -8, x^4, and y^5. The coefficient is -8.
- The coefficient of xy is 1. The expression xy can be written as $1(xy)$. The coefficient of $-xy$ is -1, because $-xy = (-1)(xy)$.

Caution: The numerical coefficient in an algebraic expression is multiplied by the product of all the other factors. Thus, $2ab$ means $2(ab)$. It does *not* mean $2(a)2(b)$.

- The expression $2a - 7b$ is a sum of two terms. The coefficient of the first term is 2. The coefficient of the second term is -7.
- The expression $-6p^2q^3$ is a product of three factors. The coefficient in the expression is -6.

- The coefficient of the second term in the expression $p^2 - 4p^2q^3 + q^5$ is -4.
- The expression $2(x - 3) - 3y$ is a sum of two terms. The coefficient of the first term is 2. The coefficient of the second term is -3.
- The expression $2a - (3 - b) + c$ is a sum of three terms. The coefficients of the terms are $2, -1$, and 1.

In algebra, we use the symbols + for addition, – for subtraction, and ÷ for division. However, for multiplication, we usually do *not* use the × symbol. We write the factors side by side or enclose them within parentheses. For example, the product of x and y is written xy. The product of 3, 5, a, and b is written as $(3)(5)ab$. It is *not*, however, written as $35ab$.

EVALUATING ALGEBRAIC EXPRESSIONS

To evaluate an algebraic expression:
- Substitute the number value for each of the variables.
- Enclose the number values within parentheses.
- Perform the arithmetic operations as you would do in arithmetic.

Exercise 11.2

Evaluate each of the algebraic expressions as indicated:

a) $a + 3b$ when $a = 2$ and $b = 3$

b) $3xy$ when $x = \dfrac{1}{2}$ and $y = \dfrac{-1}{3}$

c) $2p + q^2 - 3r$ when $p = -1$, $q = -2$, and $r = 3$

d) $mn + 4m^2 - 5n$ when $m = -2$ and $n = -3$

Solution 11.2

a) Substitute the value 2 for a and the value 3 for b in the given expression.

We have $a + 3b$

$= (2) + (3)(3)$

$= 2 + 9$

$= 11$

b) Substituting, we have $3xy$

$= 3\left(\dfrac{1}{2}\right)\left(\dfrac{-1}{3}\right)$

$= \dfrac{-3}{6}$

$= \dfrac{-1}{2}$

c) Substituting, we have $2p + q^2 - 3r$
$= 2(-1) + (-2)^2 - (3)(3)$
$= 2(-1) + (+4) - (3)(3)$
$= (-2) + (+4) - (9)$
$= (+2) - (9)$
$= -7$

d) Substituting, we have $mn + 4m^2 - 5n$
$= (-2)(-3) + 4(-2)^2 - 5(-3)$
$= (-2)(-3) + 4(+4) - 5(-3)$
$= 6 + 16 - (-15)$
$= 22 - (-15)$
$= 37$

Exercise 11.3

Evaluate each of the algebraic expressions as indicated:
a) $3u - (v - 4u) + 2v$ when $u = 0.4$ and $v = -0.3$
b) $p^2q - 5pq + 3q^2$ when $p = -0.3$ and q -0.2
c) $-3(x^2 + y^3)$ when $x = 2.3$ and $y = -5.1$

Solution 11.3

a) Substituting, we have $3u - (v - 4u) + 2v$
$= 3(0.4) - [(-0.3) - 4(0.4)] + 2(-0.3)$
$= 3(0.4) - [(-0.3) - (1.6)] + 2(-0.3)$
$= 3(0.4) - (-1.9) + 2(-0.3)$
$= 1.2 - (-1.9) + (-0.6)$
$= 3.1 + (-0.6)$
$= 2.5$

b) Substituting, we have $p^2q - 5pq + 3q^2$
$= (-0.3)^2(-0.2) - 5(-0.3)(-0.2) + 3(-0.2)^2$
$= (0.09)(-0.2) - 5(-0.3)(-0.2) + 3(0.04)$
$= (-0.018) - (0.3) + (0.12)$
$= (-0.318) + (0.12)$
$= -0.198$

c) Substituting, we have $-3(x^2 + y^3)$
$= -3[(2.3)^2 + (-5.1)^3]$
$= -3 [(5.29) + (-132.651)]$
$= -3(-127.361)$
$= 382.083$

Exercise 11.4
Evaluate the algebraic expression $\dfrac{2a^3-3b^2}{b+c}$ when $a=-3$, $b=-2$, and $c=4$.

Solution 11.4
Substituting, we have $\dfrac{2a^3-3b^2}{b+c}$

$= \dfrac{2(-3)^3-3(-2)^2}{(-2)+(4)}$

$= \dfrac{2(-27)-(12)}{(-2)+(4)}$

$= \dfrac{-54-(12)}{2}$

$= \dfrac{-66}{2}$

$= -33$

Exercise 11.5
Evaluate the algebraic expression $(a^3+b^3) \div (a+b)$ when $a=-3$ and $b=-2$.

Solution 11.5
Substituting, we have $(a^3+b^3) \div (a+b)$
$= [(-3)^3+(-2)^3] \div [(-3)+(-2)]$
$= [(-27)+(-8)] \div [(-3)+(-2)]$
$= (-35) \div (-5)$
$= 7$

Exercise 11.6
Evaluate the algebraic expression $-5x^3yz^2w$ when $x=0.2$, $y-0.3$, $z=-3$, and $w=-0.5$.

Solution 11.6
Substituting, we have
$-5x^3yz^2w$
$= -5(0.2)^3(-0.3)(-3)^2(-0.5)$
$= -5(0.008)(-0.3)(9)(-0.5)$
$= (-0.04)(-0.3)(9)(-0.5)$
$= (-0.04)(-2.7)(-0.5)$
$= (0.108)(-0.5)$
$= -0.054$

11.2 SIMPLIFYING ALGEBRAIC EXPRESSIONS

In this section, you learn to simplify algebraic expressions by combining like terms. In an algebraic expression, **like terms** are terms that contain the same variables with the same exponents. Terms that are not like terms are called **unlike terms.** All constant terms are like terms. Like terms can differ only by their numerical coefficients.

Exercise 11.7

Determine if the following expressions are like or unlike terms:

a) $3a, -4a, 11a, 0.2a$

b) $\frac{3}{5}uv, \frac{-7}{8}uv, \frac{1}{9}uv$

c) $2m, 3n$

d) $3x^3, x^3, -8x^3$

e) $w^3, 7w^5$

f) a^2b^3, a^3b^2

g) $u^5, 3$

h) $0, -2, 4.8, \sqrt{5}, \pi$

i) $-2x^2y, 3x^2y, -7x^2y$

j) $2(a+b), -3(a-b)$

Solution 11.7

a) Like terms; these are a terms.

b) Like terms; these are uv terms.

c) Unlike terms; the *variables* are different.

d) Like terms; these are x^3 terms.

e) Unlike terms; the *exponents* are different.

f) Unlike terms; the *exponents* are different.

g) Unlike terms; we have a u^5 term and a *constant* term.

h) Like terms; these are *constant* terms.

i) Like terms; these are x^2y terms.

j) Unlike terms. The *variables* are different.

An Introduction to Algebra

COMBINING LIKE TERMS

To combine like terms:

1. Add their numerical coefficients.
2. Multiply the sum by the common variable part.

Exercise 11.8

Combine the following:

a) $12a + 5a$
b) $b - 5b$
c) $6x^5 - 9x^5$
d) $-4t^3 + 6t^3 - 2t^3$
e) $7p^2q^3 - 3p^2q^3 - p^2q^3$
f) $0.5y - 2.7y + 3.1y - 1.9y$
g) $\dfrac{7}{9}w - w + \dfrac{4}{9}w + w$

Solution 11.8

a) $12a + 5a = (12 + 5)a = 17a$
b) $b - 5b = [b + (-5b)] = [1 + (-5)]b = -4b$ (Remember that b means $(1)b$.)
c) $6x^5 - 9x^5 = (6 - 9)x^5 = -3x^5$
d) $-4t^3 + 6t^3 - 2t^3 = (-4 + 6 - 2)t^3 = (0)t^3 = 0$
e) $7p^2q^3 - 3p^2q^3 - p^2q^3 = (7 - 3 - 1)p^2q^3 = 3p^2q^3$
f) $0.5y - 2.7y + 3.1y - 1.9y = (0.5 - 2.7 + 3.1 - 1.9)y = -1y = -y$
g) $\dfrac{7}{9}w - \dfrac{4}{9}w + w = \left(\dfrac{7}{9} - \dfrac{4}{9} + 1\right)w = \dfrac{12}{9}w = \dfrac{4}{3}w$

If an algebraic expression is a sum of two or more terms, we can combine all groups of like terms.

Exercise 11.9

Combine the like terms in $2p + 4q + 7p - 5q$.

Solution 11.9

The given expression is a sum of four terms. The first and third terms may be combined, because they are like terms. The second and fourth terms are like terms and may also be combined. We have $2p + 4q + 7p - 5q$

$= (2p + 7p) + (4q - 5q)$ (regroup the terms)
$= (2 + 7)p + (4 - 5)q$ (add like terms in each group)
$= 9p + (-1)q$ (simplify)
$= 9p - q$ (rewrite)

Because $9p$ and $-q$ have different variable factors, they are not like terms and cannot be combined.

Exercise 11.10
Simplify:
a) $6x^2y^3 - 3x^3y^2 + 5x^2y^3$
b) $5ab^2 - 3a^2b + 4a^2b - ab^2$
c) $3x^2y^4 - 2x^4y^2 - x^2y^4 + 6x^4y^2 - x^2y^2$
d) $3x - 4.1y - 3.7z - 3.9x - y + z - 2y + 3.1z - x$

Solution 11.10
a) $6x^2y^3 - 3x^3y^2 + 5x^2y^3$
$= (6x^2y^3 + 5x^2y^3) - 3x^3y^2$ (regroup)
$= (6 + 5)x^2y^3 - 3x^3y^2$ (add like terms)
$= 11x^2y^3 - 3x^3y^2$ (simplify)

Because $11x^2y^3$ and $-3x^3y^2$ are unlike terms (the exponents are different), they cannot be combined.

b) $5ab^2 - 3a^2b + 4a^2b - ab^2$
$= (5ab^2 - ab^2) + (-3a^2b + 4a^2b)$ (regroup)
$= (5 - 1)ab^2 + (-3 + 4)a^2b$ (add like terms)
$= 4ab^2 + a^2b$ (simplify)

c) $3x^2y^4 - 2x^4y^2 - x^2y^4 + 6x^4y^2 - x^2y^2$
$= (3x^2y^4 - x^2y^4) + (-2x^4y^2 + 6x^4y^2) - x^2y^2$
$= (3 - 1)x^2y^4 + (-2 + 6)x^4y^2 - x^2y^2$
$= 2x^2y^4 + 4x^4y^2 - x^2y^2$

d) $3x - 4.1y - 3.7z - 3.9x - y + z - 2y + 3.1z - x$
$= (3x - 3.9x - x) + (-4.1y - y - 2y) + (-3.7z + z + 3.1z)$
$= (3 - 3.9 - 1)x + (-4.1 - 1 - 2)y + (-3.7 + 1 + 3.1)z$
$= (-1.9)x + (-7.1)y + (0.4)z$
$= -1.9x - 7.1y + 0.4z$

REMOVING SYMBOLS OF GROUPING

If algebraic expressions contain symbols of grouping, we must first remove them before simplifying. Note the following rules for removing symbols of grouping:

1. To remove parentheses (or other symbols of grouping), do the following:

    a. If the parentheses are preceded by a + sign, drop the parentheses and the + sign. Leave the sign of each term within the parentheses as is.

    b. If the parentheses are preceded by a − sign, drop the parentheses and the − sign. *Change* the sign of *each term* within the parentheses.

2. If two or more grouping symbols are involved in the expression, remove the innermost symbols first and work outward. If no sign precedes a grouping symbol or a term within a grouping symbol, it is understood to be a + sign.

Exercise 11.11
Remove the parentheses: $2u + (v - w)$

Solution 11.11
$2u + (v - w)$
$= 2u + (+v - w)$ (sign preceding v is understood to be +)
$= 2u + v - w$ (remove the parentheses preceded by a + sign)

Exercise 11.12
Remove the parentheses: $3a - (2b + c)$

Solution 11.12
$3a - (2b + c)$
$= 3a - (+2b + c)$ (sign preceding $2b$ is understood to be +)
$= 3a - 2b - c$ (remove parentheses preceded by a − sign; *change* the sign of *each term* within the parentheses)

Exercise 11.13
Remove the grouping symbols: $3p - [2q + (r - 4s)]$

Solution 11.13
$3p - [2q + (r - 4s)]$ (start with innermost parentheses)
$= 3p - [2q + (+r - 4s)]$ (sign preceding r is understood to be +)
$= 3p - [2q + r - 4s]$ (remove parentheses preceded by a + sign)
$= 3p - [+2q + r - 4s]$ (sign preceding q is understood to be +)
$= 3p - 2q - r + 4s$ (remove parentheses preceded by a − sign; *change* sign of *each term* within brackets)

After removing symbols of grouping, we combine all groups of like terms, if any.

Exercise 11.14
Simplify by combining like terms:
a) $(4p - 3q) - (5r - 7p) - (q - 3r)$
b) $-(4x - 5y) + (6x + 3y)$
c) $3m - [4n - (3m - 6n)] - 7m$

Solution 11.14
a) $(4p - 3q) - (5r - 7p) - (q - 3r)$ (remove parentheses; + sign understood)
$= 4p - 3q - (5r - 7p) - (q - 3r)$ (remove parentheses preceded by − sign)
$= 4p - 3q - 5r + 7p - (q - 3r)$ (remove parentheses preceded by a − sign)
$= 4p - 3q - 5r + 7p - q + 3r$ (add the p terms)

$= 11p - 3q - 5r - q + 3r$ (add the q terms)
$= 11p - 4q - 5r + 3r$ (add the r terms)
$= 11p - 4q - 2r$

b) $-(4x - 5y) + (6x + 3y)$
$= -4x + 5y + (6x + 3y)$ (remove parentheses preceded by a $-$ sign)
$= -4x + 5y + 6x + 3y$ (remove parentheses preceded by a $+$ sign)
$= 2x + 5y + 3y$ (add the x terms)
$= 2x + 8y$ (add the y terms)

c) $3m - [4n - (3m - 6n)] - 7m$
$= 3m - [4n - 3m + 6n] - 7m$ (remove parentheses preceded by a $-$ sign)
$= 3m - [10n - 3m] - 7m$ (add the n terms)
$= 3m - 10n + 3m - 7m$ (remove parentheses preceded by a $-$ sign)
$= -m - 10n$ (add the m terms)

Exercise 11.15
Simplify by combining like terms: $-2s - \{2 + 3t + [4 - 2s - (3t - 2s) + 5] - 2\} + 5s$

Solution 11.15
$-2s - \{2 + 3t + [4 - 2s - (3t - 2s) + 5] - 2\} + 5s$
$= -2s - \{2 + 3t + [4 - 2s - 3t + 2s + 5] - 2\} + 5s$
$= -2s - \{2 + 3t + [9 - 3t] - 2\} + 5s$
$= -2s - \{2 + 3t + 9 - 3t - 2\} + 5s$
$= -2s - \{9\} + 5s$
$= -2s - 9 + 5s$
$= 3s - 9$

Exercise 11.16
Simplify by combining like terms:
a) $3x - [2y + (4x - 2y) - (5y - 3x)]$
b) $-(2p + 3q) - (4p - 2q)$

Solution 11.16
a) $3x - [2y + (4x - 2y) - (5y - 3x)]$
$= 3x - [2y + 4x - 2y - (5y - 3x)]$
$= 3x - [2y + 4x - 2y - 5y + 3x]$
$= 3x - [-5y + 4x + 3x]$
$= 3x - [-5y + 7x]$
$= 3x + 5y - 7x$
$= -4x + 5y$

b) $-(2p + 3q) - (4p - 2q)$
$= -2p - 3q - (4p - 2q)$
$= -2p - 3q - 4p + 2q$
$= -6p - 3q + 2q$
$= -6p + (-q)$
$= -6p - q$

Exercise 11.17
Simplify by combining like terms: $(0.1a + 0.2b - 0.3c) + (0.4a - 0.6c) - (0.5a - 0.4b)$

Solution 11.17
$(0.1a + 0.2b - 0.3c) + (0.4a - 0.6c) - (0.5a - 0.4b)$
$= 0.1a + 0.2b - 0.3c + (0.4a - 0.6c) - (0.5a - 0.4b)$
$= 0.1a + 0.2b - 0.3c + 0.4a - 0.6c - (0.5a - 0.4b)$
$= 0.1a + 0.2b - 0.3c + 0.4a - 0.6c - 0.5a + 0.4b$
$= (0)a + 0.2b - 0.3c - 0.6c + 0.4b$
$= 0.6b - 0.3c - 0.6c$
$= 0.6b - 0.9c$

If an algebraic expression is a sum and the sum is multiplied by a factor, *every* term of the sum *must* be multiplied by that factor.

Exercise 11.18
Multiply:
a) $5(x + y)$
b) $-7(3a - 2b + c)$
c) $(3r - 5s)(-4)$

Solution 11.18
a) $5(x + y)$
$= (5)(x) + (5)(y)$ (*each term* within the parentheses is multiplied by the factor 5)
$= 5x + 5y$ (simplify)
b) $-7(3a - 2b + c)$
$= (-7)(3a) + (-7)(-2b) + (-7)(c)$ (*each term* within the parentheses is multiplied by the factor -7)
$= -21a + 14b - 7c$ (simplify)
c) $(3r - 5s)(-4)$
$= (3r)(-4) + (-5s)(-4)$ (*each term* within the parentheses is multiplied by the factor -5)
$= -12r + 20s$ (simplify)

Exercise 11.19

Simplify the following expressions by combining like terms:

a) $4(x - 2y) - 5(3y - 2x)$

b) $\dfrac{1}{2}(6r - 2s - t) - \dfrac{1}{3}(3s - 3r + 5t)$

Solution 11.19

a) $4(x - 2y) - 5(3y - 2x)$
$= (4)(x) - (4)(2y) - 5(3y - 2x)$ (multiply each term in the parentheses by 4)
$= 4x - 8y - 5(3y - 2x)$ (simplify)
$= 4x - 8y + (-5)(3y - 2x)$ (rewrite, using the definition of subtraction)
$= 4x - 8y + (-5)(3y) + (-5)(-2x)$ (multiply each term in the parentheses by -5)
$= 4x - 8y + (-15y) + (10x)$ (simplify)
$= 14x - 8y + (-15y)$ (add the x terms)
$= 14x - 23y$ (add the y terms)

b) $\dfrac{1}{2}(6r - 2s - t) - \dfrac{1}{3}(3s - 3r + 5t)$

$= \left(\dfrac{1}{2}\right)(6r) + \left(\dfrac{1}{2}\right)(-2s) + \left(\dfrac{1}{2}\right)(-t) - \left(\dfrac{1}{3}\right)(3s - 3r + 5t)$

$= 3r - s - \left(\dfrac{1}{2}\right)t - \left(\dfrac{1}{3}\right)(3s - 3r + 5t)$

$= 3r - s - \left(\dfrac{1}{2}\right)t - \left(\dfrac{1}{3}\right)(3s) + \left(\dfrac{-1}{3}\right)(-3r) + \left(\dfrac{-1}{3}\right)(5t)$

$= 3r - s - \left(\dfrac{1}{2}\right)t - s + r + \left(\dfrac{-5}{3}\right)t$

$= 4r - s - \left(\dfrac{1}{2}\right)t - s + \left(\dfrac{-5}{3}\right)t$

$= 4r - 2s - \left(\dfrac{1}{2}\right)t + \left(\dfrac{-5}{3}\right)t$

$= 4r - 2s - \dfrac{13}{6}t$

Exercise 11.20

Simplify the following by combining like terms: $4a - \{(3a - 5b) - [2(b - 2a) + 3(4a - 3b)]\}$

Solution 11.20

$4a - \{(3a - 5b) - [2(b - 2a) + 3(4a - 3b)]\}$
$= 4a - \{(3a - 5b) - [2b - 4a + 3(4a - 3b)]\}$
$= 4a - \{(3a - 5b) - [2b - 4a + 12a - 9b]\}$
$= 4a - \{(3a - 5b) - [2b + 8a - 9b]\}$
$= 4a - \{(3a - 5b) - [-7b + 8a]\}$
$= 4a - \{3a - 5b - [-7b + 8a]\}$
$= 4a - \{3a - 5b + 7b - 8a\}$
$= 4a - \{-5a - 5b + 7b\}$
$= 4a - \{-5a + 2b\}$
$= 4a + 5a - 2b$
$= 9a - 2b$

11.3 EQUATIONS

Many applications can be solved by using an equation. An **equation** is a statement that two expressions are equal. The symbol = is used to denote the equality and is read, "is equal to." The following are examples of equations:

- $2x + 3 = 5$
- $3 + 7 = 10$
- $\dfrac{x+1}{5} = 3$
- $3(y - 1) = 3y - 3$
- $\sqrt{2t + 7} = 0$

The expression on the left-hand side of the equal sign is called the **left member** of the equation. The expression on the right-hand side of the equal sign is called the **right member.**

An equation may be true or it may be false.

- The equation $5 + 7 = 12$ is a true statement, because the left-hand member *is* equal to the right member.
- The equation $5 \times 6 = 11$ is a false statement, because the left-hand member *is not* equal to the right member.
- The equation $x + 5 = 9$ may be true or it may be false depending upon what value is substituted for x. If $x = 7$, the equation $x + 5 = 9$ becomes $7 + 5 = 9$, which *is not* a true statement. If $x = 4$, the equation $x + 5 = 9$ becomes $4 + 5 = 9$, which *is* a true statement.

If an equation contains a variable, then the equation may be true for some values of the variable and false for other values.

SOLUTIONS TO EQUATIONS

To solve an equation that contains a variable means to determine all values of the variable that make the equation a true statement. Each such value of the variable is called a **solution** of the equation. To determine whether a number is a solution to an equation, substitute the number for the variable and see whether the statement is true or false.

Exercise 11.21

Determine whether 5 is a solution of the equation $y - 2 = 7$.

Solution 11.21

Replace y in the equation by 5. We have
$$y - 2 = 7$$
$$5 - 2 \;?\; 7$$
$$3 \neq 7$$

Therefore, 5 *is not* a solution of the given equation.

Exercise 11.22

Determine whether 2 is a solution of the equation $2u + 5 = 9$.

Solution 11.22

Replace u in the equation by 2. We have
$$2u + 5 = 9$$
$$2(2) + 5 \;?\; 9 \quad \text{(remember, } 2u \text{ means 2 times } u.)$$
$$4 + 5 \;?\; 9$$
$$9 = 9 \checkmark$$

Therefore, 2 *is* a solution of the given equation.

Exercise 11.23

Determine whether −4 is a solution of the equation $\dfrac{x - 5}{3} = 2$.

Solution 11.23

Replace x in the equation by −4. We have
$$\frac{x - 5}{3} = 2$$
$$\frac{-4 - 5}{3} \;?\; 2$$
$$\frac{-9}{3} \;?\; 2$$
$$-3 \neq 2$$

Therefore, −4 *is not* a solution of the given equation.

Exercise 11.24
Determine whether 7 is a solution of the equation $2(3 - p) = -8$.

Solution 11.24
Replace p in the equation by 7. We have
$2(3 - p) = -8$
$2(3 - 7) \ ? \ -8$
$2(-4) \ ? \ -8$
$-8 = -8 \checkmark$

Therefore, 7 *is* a solution of the given equation.

LINEAR EQUATIONS

In this section, we study the solution of linear equations in a single variable. A **linear equation in a single variable** is an algebraic equation that contains only that variable, with an exponent of 1, and such that the variable does not appear in any denominators.

The following are examples of linear equations:
- $y + 9 = 5$ *is* a linear equation in the variable y.
- $0.7w - 4 = 5$ *is* a linear equation in the variable w.
- $t + 4 = 2 - 3t$ *is* a linear equation in the variable t.
- $-3(u + 4) = 2(5 - u)$ *is* a linear equation in the variable u.
- $2x - 7(x + 1) = 29$ *is* a linear equation in the variable x.

The following are examples of equations that are *not* linear equations in a single variable:
- $v^2 + 7 = 0$ *is not* a linear equation, because the variable v has an exponent 2.
- $x(y + 2) = 3$ *is not* a linear equation in a single variable, because there are two variables.
- $y^{-2} + 5 = 10$ *is not* a linear equation, because the first term has an exponent -2.
- $\dfrac{1}{s} = 10$ *is not* a linear equation, because the variable s appears in the denominator on the left-hand side.
- $3u - 5v = 7$ *is not* a linear equation in a single variable, because there are two variables.

IDENTITY

Some equations are true for *all* values of the variable involved. Such equations are called **identities.**
- The equation $9 + s = s + 9$ is an identity; it is true for *all* values of s.
- The equation $4(x - 3) = 4x - 12$ is an identity; it is true for *all* values of x.

Some equations in a single variable have *no* solutions. For example, the equation $x = x + 1$ has no solutions. No matter what value is substituted for x, the right hand side of the equation will be 1 more than that value.

In this chapter, most of the equations considered have only one solution. An equation that is true for some but not all values assigned to its variables is called a **conditional equation.**

11.4 SOLVING LINEAR EQUATIONS

In this section, you learn to solve basic linear equations in a single variable. Two linear equations in the same variable that have the same solutions are called **equivalent equations.** We solve a linear equation by changing the given equation into equivalent equations until the variable is all by itself on one side of the equation with a coefficient of 1.

ADDITION RULE

We can *add* the *same* quantity to *both* sides of a linear equation to obtain an equivalent equation. That is, if
$$x - a = b,$$
then
$$(x - a) + a = b + a$$
or
$$x = b + a$$

Exercise 11.25

Solve the equation $y - 6 = 2$ for y.

Solution 11.25

To solve the equation, we *add* 6 to *both* sides of the equation. Hence, $y - 6 = 2$ becomes
$$(y - 6) + 6 = 2 + 6 \text{ (add 6 to } both \text{ sides)}$$
$$y + (-6 + 6) = 2 + 6 \text{ (regroup)}$$
$$y + 0 = 8 \text{ (simplify)}$$
$$y = 8 \text{ (simplify)}$$

Therefore, 8 is the solution. We can check this by substituting 8 for y in the *original* equation.

$$\text{Check: If } y = 8, \text{ does } y - 6 = 2?$$
$$y - 6 = 2$$
$$8 - 6 \text{ ? } 2$$
$$2 = 2 \checkmark$$

Exercise 11.26

Solve the equation $t - 9 = 5$ for t.

Solution 11.26

To solve this equation, we *add* 9 to *both* sides of the equation. Hence, $t - 9 = 5$ becomes
$$(t - 9) + 9 = 5 + 9 \text{ (add 9 to } both \text{ sides)}$$
$$t + (-9 + 9) = 5 + 9 \text{ (regroup)}$$

$$t + 0 = 14 \text{ (simplify)}$$
$$t = 14 \text{ (simplify)}$$

Therefore, 14 is the solution.

$$\text{Check: If } t = 14, \text{ does } t - 9 = 5?$$
$$t - 9 = 5$$
$$14 - 9 \;?\; 5$$
$$5 = 5 \checkmark$$

Exercise 11.27

Solve the equation $p - \dfrac{1}{3} = \dfrac{5}{7}$ for p.

Solution 11.27

$p - \dfrac{1}{3} = \dfrac{5}{7}$ ($\dfrac{1}{3}$ is being *subtracted* from p)

$p = \dfrac{5}{7} + \dfrac{1}{3}$ (*add* $\dfrac{1}{3}$ to *both* sides)

$p = \dfrac{22}{21}$ (add fractions)

Therefore, $\dfrac{22}{21}$ is the solution.

Check: $p - \dfrac{1}{3} = \dfrac{5}{7}$

$\dfrac{22}{21} - \dfrac{1}{3} \;?\; \dfrac{5}{7}$

$\dfrac{22}{21} - \dfrac{7}{21} \;?\; \dfrac{5}{7}$

$\dfrac{15}{21} \;?\; \dfrac{5}{7}$

$\dfrac{5}{7} = \dfrac{5}{7} \checkmark$

Exercise 11.28

Solve the equation $q - 3.45 = -7.68$ for q.

Solution 11.28

$$q - 3.45 = -7.68 \text{ (3.45 is being } subtracted \text{ from } q)$$
$$q = -7.68 + 3.45 \text{ (} add \text{ 3.45 to } both \text{ sides)}$$
$$q = -4.23 \text{ (add decimals)}$$

The solution is -4.23.

$$\text{Check: } q - 3.45 = -7.68$$
$$-4.23 - 3.45 \;?\; -7.68$$
$$-4.23 + (-3.45) \;?\; -7.68$$
$$-7.68 = -7.68 \checkmark$$

SUBTRACTION RULE

We can *subtract* the *same* quantity from *both* sides of a linear equation to obtain an equivalent equation. That is, if

$$x + a = b,$$

then

$$(x + a) - a = b - a$$

or

$$x = b - a$$

Exercise 11.29
Solve the equation $x + 7 = 3$ for x.

Solution 11.29
To solve the equation, we *subtract* 7 from *both* sides. Hence, $x + 7 = 3$ becomes

$(x + 7) - 7 = 3 - 7$ (*subtract* 7 from *both* sides)
$x + (7 - 7) = 3 - 7$ (regroup)
$x + 0 = -4$ (simplify)
$x = -4$ (simplify)

Therefore, -4 is the required solution.

Check: $x + 7 = 3$
$-4 + 7 \ ? \ 3$
$3 = 3 \checkmark$

Exercise 11.30
Solve the equation $s + 3 = 8$ for s.

Solution 11.30

$s + 3 = 8$ (3 is being *added* to s)
$s = 8 - 3$ (*subtract* 3 from *both* sides)
$s = 5$ (simplify)

The solution is 5.

Check: $s + 3 = 8$
$5 + 3 \ ? \ 8$
$8 = 8 \checkmark$

Exercise 11.31
Solve the equation $p + 12 = 5$ for p.

Solution 11.31

$$p + 12 = 5 \text{ (12 is being } added \text{ to } p)$$
$$p = 5 - 12 \text{ (}subtract \text{ 12 from } both \text{ sides)}$$
$$p = -7 \text{ (simplify)}$$

The required solution is -7.

$$\text{Check: } p + 12 = 5$$
$$-7 + 12 \ ? \ 5$$
$$5 = 5 \checkmark$$

Exercise 11.32

Solve the equation $x + \dfrac{3}{2} = \dfrac{1}{5}$ for x.

Solution 11.32

$$x + \frac{3}{2} = \frac{1}{5} \ (\frac{3}{2} \text{ is being } added \text{ to } x)$$
$$x = \frac{1}{5} - \frac{3}{2} \ (subtract \ \frac{3}{2} \text{ from } both \text{ sides})$$
$$x = \frac{-13}{10} \text{ (simplify)}$$

The solution is $\dfrac{-13}{10}$

$$\text{Check: } x + \frac{3}{2} = \frac{1}{5}$$
$$\frac{-13}{10} + \frac{3}{2} \ ? \ \frac{1}{5}$$
$$\frac{-13}{10} + \frac{15}{10} \ ? \ \frac{1}{5}$$
$$\frac{2}{10} \ ? \ \frac{1}{5}$$
$$\frac{1}{5} = \frac{1}{5} \checkmark$$

Exercise 11.33

Solve the equation $x + 0.7 = 0.53$ for x.

Solution 11.33

$$x + 0.7 = 0.53 \text{ (0.7 is being } added \text{ to } x)$$
$$x = 0.53 - 0.7 \text{ (}subtract \text{ 0.7 from } both \text{ sides)}$$
$$x = -0.17 \text{ (simplify)}$$

The solution is −0.17.

$$\text{Check: } x + 0.7 = 0.53$$
$$-0.17 + 0.7 \,?\, 0.53$$
$$0.53 = 0.53 \checkmark$$

MULTIPLICATION RULE

We can *multiply both* sides of a linear equation by the *same nonzero* quantity to obtain an equivalent equation. That is, if

$$\frac{x}{a} = b \ (a \neq 0),$$

then

$$a\left(\frac{x}{a}\right) = (a)(b)$$

or

$$x = ab$$

Exercise 11.34

Solve the equation $\frac{u}{5} = 7$ for u.

Solution 11.34

To solve the equation, we *multiply both* sides of the equation by 5. Hence, $\frac{u}{5} = 7$ becomes

$$(5)\frac{u}{5} = (7)(5) \text{ (multiply } both \text{ sides by 5)}$$

or

$$u = 35 \text{ (simplify)}$$

Therefore, 35 is the solution.

$$\text{Check: } \frac{u}{5} = 7$$
$$\frac{35}{5} \,?\, 7$$
$$7 = 7 \checkmark$$

Exercise 11.35

Solve the equation $\frac{w}{2} = 3.9$ for w.

Solution 11.35

$$\frac{w}{2} = 3.9 \quad (w \text{ is being } divided \text{ by 2})$$

$$(2)\left(\frac{w}{2}\right) = (2)(3.9) \quad (multiply \text{ both sides by 2})$$

$$w = 7.8$$

The solution is 7.8.
  Check: $w \div 2 = 3.9$
    $7.8 \div 2 ? 3.9$
      $3.9 = 3.9$ ✓

Exercise 11.36

Solve the equation $\frac{x}{-3} = \frac{1}{4}$ for x.

Solution 11.36

$$\frac{x}{-3} = \frac{1}{4} \quad (x \text{ is being } divided \text{ by } -3)$$

$$(-3)\left(\frac{x}{-3}\right) = (-3)\left(\frac{1}{4}\right) \quad (multiply \text{ both sides by } -3)$$

$$x = \frac{-3}{4}$$

The solution is $\frac{-3}{4}$.

Check: $\frac{x}{-3} = \frac{1}{4}$

$$\left(\frac{-3}{4}\right) \div (-3) ? \frac{1}{4}$$

$$\frac{1}{4} = \frac{1}{4} \checkmark$$

Exercise 11.37

Solve the equation $\frac{p}{3.7} = 1.6$ for p.

Solution 11.37

$$\frac{p}{3.7} = 1.6 \quad (p \text{ is being } divided \text{ by 3.7})$$

$$(3.7)\left(\frac{p}{3.7}\right) = (3.7)(1.6) \quad (multiply \text{ both sides by 3.7})$$

$$p = 5.92$$

The solution is 5.92.

Check: $\dfrac{p}{3.7} = 1.6$

$\dfrac{5.92}{3.7} \;?\; 1.6$

$1.6 = 1.6$ ✓

Exercise 11.38

Solve the equation $\dfrac{v}{4} = -1.7$ for v.

Solution 11.38

$\dfrac{v}{4} = -1.7$ (v is being *divided* by 4)

$(4)\left(\dfrac{v}{4}\right) = (4)(-1.7)$ (*multiply both* sides by 4)

$v = -6.8$

The solution is −6.8.

Check: $\dfrac{v}{4} = -1.7$

$\dfrac{(-6.8)}{4} \;?\; -1.7$

$-1.7 = -1.7$ ✓

DIVISION RULE

We *divide both* sides of a linear equation by the *same nonzero* quantity to obtain an equivalent equation. That is, if

$$ax = b \;(a \neq 0),$$

then

$$\dfrac{ax}{a} = \dfrac{b}{a}$$

or

$$x = \dfrac{b}{a}$$

Exercise 11.39

Solve the equation $3x = 7$ for x.

Solution 11.39

To solve the equation, we *divide both* sides of the equation by 3. Hence, $3x = 7$ becomes

$$\frac{3x}{3} = \frac{7}{3} \text{ (divide both sides by 3)}$$

or

$$x = \frac{7}{3} \text{ (simplify)}$$

Therefore, $\frac{7}{3}$ is the solution.

Check: $3x = 7$

$$(3)\left(\frac{7}{3}\right) ? 7$$

$$7 = 7 \checkmark$$

Exercise 11.40

Solve the equation $-2p = 27$ for p.

Solution 11.40

$-2p = 27$ (p is being *multiplied* by -2)

$$\frac{-2p}{-2} = \frac{27}{-2} \text{ (divide both sides by } -2\text{)}$$

$$p = \frac{-27}{2}$$

The solution is $\frac{-27}{2}$.

Check: $-2p = 27$

$$-2\left(\frac{-27}{2}\right) ? 27$$

$$27 = 27 \checkmark$$

Exercise 11.41

Solve the equation $4.6q = 1.38$ for q.

Solution 11.41

$4.6q = 1.38$ (q is being *multiplied* by 4.6)

$$\frac{4.6q}{4.6} = \frac{1.38}{4.6} \text{ (divide both sides by 4.6)}$$

$$q = 0.3$$

The solution is 0.3.
Check: $4.6q = 1.38$
$(4.6)(0.3) \stackrel{?}{=} 1.38$
$1.38 = 1.38$ ✓

Exercise 11.42

Solve the equation $\frac{3}{4}y = \frac{5}{3}$ for y.

Solution 11.42

$\frac{3}{4}y = \frac{5}{3}$ (y is being *multiplied* by $\frac{3}{4}$)

$\frac{3}{4}y \div \frac{3}{4} = \frac{5}{3} \div \frac{3}{4}$ (*divide both* sides by $\frac{3}{4}$)

$\frac{3}{4}y \times \frac{4}{3} = \frac{5}{3} \times \frac{4}{3}$ (multiply by reciprocal)

$\left(\frac{3}{4} \times \frac{4}{3}\right)y = \frac{5}{3} \times \frac{4}{3}$ (rewrite)

$y = \frac{20}{9}$

The solution is $\frac{20}{9}$.

Check: $\frac{3}{4}y = \frac{5}{3}$

$\left(\frac{3}{4}\right)\left(\frac{20}{9}\right) \stackrel{?}{=} \frac{5}{3}$

$\left(\frac{\cancel{3}^{1}}{\cancel{4}_{1}}\right)\left(\frac{\cancel{20}^{5}}{\cancel{9}_{3}}\right) \stackrel{?}{=} \frac{5}{3}$

$\frac{5}{3} = \frac{5}{3}$ ✓

Exercise 11.43

Solve the equation $\frac{2}{3}x = \frac{3}{4}$ for x.

Solution 11.43

$$\frac{2}{3}x = \frac{3}{4} \quad (x \text{ is being } \textit{multiplied} \text{ by } \frac{2}{3})$$

$$\frac{2}{3}x \div \frac{2}{3} = \frac{3}{4} \div \frac{2}{3} \quad (\text{divide both sides by } \frac{2}{3})$$

$$\left(\frac{2}{3}x\right)\left(\frac{3}{2}\right) = \left(\frac{3}{4}\right)\left(\frac{3}{2}\right) \quad (\text{multiply by reciprocal})$$

$$\left(\frac{2}{3}\right)\left(\frac{3}{2}\right)x = \left(\frac{3}{4}\right)\left(\frac{3}{2}\right) \quad (\text{rewrite})$$

$$x = \frac{9}{8}$$

The solution is $\frac{9}{8}$.

Check: $\frac{2}{3}x = \frac{3}{4}$

$$\frac{2}{3}\left(\frac{9}{8}\right) \stackrel{?}{=} \frac{3}{4}$$

$$\frac{3}{4} = \frac{3}{4} \checkmark$$

COMBINED OPERATIONS

Sometimes more than one operation is necessary to solve an equation.

Exercise 11.44

Solve the equation $4x - 8 = 6$ for x.

Solution 11.44

First, add 8 to both sides of the equation to get the x-term by itself on the left-hand side of the equation. We have

$$(4x - 8) + 8 = 6 + 8$$

or

$$4x = 14$$

Next, we *divide* both sides of the equation by 4. Therefore,

$$4x = 14$$

becomes

$$\frac{4x}{4} = \frac{14}{4}$$

or
$$x = \frac{7}{2}$$

Therefore, $\frac{7}{2}$ is the solution.

$$\text{Check: } 4x - 8 = 6$$
$$4\left(\frac{7}{2}\right) - 8 \,?\, 6$$
$$14 - 8 \,?\, 6$$
$$6 = 6 \checkmark$$

Exercise 11.45
Solve the equation $5p + 1 = 9$ for p.

Solution 11.45
$(5p + 1) - 1 = 9 - 1$ (subtract 1 from both sides)
$$5p = 8$$
$$\frac{5p}{5} = \frac{8}{5} \text{ (divide both sides by 5)}$$
$$p = \frac{8}{5}$$

The solution is $\frac{8}{5}$.

$$\text{Check: } 5p + 1 = 9$$
$$5\left(\frac{8}{5}\right) + 1 \,?\, 9$$
$$8 + 1 \,?\, 9$$
$$9 = 9 \checkmark$$

Exercise 11.46
Solve each of the following equations as indicated. The checks are left to the student.

a) $\frac{2-q}{4} = 2$; solve for q

b) $3(t - 1) = 11$; solve for t

c) $0.5x - 0.1 = 1.7$; solve for x

d) $4(3 - 5r) = -13$; solve for r

Solution 11.46

a) $\dfrac{2-q}{4} = 2$

$\left(\dfrac{2-q}{4}\right) = 4(2)$ (multiply both sides by 4)

$2 - q = 8$

$-q = 6$ (subtract 2 from both sides)

$q = -6$ (multiply both sides by -1)

The solution is -6.

b) $3(t - 1) = 11$

$3t - 3 = 11$ (multiply $(t - 1)$ by 3)

$3t = 14$ (add 3 to both sides)

$t = \dfrac{14}{3}$ (divide both sides by 3)

The solution is $\dfrac{14}{3}$.

c) $0.5x - 0.1 = 1.7$

$0.5x = 1.8$ (add 0.1 to both sides)

$x = \dfrac{1.8}{0.5}$ (divide both sides by 0.5)

$x = 3.6$ (simplify)

The solution is 3.6.

d) $4(3 - 5r) = -13$

$12 - 20r = -13$ (multiply $(3 - 5r)$ by 4)

$-20r = -25$ (subtract 12 from both sides)

$r = \dfrac{-25}{-20}$ (divide both sides by -20)

$r = \dfrac{5}{4}$ (simplify)

The solution is $\dfrac{5}{4}$.

Exercise 11.47
Solve each of the following equations as indicated. The checks are left to the student.

a) $\dfrac{1-3u}{5} = 1$ for u

b) $1.7q + 2.5 = -0.7$ for q

c) $\dfrac{-3y}{4} + 4 = \dfrac{1}{3}$ for y

d) $\dfrac{3}{7}(5x - 3) = \dfrac{-1}{3}$ for x

Solution 11.47

a) $\quad \dfrac{1-3u}{5} = 1$

$5\left(\dfrac{1-3u}{5}\right) = 5(1)$ (multiply both sides by 5)

$1 - 3u = 5$

$\quad -3u = 4$ (subtract 1 from both sides)

$\quad\quad u = \dfrac{-4}{3}$ (divide both sides by -3)

The solution is $\dfrac{-4}{3}$.

b) $1.7q + 2.5 = -0.7$

$\quad 1.7q = -3.2$ (subtract 2.5 from both sides)

$\quad \dfrac{1.7q}{1.7} = \dfrac{-3.2}{1.7}$ (divide both sides by 1.7)

$\quad\quad q = \dfrac{-32}{17}$ (simplify)

The solution is $\dfrac{-32}{17}$.

c) $\quad \dfrac{-3y}{4} + 4 = \dfrac{1}{3}$

$12\left(\dfrac{-3y}{4} + 4\right) = 12\left(\dfrac{1}{3}\right)$ (multiply both sides by 12)

$-9y + 48 = 4$ (simplify)

$\quad -9y = -44$ (subtract 48 from both sides)

$\quad \dfrac{-9y}{-9} = \dfrac{-44}{-9}$ (divide both sides by -9)

$$y = 4\frac{8}{9} \text{ (simplify)}$$

The solution is $4\frac{8}{9}$.

d) $\quad \dfrac{3}{7}(5x-3) = \dfrac{-1}{3}$

$$\frac{15x}{7} - \frac{9}{7} = \frac{-1}{3} \text{ (multiply } (5x-3) \text{ by } \frac{3}{7}\text{)}$$

$$21\left(\frac{15x}{7} - \frac{9}{7}\right) = 21\left(\frac{-1}{3}\right) \text{ (multiply both sides by 21)}$$

$$45x - 27 = -7 \text{ (simplify)}$$
$$45x = 20 \text{ (add 27 to both sides)}$$
$$x = \frac{4}{9} \text{ (divide both sides by 45)}$$

The solution is $\dfrac{4}{9}$.

11.5 THE LANGUAGE OF ALGEBRA

In this section, we translate word expressions into algebraic expressions.

Exercise 11.48

Write the following word expressions as algebraic expressions:

a) the sum of a and b

b) n increased by 8

c) y decreased by 7

d) the total of t and 11

e) subtract x from y

f) from x, subtract y

g) the difference of y and 6

h) 15 less p

i) 11 less than p

j) 5 diminished by q

k) r exceeds s

Solution 11.48

a) $a + b$
b) $n + 8$
c) $y - 7$
d) $t + 11$
e) $y - x$
f) $x - y$
g) $y - 6$
h) $15 - p$
i) $p - 11$
j) $5 - q$
k) $r > s$

Exercise 11.49

Write the following word expressions as algebraic expressions:

a) one-fifth of x
b) the product of x and y
c) the quotient of x and 3
d) the square of u
e) the cube of w
f) twice z
g) half of p

Solution 11.49

a) $\frac{1}{5}x$
b) xy
c) $\frac{x}{3}$ or $x \div 3$
d) u^2
e) w^3
f) $2z$
g) $\frac{p}{2}$

Exercise 11.50

Write the following word expressions as algebraic expressions:
- a) the sum of x and y minus their product
- b) seven more than twice the cube of y
- c) the product of the squares of m and n
- d) the square of the product of m and n
- e) the sum of twice x and four times y
- f) the square of twice the sum of w and 3
- g) the sum of the squares of x, y, and z

Solution 11.50

- a) $x + y - xy$
- b) $2y^3 + 7$
- c) m^2n^2
- d) $(mn)^2$
- e) $2x + 4y$
- f) $[2(w + 3)]^2$
- g) $x^2 + y^2 + z^2$

11.6 WORD PROBLEMS

We now use linear equations in one variable to solve some word problems. Use the following procedure to solve a word problem:

1. Read (and, if necessary, reread) the word problem carefully and determine what it is all about.
2. Identify all unknown quantities and assign a variable to each.
3. Draw a diagram or set up a chart that summarizes all the given information.
4. Translate (rewrite) the word statement(s) into an equation.
5. Solve the equation.
6. Check your results.
7. Carefully study the examples of the word problems that are given, because most problems are likely to have some similarity to these.

Exercise 11.51

The sum of 15 and twice a certain number is 33. What is the number?

Solution 11.51

In this problem, we are adding two numbers, 15 and twice a certain number. We could let n = the number. Then twice the number would be $2n$ and the sum would be $15 + 2n$. Because the sum is 33, we form and solve the equation

$$15 + 2n = 33$$
$$2n = 18$$
$$n = 9$$

The certain number is 9. Check to see that 15 plus twice 9 is equal to 33.

Exercise 11.52

A rectangle has a length of 18 inches. If its perimeter is 62 inches, what is the width of the rectangle?

Solution 11.52

Consider the rectangle below.

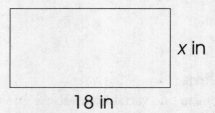

We could let x = the width (in inches). Because the perimeter P of the rectangle is the distance around, we have P = twice the length plus twice the width. Substituting x inches for width, 18 inches for length, and 62 inches for P, we have

$$P = 2(\text{length}) + 2(\text{width})$$
$$62 \text{ in.} = 2(18 \text{ in.}) + 2(x \text{ in.})$$
$$62 = 36 + 2x$$
$$26 = 2x$$
$$13 = x$$

Therefore, the width of the rectangle is 13 inches.

Exercise 11.53

The sum of three consecutive whole numbers is 168. What are the numbers?

Solution 11.53

Consecutive whole numbers follow one right after the other as you count. We could let n = the smallest number.

Then $n + 1$ = the next larger one and $n + 2$ = the largest one.

Because the sum of the three numbers is 168, we have

$$n + (n + 1) + (n + 2) = 168$$
$$n + (n + 1) + (n + 2) = 168$$
$$n + n + 1 + n + 2 = 168$$
$$3n + 3 = 168$$
$$3n = 165$$
$$n = 55$$

Hence, the solution is

$$n = 55$$
$$n + 1 = 56$$
$$n + 2 = 57$$

and the three consecutive whole numbers are 55, 56, and 57.

Exercise 11.54

Mary has twenty-five coins in her purse for a total of $3.70. If the coins are dimes and quarters, how many quarters does Mary have in her purse?

Solution 11.54

For this problem, we could let y = the number of quarters. Then $25 - y$ = the number of dimes (because there are twenty-five coins). Summarizing our data, we have

	Number of coins	Value per coin	Value of coins
quarters	y	$.25	$.25y
dimes	$25 - y$	$.10	$.10(25 - y)

Because the value of the quarters plus the value of the dimes is equal to the value of the 25 coins, we have

$$\$.25y + \$.10(25 - y) = \$3.70$$
$$25y + 10(25 - y) = 370$$
$$25y + 250 - 10y = 370$$
$$15y + 250 = 370$$
$$15y = 120$$
$$y = 8$$

Therefore, Mary has 8 quarters in her purse. She also has $(25 - y)$ or 17 dimes. Check that the total is $3.70.

Exercise 11.55

A triangle has a perimeter of 350 centimeters. The longest side of the triangle is twice the length of the shortest side. The third side is 10 centimeters longer than the shortest side. What are the lengths of the sides of the triangle?

Solution 11.55

We are to determine the lengths of the three sides of the triangle.

Let s = the length of the shortest side (in cm).
Then $s + 10$ = the length of the third side (in cm).
And $2s$ = the length of the longest side (in cm).

The perimeter of the triangle is the sum of the lengths of its sides. Hence,

$$s \text{ cm} + (s + 10) \text{ cm} + 2s \text{ cm} = 350 \text{ cm}$$
$$s + (s + 10) + 2s = 350$$
$$s + s + 10 + 2s = 350$$
$$4s + 10 = 350$$
$$4s = 340$$
$$s = 85$$

The solution is
$$s = 85$$
$$s + 10 = 95$$
$$2s = 170$$

The sides are 85 cm, 95 cm, and 170 cm. Check to determine that the perimeter is 350 cm.

Exercise 11.56

In a college general mathematics course, there are freshmen, sophomores, and juniors enrolled. There are twice as many freshmen as there are juniors. The number of sophomores is thirteen less than twice the number of juniors. In addition, there are seven noncredit students enrolled. If the total enrollment in the course is 339, how many freshmen are enrolled?

Solution 11.56

We want to determine the number of freshmen enrolled. However, this number depends upon the number of juniors enrolled. Hence, we let n = the number of juniors enrolled. Then $2n$ = the number of freshmen enrolled and $2n - 13$ = the number of sophomores enrolled. The total number of students enrolled is equal to the number of freshmen plus the number of sophomores plus the number of juniors plus the number of noncredit students enrolled. We have

$$2n + (2n - 13) + n + 7 = 339$$
$$2n + 2n - 13 + n + 7 = 339$$
$$5n - 6 = 339$$
$$5n = 345$$
$$n = 69$$
$$2n = 138$$

Hence, there are 138 freshmen enrolled in the course. There are also $n = 69$ juniors, $2n - 13 = 125$ sophomores, and 7 noncredit students enrolled. Check that $138 + 125 + 69 + 7$ is equal to 339.

11.7 FORMULAS

A mathematical **formula** is a relationship between two or more quantities or amounts. It states a rule or method for doing something. In this section, we evaluate formulas. To evaluate a formula:

1. Replace all variables with the given or known values.
2. Simplify the result.

DISTANCE-RATE-TIME FORMULA

The formula $d = rt$ expresses the relationship between the distance (d) traveled by an object in uniform motion, and the rate (r) and time (t).

Exercise 11.57

What distance does Isabel travel if she drives at a rate of 65 mph for 4.5 hours?

Solution 11.57

We must evaluate d (in miles) given $r = 65$ mph and $t = 4.5$ hours. Hence, $d = rt$
$$d = (65 \text{ mph})(4.5 \text{ hr})$$
$$d = \left(\frac{65 \text{ mi}}{1 \text{ hr}}\right)(4.5 \text{ hr})$$
$$d = 292.5 \text{ mi.}$$
Therefore, Isabel travels 292.5 miles in 4.5 hours at 65 mph.

Exercise 11.58

How long will it take Brice to run 10 miles at the rate of 4 miles per hour?

Solution 11.58

We must evaluate t (in hours) given $d = 10$ miles and $r = 4$ mph. Hence, $d = rt$
$$10 \text{ mi} = (4 \text{ mph})(t \text{ hr})$$
$$10 \text{ mi} = 4t \text{ mi}$$
$$\frac{10}{4} = t$$
$$2.5 = t.$$
Therefore, it will take Brice 2.5 hours to run 10 miles at 4 mph.

TEMPERATURE

The relationship between degrees Celsius (C) and degrees Fahrenheit (F) is given by the formula $C = \frac{5}{9}(F - 32°)$.

Exercise 11.59
Determine the Celsius temperature if the Fahrenheit temperature is 87°.

Solution 11.59
Using the given formula with F = 87°, we have

$$C = \frac{5}{9}(F - 32°)$$

$$C = \frac{5}{9}(87° - 32°)$$

$$C = \frac{5}{9}(55°)$$

$$C = 30\frac{5}{9}°$$

Therefore, a Fahrenheit temperature of 87° is equivalent to a Celsius temperature of $30\frac{5}{9}°$.

Exercise 11.60
Determine the Fahrenheit temperature if the Celsius temperature is 26°.

Solution 11.60
Using the formula with C = 26°, we have $C = \frac{5}{9}(F - 32°)$

$$26° = \frac{5}{9}(F - 32°)$$
$$234° = 5(F - 32°)$$
$$234° = 5F - 160°$$
$$394° = 5F$$
$$78.8° = F$$

Therefore, a Celsius temperature of 26° is equivalent to 78.8° Fahrenheit.

IQ

Intelligence quotient (IQ) is the relationship between mental age (M) and chronological age (C) and is given by the formula $IQ = \frac{100M}{C}$.

Exercise 11.61
Determine the IQ of a 14-year-old boy who has a mental age of 16 years.

Solution 11.61
We must evaluate IQ given that M = 16 yr and C = 14 yr. Hence,

$$IQ = \frac{100M}{C}$$

$$IQ = \frac{(100)(16 \text{ yr})}{14 \text{ yr}}$$

$$IQ = \frac{1,600}{14}$$

$$IQ = 114\frac{2}{7}$$

Therefore, the boy has an IQ of $114\frac{2}{7}$.

Exercise 11.62
Determine the mental age of a 16-year-old girl who has an IQ of 130.

Solution 11.62
Using the formula $IQ = \frac{100M}{C}$, we determine M given that C = 16 years and IQ = 130 as follows:

$$IQ = \frac{100M}{C}$$

$$130 = \frac{100M}{16 \text{ yr}}$$

$$(130)(16 \text{ yr}) = 100M$$

$$2,080 \text{ yr} = 100M$$

$$20.8 \text{ yr} = M$$

Therefore, the girl's mental age is 20.8 years.

AVERAGES

The average of two or more quantities is the result obtained by dividing the sum of the quantities by the number of quantities. It is also called the arithmetical mean.

Exercise 11.63
The average for five test scores is given by the formula $A = \frac{a+b+c+d+e}{5}$, where a = the first test score, b = the second test score, c = the third test score, d = the fourth test score, and e = the fifth score. Evaluate A if $a = 85$, $b = 69$, $c = 77$, $d = 81$, and $e = 78$.

Solution 11.63

Substituting the test scores in the given formula, we have:

$$A = \frac{a+b+c+d+e}{5}$$

$$A = \frac{85+69+77+81+78}{5}$$

$$A = \frac{390}{5}$$

$$A = 78$$

Therefore, the average of the five test scores is 78.

Exercise 11.64

Stacey has three exam scores of 88, 79, and 80. What score must she receive on the fourth exam to have an average of 84?

Solution 11.64

Using the formula $A = \frac{a+b+c+d}{4}$ with A = 84, $a = 88$, $b = 79$, and $c = 80$, we must solve for d as follows:

$$A = \frac{a+b+c+d}{4}$$

$$84 = \frac{88+79+80+d}{4}$$

$$84 = \frac{247+d}{4}$$

$$336 = 247 + d$$

$$89 = d$$

Therefore, Stacey must receive a score of 89 on her fourth test.

GRAVITY

The formula $s = \frac{1}{2}gt^2$ represents the distance (s) that a free-falling body travels in a given amount of time (t) (measured in seconds) under the effects of gravity (g), where g = 32 ft/sec^2.

Exercise 11.65

A student's billfold is dropped from the top of a 700-ft dormitory building. How far will the billfold fall in 5.8 seconds?

Solution 11.65

We are asked to determine the value of s given $g = 32$ ft/sec² and $t = 5.8$ sec. Using the given formula, we have

$$s = \frac{1}{2} gt^2$$

$$s = \frac{1}{2}(32 \text{ ft/sec}^2)(5.8 \text{ sec})^2$$

$$s = \frac{1}{2}(\frac{32 \text{ ft}}{\text{sec}^2})(5.8)^2 (\text{sec})^2$$

$$s = \frac{1}{2}(32)(5.8)^2 \text{ ft}$$

$$s = \frac{1}{2}(32)(33.64) \text{ ft}$$

$$s = 538.24 \text{ ft}$$

Therefore, the billfold will fall 538.24 feet in 5.8 seconds.

Exercise 11.66

A ring is dropped from the top of a 300-ft platform. How close to the ground will the ring be after 3.5 seconds?

Solution 11.66

There are two parts to this problem.

Step 1: We first must determine how far the ring will drop in 3.5 seconds. Using the formula $s = \frac{1}{2} gt^2$ with $g = 32$ ft/sec² and $t = 3.5$ sec, we determine s as follows:

$$s = \frac{1}{2} gt^2$$

$$s = \frac{1}{2}(\frac{32 \text{ ft}}{\text{sec}^2})(3.5 \text{ sec})^2$$

$$s = \frac{1}{2}(\frac{32 \text{ ft}}{\text{sec}^2})(3.5)^2 (\text{sec})^2$$

$$s = \frac{1}{2}(32)(3.5)^2 \text{ ft}$$

$$s = \frac{1}{2}(32)(12.25) \text{ ft}$$

$$s = 196 \text{ ft}$$

Therefore, the ring will drop 196 feet in 3.5 seconds.

Step 2: We must now determine how close the ring is to the ground. Because the platform is 300 feet high and the ring drops 196 feet, the ring will be 300 − 196 = 104 feet from the ground. Therefore, after 3.5 seconds, the ring will be 104 feet from the ground.

AREA OF A TRAPEZOID

The area (A) of a trapezoid is given by the formula $A = \frac{1}{2}h(b + B)$, where h = the height, b = length of the shorter base, and B = the length of the longer base (h, b, and B are all measured in the same units; A is measured in the unit squared).

Exercise 11.67
Determine the value of h given that $b = 11$ cm, $B = 17$ cm, and $A = 196$ cm².

Solution 11.67
Using the given formula, we determine the value of h as follows:

$$A = \frac{1}{2}h(b + B)$$

$$196\,cm^2 = \frac{1}{2}h(11\,cm + 17\,cm)$$

$$196\,cm^2 = \frac{1}{2}h(28\,cm)$$

$$196\,cm^2 = 14h\,cm$$

$$14\,cm = h$$

Therefore, the height of the trapezoid is 14 cm.

In this chapter, we discussed algebraic expressions and their evaluation. We also discussed simplifying algebraic expressions. Linear equations in one variable, together with their solutions, were introduced. We also introduced the language of algebra and how to change word expressions to algebraic expressions. Finally, we considered word problems that can be solved by using linear equations.

TEST YOURSELF

In Exercises 1–3, evaluate the given expressions for the indicated values of the variables.

1) $4x - 7y$, if $x = -3$ and $y = 1.2$
2) $2p - (3p - q)$, if $p = -2.1$ and $q = 3.2$
3) $-3u(2v + 4)$, if $u = -0.3$ and $v = -0.7$

In Exercises 4–6, simplify the given expressions by combining like terms.

4) $5a - 3b + 2 - 3a + 7b$
5) $-4x^2y + 5xy^2 - 6x^2y^2 - x^2y^2 - x^2y + xy^2$
6) $2p - 3q + 0.7p - 1.6q - 4q - 5.1p + 6.7q$

In Exercises 7–10, write the corresponding algebraic expression for each of the given word expressions.

7) The sum of u, v, and the square of w.
8) Four times the square of the sum of x and y.
9) Eight less than six times p.
10) The sum of p and the product of q and r.

In Exercises 11–18, solve each of the given equations for the indicated variable. Check your results.

11) $x + 11 = 4$
12) $3y - 1.2 = 5.7$
13) $5t - 13 = -4$
14) $7 - 9u = 10$
15) $\dfrac{y-3}{-2} = -1$
16) $\dfrac{5-2x}{2} = 0.5$
17) $v - (-3.8) = -8.2$
18) $-23 = 4q + 5$

19) A parking lot is in the form of a rectangle. Determine its area if its length is 123 yards and its width is 248 feet.

20) Jay is older than Heather. The sum of their ages is sixty-four years. Five years from now, Jay will be twice as old as Heather was nine years ago. How old is each now?

21) The sum of four consecutive whole numbers is 154. What are the numbers?

22) A triangle has a perimeter of 47 feet. The longest side of the triangle is three times the length of the shortest side. The third side is 12 feet longer than the shortest side. What are the lengths of the sides of the triangle?

23) Determine the Celsius temperature if the Fahrenheit temperature is 78 degrees.

24) A price comparison of five stores was made for the price of a video game. In the first store, the price was $19.97; in the second store, the price was $21.95; in the third store, the price was $18.99; in the fourth store, the price was $22.75; in the fifth store, the price was $20.59. The video games in all five stores were identical. What was the average price for the video game for the five stores?

25) For the first three weeks of a particular month, Breana, working at a hardware store, received commissions of $391, $376, and $407. What must her commission be for the fourth week to have an average weekly commission of $382 for the four weeks?

TEST YOURSELF WITH THE HANDHELD CALCULATOR

For each of the following, evaluate the given expression for the indicated value of the variable.

1) $2.5u - 3.2v$, if $u = -1.7$ and $v = 4.3$
2) $2.4t - (5.6t - 7.2s)$, if $s = 0.9$ and $t = -3.6$
3) $-7.9r(2.8s - 4)$, if $r = -0.3$ and $s = -4.9$
4) $2a + b^2 - 4c$, if $a = -1.6$, $b = -2.4$, and $c = 4.5$
5) $pq + 4p^2 - 5q$, if $p = -2.9$ and $q = -3.8$
6) $p^2q - 7pq + 5q^2$, if $p = -0.4$ and $q = -0.7$
7) $(x^3 - y^3) \div (x - y)$, if $x = -13$ and $y = -12$
8) $-4r^3st^2u$, if $r = 0.3$, $s = -0.2$, $t = -4$, and $u = -0.9$
9) $(3a^4 - 2b^3) \div (b + 2c)$, if $a = -3.2$, $b = -2.6$, and $c = 4.3$
10) $6x^2y^3 - 3x^3y^2 + 5x^2y^2$, if $x = -0.3$ and $y = 0.4$

TEST YOURSELF ANSWERS

1) $4x - 7y = 4(-3) - 7(1.2) = -12 - 8.4 = -20.4$
2) $2p - (3p - q) = 2(-2.1) - [3(-2.1) - 3.2] = -4.2 - (-6.3 - 3.2) = -4.2 - (-9.5) = -4.2 + 9.5 = 5.3$
3) $-3u(2v + 4) = -3(-0.3)[2(-0.7) + 4] = +0.9(-1.4 + 4) = +0.9(2.6) = +2.34$
4) $\mathbf{5a} - 3b + 2 - \mathbf{3a} + 7b = 2a - 3b + 2 + 7b = 2a + 4b + 2$
5) $\mathbf{-4x^2y} + 5xy^2 - 6x^2y^2 - x^2y^2 - \mathbf{x^2y} + xy^2 = -5x^2y + \mathbf{5xy^2} - 6x^2y^2 - x^2y^2 + \mathbf{xy^2} = -5x^2y + 6xy^2 - \mathbf{6x^2y^2} - x^2y^2 = -5x^2y + 6xy^2 - 7x^2y^2$
6) $\mathbf{2p} - 3q + \mathbf{0.7p} - 1.6q - 4q - \mathbf{5.1p} + 6.7q = -2.4p - 3q - 1.6q - 4q + 6.7q = -2.4p - 1.9q$
7) $u + v + w^2$
8) $4(x + y)^2$
9) $6p - 8$
10) $p + qr$
11) $x + 11 = 4$
    $x = -7$
    check:
    $-7 + 11 \stackrel{?}{=} 4$
    $4 = 4 \checkmark$
12) $3y - 1.2 = 5.7$
    $3y = 6.9$
    $y = 2.3$

check:
$$3(2.3) - 1.2 \;?\; 5.7$$
$$6.9 - 1.2 \;?\; 5.7$$
$$5.7 = 5.7 \checkmark$$

13) $5t - 13 = -4$
$$5t = 9$$
$$t = \frac{9}{5}$$
check:
$$5\left(\frac{9}{5}\right) - 13 \;?\; -4$$
$$9 - 13 \;?\; -4$$
$$-4 = -4 \checkmark$$

14) $7 - 9u = 10$
$$-9u = 3$$
$$u = \frac{-1}{3}$$
check:
$$7 - 9\left(\frac{-1}{3}\right) \;?\; 10$$
$$7 + 3 \;?\; 10$$
$$10 = 10 \checkmark$$

15) $\dfrac{(y-3)}{(-2)} = -1$
$$y - 3 = 2$$
$$y = 5$$
check:
$$\frac{(5-3)}{(-2)} \;?\; -1$$
$$\frac{2}{-2} \;?\; -1$$
$$-1 = -1 \checkmark$$

16) $\dfrac{(5-2x)}{2} = 0.5$

$5 - 2x = 1$

$-2x = 4$

$x = 2$

check:

$\dfrac{[5-2(2)]}{2} ? 0.5$

$\dfrac{(5-4)}{2} ? 0.5$

$\dfrac{1}{2} ? 0.5$

$0.5 = 0.5$ ✓

17) $v - (-3.8) = -8.2$

$v + 3.8 = -8.2$

$v = -12$

check:

$-12 - (-3.8) ? -8.2$

$-12 + 3.8 ? -8.2$

$-8.2 = -8.2$ ✓

18) $-23 = 4q + 5$

$-28 = 4q$

$-7 = q$

check:

$-23 ? 4(-7) + 5$

$-23 ? -28 + 5$

$-23 = -23$ ✓

19) $A = (123 \text{ yd})(248 \text{ ft})$

$A = (369 \text{ ft})(248 \text{ ft})$

$A = 91,512 \text{ sq ft}$

20)

	Age		
	Now	5 years from now	9 years ago
Heather	x years	$(x+5)$ years	$(x-9)$ years
Jay	$(64-x)$ years	$(64-x)+5$ years	$(64-x)-9$ years

$(64 - x) + 5 = 2(x - 9)$
$69 - x = 2x - 18$
$87 - x = 2x$
$87 = 3x$
$29 = x$
$35 = 64 - x$

Jay is 35 yeas old, and Heather is 29 years old.

21) Let n = smallest whole number
$n + 1$ = next larger whole number
$n + 2$ = next larger whole number
$n + 3$ = largest whole number
Then $n + (n + 1) + (n + 2) + (n + 3) = 154$
$4n + 6 = 154$
$4n = 148$
$n = 37$
$n + 1 = 38$
$n + 2 = 39$
$n + 3 = 40$

The numbers are 37, 38, 39, and 40.

22) Let x = length of shortest side (in feet)
$3x$ = length of longest side (in feet)
$x + 12$ = length of third side (in feet)
Then x ft $+ 3x$ ft $+ (x + 12)$ ft $= 47$ ft
$x + 3x + x + 12 = 47$
$5x + 12 = 47$
$5x = 35$
$x = 7$
$3x = 21$
$(x + 12) = 19$

Therefore, the triangle has sides of length 7 ft, 21 ft, and 19 ft. The perimeter is 47 ft.

23) $C = \dfrac{5}{9}(F - 32°)$

$C = \dfrac{5}{9}(78° - 32°)$

$C = \dfrac{5}{9}(46°)$

$C = \left(\dfrac{230}{9}\right)°$

$C = \left(25\dfrac{5}{9}\right)°$

24) $\text{Avg} = \dfrac{\$19.97 + \$21.95 + \$18.99 + \$22.75 + \$20.59}{5}$

$= \dfrac{\$104.25}{5}$

$\$20.85$

25) $\text{Avg } \$382 = \dfrac{\$391 + \$376 + \$407 + x}{4}$

$\$382 = \dfrac{\$1{,}174 + x}{4}$

$\$1{,}528 = \$1{,}174 + x$

$\$354 = x$

Therefore, Breana's commission for the fourth week should be $354 to have an average commission for the four weeks of $382.

TEST YOURSELF ANSWERS WITH THE HANDHELD CALCULATOR

1) $2.5u - 3.2v = (2.5)(-1.7) - (3.2)(4.3) = -18.01$
2) $2.4t - (5.6t - 7.2s) = (2.4)(-3.6) - [(5.6)(-3.6) - (7.2)(0.9)] = 18$
3) $-7.9(-0.3)(2.8s - 4) = (-7.9)(-0.3)[(2.8)(-4.9) - 4] = -41.9964$
4) $2a + b^2 - 4c = 2(-1.6) + (-2.4)^2 - 4(4.5) = -15.44$
5) $pq + 4p^2 - 5q = (-2.9)(-3.8) + 4(-2.9)^2 - 5(-3.8) = 63.66$
6) $p^2q - 7pq + 5q^2 = (-0.4)^2(-0.7) - 7(-0.4)(-0.7) + 5(-0.7)^2 = 0.378$
7) $(x^3 - y^3) \div (x - y) = [(-13)^3 - (-12)^3] \div [-13 - (-12)] = 469$
8) $-4r^3st^2u = (-4)(0.3)^3(-0.2)(-4)^2(-0.9) = 311.04$
9) $(3^4a^4 - 2b^3) \div (b + 2c) = [3(-3.2)^4 - 2(-2.6)^3] \div [-2.6 + 2(4.3)] \approx 58.28746$
10) $6x^2y^3 - 3x^3y^2 + 5x^2y^2 = 6(-0.3)^2(0.4)^3 - 3(-0.3)^3(0.4)^2 + 5(-0.3)^2(0.4)^2 = 0.11952$

CHAPTER 12
Statistics

In this chapter, we introduce the most basic concepts involved in statistics and indicate how statistics can be used. You will gain a better understanding of the use of statistics and statistical terms encountered in newspapers, magazines, on television, and in other media. Although the branch of statistics known as inferential statistics is very interesting, we emphasize the branch known as descriptive statistics.

12.1 INTRODUCTION

The branch of mathematics concerned with collecting and analyzing data is called **statistics**. There are two basic branches of statistics: descriptive statistics and inferential statistics. **Descriptive statistics** is concerned with collecting, tabulating, summarizing, and presenting information known about some particular situation. From this information, we may attempt to infer something about a larger situation about which we do not have complete information. This use of statistics is known as **statistical inference** or **inferential statistics.**

Throughout this chapter, we will be concerned with some basic concepts of descriptive statistics.

12.2 FREQUENCY DISTRIBUTIONS

In this section, we examine sets of data and frequency distributions associated with them. Histograms and frequency polygons are discussed.

FREQUENCY DISTRIBUTIONS

Suppose that we have a group of fifty people and we want to know whether their birthdays for the year fall on one day of the week more often than on other days of the week. One way to find out is to list the birthdays according to the day of the week, using 1 for Sunday, 2 for Monday, 3 for Tuesday, 4 for Wednesday, 5 for Thursday, 6 for Friday, and 7 for Saturday. (See Table 12.1.)

Table 12.1

7	7	5	1	4	3	2	2	2	1
6	4	7	1	3	3	4	4	6	5
7	3	7	7	3	1	6	3	2	5
4	4	2	7	5	1	6	6	3	4
1	1	7	6	3	5	5	5	2	7

The entries in Table 12.1 are the "raw" data. They are not in any particular order or ranking. We could now list all the entries from 1 through 7 and record alongside each entry the number of times it occurs in the table. The number of times an entry appears is called the **frequency** and is denoted by f. When this listing has been completed, the resulting table is called a **frequency distribution**. In Table 12.2, we show the frequency distribution for the ungrouped data given in Table 12.1. The symbol X represents the individual entries.

Table 12.2

X	f
1	7
2	6
3	8
4	7
5	7
6	6
7	9

In this frequency distribution, we see that 7 (Saturday) occurs more frequently than any of the other entries. Hence, more birthdays for this group of fifty people will occur on Saturday for the year than for any other day of the week.

Exercise 12.1

Construct a frequency distribution for the following set of data:

```
3  5  8  1  2  2  6  7  6  6
9  2  4  4  3  7  9  9  8  8
1  1  1  4  7  4  6  5  5  9
6  5  7  3  3  1  1  1  6  3
4  4  9  8  2  6  4  4  8  5
5  2  2  3  1  4  6  6  9  2
```

Solution 12.1

There are sixty entries listed above. The frequency distribution is given in Table 12.3.

Table 12.3

X	f
1	8
2	7
3	6
4	9
5	6
6	9
7	4
8	5
9	6

Exercise 12.2

Consider the following set of test grades for a group of twenty students:

```
94   100   76   88   63   67   91   88   88   59
98    58   88   83  100   89   95   90   97   90
```

Construct a frequency distribution for this set of grades.

Solution 12.2

The highest grade is 100, and the lowest grade is 58. The frequency distribution lists all the grades from 100 through 58. If a grade in between is not listed, a 0 is listed for its frequency. The frequency distribution is given in Table 12.4.

Table 12.4

X	f	X	f	X	f
100	2	85	0	70	0
99	0	84	0	69	0
98	1	83	1	68	0
97	1	82	0	67	1
96	0	81	0	66	0
95	1	80	0	65	0
94	1	79	0	64	0
93	0	78	0	63	1
92	0	77	0	62	0
91	1	76	1	61	0
90	2	75	0	60	0
89	1	74	0	59	1
88	4	73	0	58	1
87	0	72	0		
86	0	71	0		

GROUPED FREQUENCY DISTRIBUTIONS

The scores listed in Table 12.4 are widely spread, and many have a 0 frequency associated with them. In situations of this type, it is sometimes useful to group the scores into class intervals such that any particular score will belong to exactly one such class interval. When this is done, the resulting frequency distribution is called a **grouped frequency distribution.**

How many class intervals should there be for a grouped frequency distribution? Although no particular number is universally prescribed, statisticians do agree that most data of the type we are concerned with in this chapter can be accommodated by from ten to twenty class intervals, aiming for approximately fifteen class intervals.

Now that we know how many class intervals we should have for a grouped frequency distribution, how do we go about determining what they are? The first step is to determine the difference between the highest and lowest scores, which for the data in Table 12.4 is $100 - 58 = 42$. We then add 1 to obtain the total number of scores or potential scores. Hence, we have $42 + 1 = 43$. The next step is to divide 43 by 15 (in this case) to obtain the number of scores or potential scores in each class interval and round the result to the nearest whole number. We have $43 \div 15 = 2\frac{13}{15}$ or 3, rounded to the nearest whole number. With three scores to an interval and a total of forty-three scores or potential scores, we need fifteen class intervals for this distribution.

Next, consider the highest score in the distribution and let that be the largest number in the highest class interval. Hence, in this case 100 is the largest number in the highest class interval. Subtract 2 from this to obtain the lowest number in this class interval. Hence, the highest class interval would be 100–98, which does contain three scores or potential scores. The next lower class interval is 97–95. The next lower class interval would be 94–92. Continue this procedure to obtain each lower class interval until all the scores in the distribution have been included.

The resulting grouped frequency distribution for the data in Table 12.4 is given in Table 12.5.

Table 12.5

Class Interval	f	Class Interval	f
100–98	3	76–74	1
97–95	2	73–71	0
94–92	1	70–68	0
91–89	4	67–65	1
88–86	4	64–62	1
85–83	1	61–59	1
82–80	0	58–56	1
79–77	0		

It should be noted that in an ungrouped frequency distribution (such as in Table 12.4), the individual scores are preserved. However, in a grouped frequency distribution this is not so. For example, in the class interval 91–89 there are four scores, but the identity of the individual scores is not known unless we refer back to the original data.

Exercise 12.3

The following scores were recorded for sixty students taking an English examination:

82	79	63	42	57	86	98	79	90	75
96	87	72	71	68	84	90	89	84	91
61	83	55	74	77	89	51	71	73	84
97	90	92	80	81	63	59	68	74	77
71	77	88	90	96	85	84	65	69	85
84	79	77	75	60	58	49	88	83	92

Construct a grouped frequency distribution for the above data taking the lowest class interval as 45–42.

Solution 12.3

The highest score is 98, and the lowest score is 42. With the lowest class interval as 45–42, we would need fifteen class intervals to include all the scores or potential scores in the above set of data. The grouped frequency distribution is given in Table 12.6.

Table 12.6

Class Interval	f	Class Interval	f
101–98	1	69–66	3
97–94	3	65–62	3
93–90	7	61–58	4
89–86	6	57–54	2
85–82	10	53–50	1
81–78	5	49–46	1
77–74	8	45–42	1
73–70	5		

Exercise 12.4

Purchases rounded to the nearest dollar for the first forty customers in a food store on a particular day are recorded below:

19	2	6	17	13	23	36	1	16	42
30	25	19	7	11	9	31	57	21	6
11	8	17	21	20	30	13	26	22	10
2	10	22	29	17	14	32	63	15	5

Construct a) a frequency distribution for the above data, and b) a grouped frequency distribution for the above data, taking the highest class interval to be 65–62.

Solution 12.4

a) Table 12.7

X	f	X	f	X	f	X	f
63	1	47	0	31	1	15	1
62	0	46	0	30	2	14	1
61	0	45	0	29	1	13	2
60	0	44	0	28	0	12	0
59	0	43	0	27	0	11	2
58	0	42	1	26	1	10	2
57	1	41	0	25	1	9	1
56	0	40	0	24	0	8	1
55	0	39	0	23	1	7	1
54	0	38	0	22	2	6	2
53	0	37	0	21	2	5	1
52	0	36	1	20	1	4	0
51	0	35	0	19	2	3	0
50	0	34	0	18	0	2	2
49	0	33	0	17	3	1	1
48	0	32	1	16	1		

b) Table 12.7

Class Interval	f	Class Interval	f
65–62	1	29–26	2
61–58	0	25–22	4
57–54	1	21–18	5
53–50	0	17–14	6
49–46	0	13–10	6
45–42	1	9–6	5
41–38	0	5–2	3
37–34	1	1–(–2)	1
33–30	4		

Exercise 12.5

Scores obtained by thirty high school seniors on the verbal portion of the CEEB examinations are given below:

```
420   560   550   490   610   620   480   600
590   570   500   530   590   490   390   560
540   490   510   720   600   630   480   610
690   610   510   490   600   460
```

Construct a grouped frequency distribution for the above data taking twenty-two scores or potential scores to a class interval and such that the lowest class interval has a lower number of 380.

Solution 12.5

Table 12.8

Class Interval	f	Class Interval	f
731–710	1	555–534	2
709–688	1	533–512	1
687–666	0	511–490	7
665–644	0	489–468	2
643–622	1	467–446	1
621–600	7	445–424	0
599–578	2	423–402	1
577–556	3	401–380	1

12.3 HISTOGRAMS AND FREQUENCY POLYGONS

Frequency distributions can also be represented graphically. The graph contains a horizontal axis that reflects the entries in the set of data. The graph also contains a vertical axis that reflects the frequencies associated with each entry. One such graph is the **histogram**.

HISTOGRAMS

A **histogram** is a set of vertical rectangles constructed side by side. The height of each rectangle is the frequency associated with the particular entry from the data. In the following figure, we illustrate a histogram for the data given in Table 12.2 in Section 12.2. The horizontal axis is labeled X and the vertical axis is labeled f.

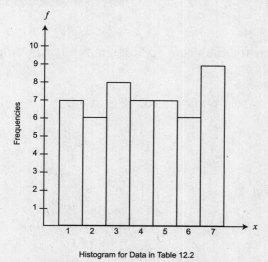

Histogram for Data in Table 12.2

Figure 12-1

Exercise 12.6

Construct a histogram for the data given in Table 12.3 in the preceding section.

Solution 12.6

The required histogram is given in Figure 12-2.

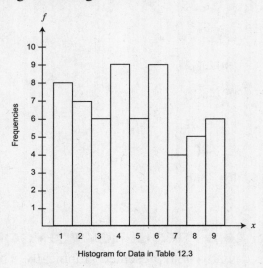

Histogram for Data in Table 12.3

Figure 12-2

Histograms can also be constructed for grouped frequency distributions. The vertical axis still reflects the frequencies, but the horizontal axis now reflects the class intervals. For example, in the class interval 164–167, the lower limit is 164 and the upper limit is 167. These upper and lower limits are referred to as the **apparent limits** of the class interval. The values recorded in this interval are given in whole numbers, so the interval would actually contain all values from 163.5 to 167.5. The **real limits,** then, for the interval 164–167 are 163.5 and 167.5. We label the real limits on the horizontal graph for such a histogram.

Exercise 12.7

Construct a histogram for the grouped frequency distribution given in Table 12.5 of the preceding section.

Solution 12.7

The required histogram is given in Figure 12-3.

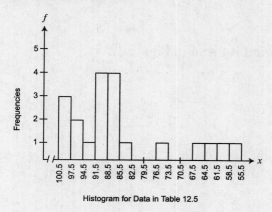

Figure 12-3

In Figure 12.3, the horizontal axis has a break in it, denoted by the symbol —//—, because we start with values far from 0. This is a customary practice that you should follow.

Exercise 12.8

Construct a histogram for the grouped frequency distribution given in Table 12.6.

Solution 12.8

The required histogram is given in Figure 12-4.

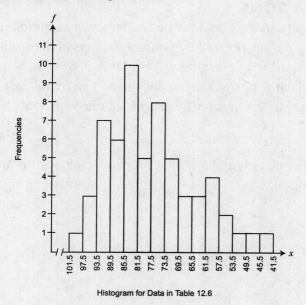

Figure 12-4

Exercise 12.9
Construct a histogram for the grouped frequency distribution given in Table 12.8.

Solution 12.9
The required histogram is given in Figure 12-5.

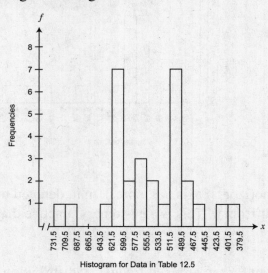

Figure 12-5

FREQUENCY POLYGONS

Another type of graph used to illustrate the frequency distribution is a **frequency polygon.** To construct a frequency polygon, the middle point of the tops of the rectangles in the corresponding histogram are connected by line segments. Because a frequency polygon always starts and terminates on the horizontal axis, it is necessary to add an extra class interval at each end of the grouped distribution. Of course, each of these added class intervals has a 0 frequency.

Exercise 12.10
Construct a frequency polygon for the grouped frequency distribution given in Table 12.5 in the preceding section.

Solution 12.10

The required frequency polygon is given in Figure 12-6.

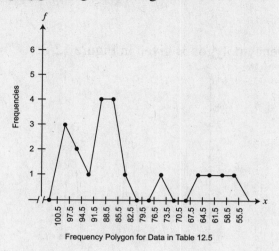

Figure 12-6

Exercise 12.11

Construct a frequency polygon for the grouped frequency distribution given in Table 12.6 in the preceding section.

Solution 12.11

The required frequency polygon is given in Figure 12-7.

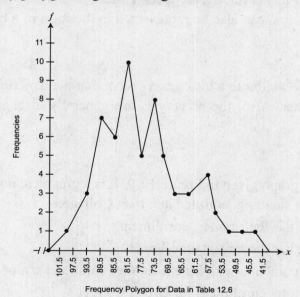

Figure 12-7

Exercise 12.12
Construct a frequency polygon for the grouped frequency distribution given in Table 12.8.

Solution 12.12
The required frequency polygon is given in Figure 12-8.

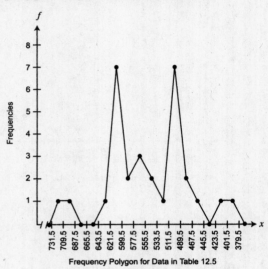

Figure 12-8

12.4 BAR GRAPHS AND LINE GRAPHS
A frequency distribution can also be represented in the form of a bar graph.

BAR GRAPHS
A **bar graph** is very similar to a histogram except that the bars (that is, rectangles) generally do not touch each other. Also, the rectangles are generally arranged horizontally instead of vertically.

Exercise 12.13
Consider the bar graph given in Figure 12-9. It is a graphical representation of the number of students, by class standing, enrolled at Urban College.
    a) Which class has the smallest enrollment?
    b) Which class has the largest enrollment?
    c) What is the combined enrollment for the junior and senior classes?
    d) What is the total enrollment for Urban College?

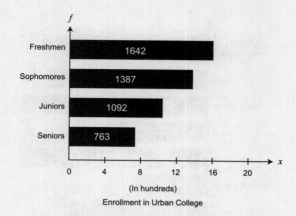

Figure 12-9

Solution 12.13

a) The senior class.
b) The freshman class.
c) 1,092 + 763 = 1,855
d) 1,642 + 1,387 + 1,092 + 763 = 4,884

Exercise 12.14

For 1985, the approximate number of births in the nine major areas of the United States is represented by the bar graph in Figure 12-10.

   a) Which region had the least number of births?
   b) Which region had the largest number of births?
   c) Which region(s) had more than 500,000 births?
   d) Which region(s) had less than 300,000 births?
   e) Which region(s) had more than 225,000 but less than 550,000 births?
   f) Using the data graphed in Figure 12-10, what was the total number of births in the United States in 1985?

Solution 12.14

a) New England.
b) East North Central.
c) Middle Atlantic, East North Central, South Atlantic, and Pacific.
d) New England, West North Central, East South Central, and Mountain.
e) Middle Atlantic, West North Central, West South Central, and Mountain.
f) 173,000 + 517,000 + 627,000 + 275,000 + 602,000 + 223,000 + 483,000 + 236,000 + 597,000 = 3,733,000

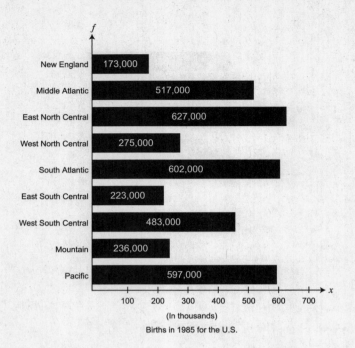

Births in 1985 for the U.S.

Figure 12-10

Exercise 12.15

The Dawson family has a budget of expenses as shown in Figure 12-11. Their total net annual income is $47,000.

a) How much is budgeted for food and rent?

b) How much is budgeted for education, entertainment, and other?

c) What percent of the Dawson family annual net income is budgeted for food?

d) What percent of the Dawson family annual net income is budgeted for education?

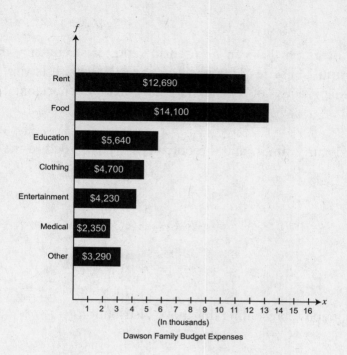

Figure 12-11

Solution 12.15

a) $14,100 + $12,690 = $26,790

b) $5,640 + $4,230 + $3,290 = $13,160

c)
$$\frac{\$14,100}{\$47,000} \times 100$$
$$= 0.3 \times 100$$
$$= 30$$

30% of the net annual income is budgeted for food.

d)
$$\frac{\$5,640}{\$47,000} \times 100$$
$$= 0.12 \times 100$$
$$= 12$$

12% of the net annual income is budgeted for education.

Line Graphs

Another type of graph that can be used to represent frequency distributions is the line graph. A **line graph** is a series of connected line segments. It is very similar to a frequency polygon except that it does not have to start and end on the horizontal axis.

Exercise 12.16

A die is tossed thirty times, and the outcomes are recorded in the frequency distribution given below:

X	f	X	f
1	3	4	5
2	5	5	6
3	7	6	4

Construct a line graph representing the above distribution.

Solution 12.16

The required line graph is given in Figure 12-12.

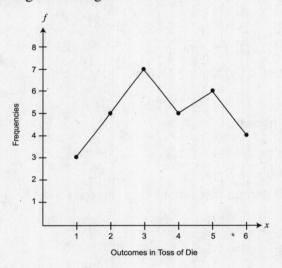

Figure 12-12

Two or more line graphs can be constructed on the same set of axes. This is especially convenient for comparison of data. We illustrate this in the following exercise.

Exercise 12.17

Consider the line graph given in Figure 12-13, which shows the comparison between monthly fatal automobile accidents for City A and City B for a recent year.

a) In what month was there the greatest number of fatal accidents?

b) In what month was there the least number of fatal accidents?

c) How many fatal accidents were there in both cities in the month of May? In the month of November?

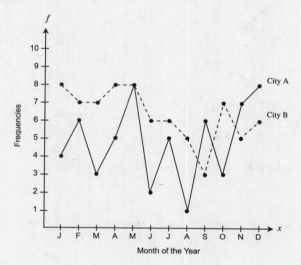

Figure 12-13

Solution 12.17

a) May. There were eight accidents in each city.

b) August. There was one accident in City A and five accidents in City B for a total of six accidents for the month.

c) Sixteen in May and twelve in November.

12.5 CIRCLE GRAPHS

The last type of graph that we look at in this chapter is the **circle graph,** which is also called a **pie chart,** because the sectors of the circle look like pieces of pie. The entries in a circle graph are given as percents of the total amount. A circle is divided into 360° (which is read, "360 degrees"). To determine the number of degrees for each entry, multiply 360 by the percent associated with the entry.

Exercise 12.18

Referring to Exercise 12.15 in the preceding section, the Dawson family has a budget of expenses as follows:

Rent	$12,690
Food	$14,100
Education	$5,640
Clothing	$4,700
Entertainment	$4,230
Medical	$2,350
Other	$3,290
Net annual income	$47,000

Construct a circle graph showing these budgeted expenses.

Solution 12.18

Begin this solution by rewriting each of the above budgeted expenses as a percent of the total net annual income:

Rent: $\dfrac{\$12,690}{\$47,000} = 27\%$

Food: $\dfrac{\$14,100}{\$47,000} = 30\%$

Education: $\dfrac{\$5,640}{\$47,000} = 12\%$

Clothing: $\dfrac{\$4,700}{\$47,000} = 10\%$

Entertainment: $\dfrac{\$4,230}{\$47,000} = 9\%$

Medical: $\dfrac{\$2,350}{\$47,000} = 5\%$

Other: $\dfrac{\$3,290}{\$47,000} = 7\%$

Next, multiply each of the above percents by 360° to determine the number of degrees in each sector (pie wedge) of the circle.

Rent: .27 × 360° = 97.2°
Food: .30 × 360° = 108°
Education: .12 × 360° = 43.2°
Clothing: .10 × 360° = 36°

Entertainment: $.09 \times 360° = 32.4°$
Medical: $.05 \times 360° = 18°$
Other: $.07 \times 360° = 25.2°$

Finally, construct the circle graph as given in Figure 12-14. A protractor is used in this construction.

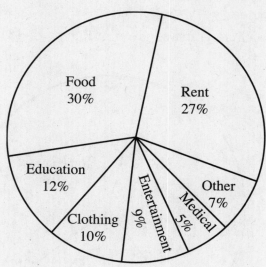

Figure 12-14

Exercise 12.19

Southern Chevrolet Dealer had a six-day marathon sale and sold the following numbers of cars:

Subcompact	112
Compact	98
Full-size	56
Luxury	14

Construct a circle graph to represent the above sales.

Solution 12.19

First, we compute the total number of cars sold in the six-day sale. This is 280 cars. Second, rewrite each of the numbers of vehicle sales as a percent of the total number of sales:

Subcompact: $\dfrac{112}{280} = 40\%$

Compact: $\dfrac{98}{280} = 35\%$

Full-size: $\dfrac{56}{280} = 20\%$

Luxury: $\dfrac{14}{280} = 5\%$

Next, multiply each of the above percents by 360° to determine the number of degrees in each sector (pie wedge) of the circle.

Subcompact: .40 × 360° = 144°
Compact: .35 × 360° = 126°
Full-size: .20 × 360° = 72°
Luxury: .05 × 360° = 18°

Finally, construct the circle graph as given in Figure 12-15. A protractor is used in this construction.

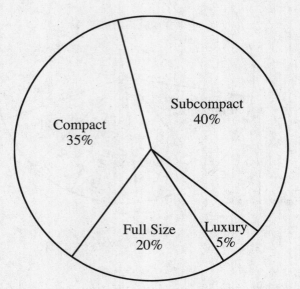

Figure 12-15

12.6 MEAN, MEDIAN, MODE

Consider a set of test grades for a particular group of twenty students:

| 100 | 100 | 98 | 97 | 95 | 94 | 91 | 90 | 90 | 89 |
| 88 | 88 | 88 | 88 | 83 | 76 | 63 | 58 | 47 | 39 |

Suppose you received a grade of 90 on the test. How well did you perform? Clearly, 90 was not the highest grade nor was it the lowest grade. There were eight students who performed at least as well as or better than you and eleven students who did not perform as well. From the distribution above we also note that 90 was not the most common grade, because that distinction belongs to 88. Finally, note that when all twenty grades are added together and the sum divided by 20, the result is 83.1, which is called the mean or the arithmetic average. Notice that only six of the grades in this distribution fall below the mean.

In the preceding discussion, we are referring to **measures of central tendency**.

THE MEAN

The **mean** or **arithmetic average** of a set of n scores is the sum of the scores divided by n, the number of scores. If the n scores are denoted by $X_1, X_2, X_3, \ldots, X_n$, the mean, denoted by $\overline{X}$, is given by the formula

$$\overline{X} = \frac{X_1 + X_2 + X_3 + \cdots + X_n}{n}.$$

Exercise 12.20

Determine the mean of each of the following sets of data:
a) 12, 16, 20, 30, and 42
b) 20, 20, 25, 30, 30, 30, 36, 40, 40, and 40
c) 12, 10, 7, 10, 9, 9, 13, 11, 8, 9, 12, 8
d) 101, 101, 98, 89, 67, 110, 98, 69
e) 33, 36, 37, 35, 36, 42, 30

Solution 12.20

a) $\overline{X} = \dfrac{12+16+20+30+42}{5} = \dfrac{120}{5} = 24$

b) $\overline{X} = \dfrac{20+20+25+30+30+30+36+40+40+40}{10} = \dfrac{311}{10} = 31.1$

c) $\overline{X} = \dfrac{12+10+7+10+9+9+13+11+8+9+12+8}{12} = \dfrac{118}{12} = 9\dfrac{5}{6}$

d) $\overline{X} = \dfrac{101+101+98+89+67+110+98+69}{8} = \dfrac{733}{8} = 91.625$

e) $\overline{X} = \dfrac{33+36+37+35+36+42+30}{7} = \dfrac{249}{7} = 35\dfrac{4}{7}$

THE MEDIAN

Consider a set of scores arranged in order of either increasing or decreasing magnitude. If the number of scores is odd, the **median** of the set is the middle score in the arrangement. If the number of scores is even, the median of the set is halfway between the two middle scores.

Exercise 12.21

Determine the median for each set of data given below:
a) 8, 10, 4, 1, 9, 7, 11
b) 93, 89, 68, 95, 88, 72, 79, 82, 69, 96
c) 2, 3, 5, 2, 6, 3, 2, 4, 5, 2, 3
d) 12, 10, 7, 10, 9, 9, 13, 11, 8, 9, 12, 8
e) 101, 101, 98, 89, 67, 110, 98, 69

Solution 12.21

a) Arranging these scores in increasing order of magnitude, we have:
1, 4, 7, **8**, 9, 10, 11
We have an odd number of scores; the median is 8.

b) Arranging these scores in increasing order of magnitude, we have:
68, 69, 72, 79, **82, 88,** 89, 93, 95, 96
We have an even number of scores; the median is 85, which is halfway between 82 and 88.

c) Arranging these scores in increasing order of magnitude, we have:
2, 2, 2, 2, 3, **3**, 3, 4, 5, 5, 6
We have an odd number of scores; the median is 3.

d) Arranging these scores in decreasing order of magnitude, we have:
13, 12, 12, 11, 10, **10, 9,** 9, 9, 8, 8, 7
We have an even number of scores; the median is 9.5, which is halfway between 10 and 9.

e) Arranging these scores in decreasing order of magnitude, we have:
110, 101, 101, **98, 98,** 89, 69, 67

We have an even number of scores; the median is halfway between the two middle scores. However, both of the middle scores are 98. Hence, the median is 98.

THE MODE

The **mode** of a set of scores is the score that occurs the greatest number of times in the distribution.

Exercise 12.22

Determine the mode for each set of scores given below:
a) 2, 4, 8, 4, 9, 6, 6, 5, 4, 9, 1
b) 21, 31, 51, 21, 61, 31, 21, 41, 51, 21, 31
c) 125, 105, 75, 105, 95, 95, 135, 115, 85, 95, 125, 85
d) 33, 36, 37, 35, 36, 42, 30
e) 101, 101, 98, 89, 67, 110, 98, 69

Solution 12.22

a) The mode is 4, because 4 occurs three times and all other scores occur fewer times.

b) The mode is 21, because 21 occurs four times and all other scores occur fewer times.

c) The mode is 95, because 95 occurs three times and all other scores occur fewer times.

d) The mode is 36, because 36 occurs twice and all other scores occur only once.

e) Both 98 and 101 occur twice and all other scores occur only once. Hence, both 98 and 101 are the mode. We say that the distribution is **bimodal.**

Now consider the following salaries of a group of workers:
$8,000; $10,000; $14,000; $14,000; $100,000

The mean of these salaries is $29,200; the median salary is $14,000, and the mode is also $14,000. Which of these measures of central tendency is most meaningful? To claim that the mean salary of $29,200 is representative of the data given is misleading, due to inclusion of the extreme value of $100,000 in the computation of the mean. Because each measurement is used in the computation of the mean, we see readily that the mean becomes very sensitive to extreme measurements unless, of course, these extreme measurements are balanced on both sides of the mean.

The median, however, is not sensitive to extreme values. Hence, this characteristic makes the median more acceptable for describing central tendency in a distribution in which the mean is not favored due to the presence of extreme values.

Of the three measures of central tendency discussed, the mean is generally the preferred statistic because of its value in more advanced statistical analysis.

Exercise 12.23

During a labor negotiation session the union representative made the statement, "The median annual wage for my membership is $28,300." The management representative, using the same data, made the statement, "The mean annual wage for your membership is $29,100." Assuming that both statements were true, what could account for the apparent discrepancy?

Solution 12.23

There were probably a few highly paid members whose salaries distorted the mean.

In this chapter, we discussed some very basic concepts of descriptive statistics. Ungrouped and grouped frequency distributions were introduced. Measures of central tendency—the mean, median, and mode—were also introduced and discussed.

TEST YOURSELF

Whenever possible, use your handheld calculator to arrive at your solutions.
In Exercises 1–15, identify each of the given statements as being true or false.

1) A general rule to follow when grouping data into class intervals is to strive for no more than six class intervals.

2) When we arrange a set of scores in increasing order of magnitude and indicate the frequency associated with each score, we have a frequency distribution.

3) The real limits for the class interval 72–76 are 71.5 and 75.5.

4) A score taken by itself is meaningless.

5) The mean, median, and mode are all measures of central tendency.

6) The median is a score or potential score that balances all the scores on either side of it.

7) Consider a distribution of scores with a mean of 50. If 10 points are added to each score in the distribution, the new mean is 60.

8) If the median of a particular distribution is 43 and 5 points are subtracted from each score in the distribution, the new median is 43.

9) The individual scores are not identified in a grouped frequency distribution.

10) If a class interval for a grouped frequency distribution is 63–66, there are three scores or potential scores represented in that interval.

11) In a grouped frequency distribution there cannot be class intervals with 0 frequency.

12) In a set of data, if the highest score is 93 and the lowest score is 17, there are 77 scores or potential scores represented among the data.

13) In an ungrouped frequency distribution the individual scores or measurements are preserved.

14) A histogram is a graphical representation for a grouped frequency distribution.

15) For a set of test scores, the median score can be higher than the mean score.

In Exercises 16–17, determine the mean, correct to the nearest tenth, for each set of numbers.

16) 17.9; 13.2; 21.0; 14.8; 20.9

17) 78.7; 66.7; 52.4; 87.3; 93.2; 36.8

18) Joshua sold seven used cars last week. The prices for the seven cars were: $1,980; $2,300; $1,185; $795; $1,075; $999; and $2,150. What was the median price of the cars sold?

19) Cindy's sales, working in a jewelry store, for the first six months of the year were: $79,122; $68,950; $73,356; $57,465; $62,355; and $80,215. What were the median sales for the six months?

20) Determine the mode for the following set of numbers:
13, 9, 26, 67, 19, 22, 54, 89

21) Determine the mode for the following set of numbers:
36, 24, 10, 36, 19, 7, 42, 24, 36, 59

22) Determine the mode for the following set of numbers:
1,024, 986, 789, 1,119, 1,024, 862, 789, 632

23) Over a period of time, a college student had $2,200 in expenses, as indicated in the following table. Fill in the blanks below.

Expenses	Amount	Percent of total	Degrees of a Circle
Food	$550		
Rent	$440		
Clothing	$330		
Books	$396		
Entertainment	$220		
Other	$264		

24) Construct a frequency distribution for the following set of data.
8 5 6 2 9 1 6 4 8 9
6 4 3 2 8 7 5 9 1 6
7 5 9 1 6 3 8 6 4 7

25) Bette has scores of 82, 94, 89, and 88 for one of her courses. To get a grade of A in the course, she needs a mean average score of 92 or better on five tests. To get a grade of B, she needs a mean average score of 84 on five tests.

   a) Bette received a grade of 93 on the fifth test. Should she receive a grade of A for the course? Why (not)?

   b) Bette received a grade of 78 on the fifth test. Should she receive a grade of B for the course? Why (not)?

TEST YOURSELF ANSWERS

1) False
2) True
3) False
4) True
5) True
6) False
7) True
8) False
9) True
10) False
11) False
12) True
13) True
14) True
15) True

16) Mean = $\dfrac{17.9+13.2+21.0+14.8+20.9}{5}$ = 17.6 (to the nearest tenth)

17) Mean = $\dfrac{78.7+66.7+52.4+87.3+93.2+36.8}{6}$ = 69.2 (to the nearest tenth)

18) First, arrange the prices from lowest to highest as follows:
    $795; $999; $1,075; $1,185; $1,980; $2,150; $2,300
    Next, select the middle number ($1,185), because 7 (the number of prices) is odd. Therefore, the median price of the cars sold is $1,185.

19) First, arrange the sales from lowest to highest as follows:
    $57,465, $62,355, $68,950, $73,356, $79,122, $80,215
    Because there is an even number (6) of sales, the median of the set of numbers is the mean of the two middle numbers, or $\dfrac{\$68,950+\$73,356}{2}$ = $71,153.

20) There is no mode for the set of numbers, because each number occurs only once.

21) The mode for the set of numbers is 36, because it occurs more times (3) than any other number in the set.

22) For this set of numbers, there are two modes, 789 and 1,024. Each of these numbers occurs more times (2) than any other number in the set.

23) Food: $\dfrac{\$550}{\$2,200} = 25\%$ $25\% \times 360° = 90°$

    Rent: $\dfrac{\$440}{\$2,200} = 20\%$ $20\% \times 360° = 72°$

    Clothing: $\dfrac{\$330}{\$2,200} = 15\%$ $15\% \times 360° = 54°$

    Books: $\dfrac{\$396}{\$2,200} = 18\%$ $18\% \times 360° = 64.8°$

    Entertainment: $\dfrac{\$220}{\$2,200} = 10\%$ $10\% \times 360° = 36°$

    Other: $\dfrac{\$264}{\$2,200} = 12\%$ $12\% \times 360° = 43.2°$

We now have:

Expenses	Amount	Percent of Total	Degrees of a Circle
Food	$550	25%	90°
Rent	$440	20%	72°
Clothing	$330	15%	54°
Books	$396	18%	64.8°
Entertainment	$220	10%	36°
Other	$264	12%	43.2°

24)

x	f
1	3
2	2
3	2
4	3
5	3
6	6
7	3
8	4
9	4

25) a) The mean average score for the five tests = $\dfrac{82+94+89+88+93}{5} = 89.2$. Because $89.2 < 92$, Bette should not receive a grade of A for the course.

b) The mean average score for the five tests = $\dfrac{82+94+89+88+78}{5} = 86.2$. Because $86.2 > 84$, Bette should receive a grade of B for the course. (Note, however, that the mean average score is not high enough for a grade of A in the course.)

CHAPTER 13

Consumer Mathematics

In Chapter 6, we discussed percents. There are many real-life applications that use percents. In this chapter, we examine some of them, including commissions, discounts, simple interest, purchase of property, and consumer credit.

13.1 COMMISSIONS

Some people work and are paid a **salary.** That is, they are paid the same amount of money each pay period. Other people, such as those who work in retail stores, sell real estate, or sell automobiles, are paid a commission. That is, they are paid a **percentage** of the amount of their sales. If a person's pay comes entirely from commission, it is called **straight commission.** Some people receive a combination of a salary and a commission. In this section, we discuss commissions.

To compute the amount of commission, you must determine the commission rate and the amount of sales. Thus,

$$\text{Commission} = \text{Commission Rate} \times \text{Amount of Net Sales}$$

Exercise 13.1

Bonnie works for straight commission and receives a commission rate of 5% on the total amount of sales. If her total sales for a week amounted to $4,350, what was her commission?

Solution 13.1

$$\text{Commission} = \text{Commission Rate} \times \text{Amount of Sales}.$$

Hence,

$$\begin{aligned}\text{Commission} &= 5\% \times \$4{,}350 \\ &= (0.05)(\$4{,}350) \\ &= \$217.50\end{aligned}$$

Bonnie's commission was $217.50.

Exercise 13.2
Barbara sells boats and receives a commission of 8% of the sale price. If Barbara sells three boats for a total of $133,500, what is her commission?

Solution 13.2

$$\text{Commission} = \text{Commission Rate} \times \text{Amount of Sales}.$$

Hence,
$$\begin{aligned}\text{Commission} &= 8\% \times \$133{,}500 \\ &= (0.08)(\$133{,}500) \\ &= \$10{,}680.00\end{aligned}$$

Barbara's commission is $10,680.

NET SALES

Returns are sometimes made after the sale of merchandise. When this happens, the amount of the returns must be subtracted from the total sales before the commission is computed. The total amount of sales is called the **gross sales.** If you subtract the amount of the returns from the amount of the gross sales, the difference is called the **net sales.** Net sales are used to determine commission.

Exercise 13.3
During the month of August, Barinem had gross sales of $79,925. His sales returns amounted to $1,854. If he received 6.5% commission on his net sales, what was Barinem's commission for the month?

Solution 13.3
We must do two things.

Step 1: Find the net sales:

$79,925	gross sales
−1,854	returns
$78,071	net sales

Step 2: Find the commission:

$ 78,071	net sales
× 0.065	6.5% commission rate
$5,074.615	commission

Barinem's commission was $5,074.62 for the month of August.

Exercise 13.4

Chris sells sportswear at the Fitness Shop. She receives a salary of $95 per week plus a commission of 5.5% of gross sales. If Chris sells $1,312 worth of merchandise during a week, what is the total amount of her pay for that week?

Solution 13.4

We must do two things.

Step 1: Determine her commission.

$$\text{Commission} = \text{Commission Rate} \times \text{Amount of Gross Sales}.$$

Hence,

$$\begin{aligned}\text{Commission} &= 5.5\% \times \$1{,}312 \\ &= (0.055)(\$1{,}312) \\ &= \$72.16\end{aligned}$$

Step 2: Add the commission to the salary:

Salary	$95.00
Commission	+ 72.16
Total pay	$167.16

Chris's total pay for the week is $167.16.

Exercise 13.5

Kelly works in an office supply store. She receives a salary of $115 per week plus a commission of 7% on her net sales. If Kelly has gross sales of $1,682 and returns of $123, what is her total pay for the week?

Solution 13.5

We must do three things.

Step 1: Determine the net sales:

$1,682	gross sales
− 123	returns
$1,559	net sales

Step 2: Determine the commission:

$1,559	net sales
× 0.07	7% commission rate
$109.13	commission

Step 3: Determine the total pay.

$115.00	salary
+109.13	commission
$224.13	total pay

Kelly's total pay for the week is $224.13.

AMOUNT OF NET SALES

If you know the commission and the commission rate, you can also determine the amount of net sales as follows:

$$\text{Amount of Net Sales} = \frac{\text{Amount of Commission}}{\text{Commission Rate}}$$

Exercise 13.6

Jennie Rose works on a straight commission of 6% of her net sales. If her commission for a given week was $288, what was the amount of her net sales?

Solution 13.6

Hence,
$$\text{Amount of Net Sales} = \frac{\text{Amount of Commission}}{\text{Commission Rate}}.$$

$$\text{Amount of Net Sales} = \frac{\$288}{6\%}$$
$$= \frac{\$288}{0.06}$$
$$= \$4,800.00$$

Jennie Rose's net sales were $4,800 for the week.

Exercise 13.7

Laura works in a gift shop and receives a salary of $125 per week plus a commission of 8% of net sales. If her total pay for the week amounted to $196.36, what was the amount of her net sales?

Solution 13.7

We must do two things.

Step 1: Determine the amount of commission:

$196.36   total pay
−125.00   salary
$ 71.36   commission

Step 2: Determine the amount of net sales:

$$\text{Amount of Net Sales} = \frac{\text{Amount of Commission}}{\text{Commission Rate}}.$$
$$= \frac{\$71.36}{8\%}$$
$$= \frac{\$71.36}{0.08}$$

Laura's net sales for the week were $892.

Exercise 13.8

Christian sells small appliances at the Big Discount Store. He earns a salary of $1,400 per month plus a commission. If his total pay for a month was $3,680 and his commission rate is 12%, what were his net sales for the month?

Solution 13.8

We must do two things.

Step 1: Determine the amount of commission:

$3,680   total pay
−1,400   salary
$2,280   commission

Step 2: Determine the amount of net sales:

$$\text{Amount of Net Sales} = \frac{\text{Amount of Commission}}{\text{Commission Rate}}.$$

$$= \frac{\$2,280}{12\%}$$

$$= \frac{\$2,280}{0.12}$$

$$= \$19,000$$

Christian's net sales for the month were $19,000.

Exercise 13.9

Andrea works and is paid a straight commission of sales. Her commission is 3.5% of the first $4,000 in sales, 5% of the next $5,000 in sales, and 6.5% for all sales over $9,000. If her total net sales amounted to $14,700, what was her commission?

Solution 13.9

There are several steps to this solution.

Step 1: Determine the commission for the first $4,000 in sales:

$4,000   in sales
× 0.035 3.5% commission rate
$140   commission

Step 2: Determine the commission for the next $5,000 in sales:

$5,000   in sales
× 0.05   5% commission rate
$250   commission

Step 3: Determine the commission for sales over $9,000:

$14,700   total sales
− 9,000   first $9,000 in sales
$ 5,700   amount over $9,000
× 0.065   6.5% commission rate
$370.50   commission

Step 4: Determine the total commission:

$140.00   commission for the first $4,000 in sales
250.00   commission for the next $5,000 in sales
+370.50   commission for sales over $9,000
$760.50   total commission

Andrea's total commission was $760.50.

Exercise 13.10

Liz sells real estate for a commission of 7.5% and undeveloped land for a commission of 10.5%. If she sells two houses for $113,900 and $92,500 and a tract of land for $147,000, what will be her total commission for the three sales?

Solution 13.10

Step 1: Determine the commission for the sale of the houses:

$113,900   sale price for one house
+ 92,500   sale price for the other house
$ 206,400   total sale price for the houses
× 0.075   7.5% commission rate
$15,480.00   commission for the houses

Step 2: Determine the commission for the sale of the undeveloped land:

$147,000   sale price for the land
× 0.105   10.5% commission rate
$15,435.00   commission for the land

Step 3: Determine the total commission:

$15,480   commission for the houses
+ 15,435   commission for the land
$30,915   total commission

Liz's commission for the three sales will be $30,915.

13.2 DISCOUNT

Some stores regularly advertise merchandise on sale, usually indicated as a certain percent off the regular price. The regular price of an item is called the **marked price.** The percent off the marked price is called the **rate of discount.** The amount off is called the **discount.** In this section, we discuss discount. Thus,

$$\text{Discount} = \text{Rate of Discount} \times \text{Marked Price}$$
$$\text{Sale Price} = \text{Marked Price} - \text{Discount}$$

Exercise 13.11

An ATV is advertised at 25% off the regular price. If the ATV is marked at $319, determine the discount and the sale price.

Solution 13.11

Step 1: Discount = Rate of discount × Marked Price
= 25% × $319
= (0.25)($319)
= $79.75

The discount is $79.75.

Step 2: Sale Price = Marked Price − Discount
= $319 − $79.75
= $239.25

The sale price of the ATV is $239.25.

Exercise 13.12

During an end-of-season sale, women's clothing is offered at 45% off the marked price. If the marked price of a particular coat is $290, what is the sale price?

Solution 13.12

We must do two things.

Step 1: Determine the discount:
Discount = Rate of Discount × Marked Price
= 45% × $290
= (0.45)($290)
= $130.50

Step 2: Determine the sale price:
Sale Price = Marked Price − Discount
= $290 − $130.50
= $159.50

The sale price for the coat is $159.50.

Exercise 13.13

A riding mower is advertised at $\frac{1}{4}$ off the marked price of $640. What is the sale price?

Solution 13.13

We must do two things.

Step 1: First, determine the discount:

Discount = Rate of discount × Marked price

$$= \frac{1}{4} \times \$640$$

$$= \$160$$

(Note that 25% could have been used for the rate of discount, because $\frac{1}{4} = 25\%$.)

Step 2: Then determine the sale price:

Sale = Marked Price − Discount
= $640 − $160
= $480

The sale price for the mower is $480.

RATE OF DISCOUNT

If you know the marked price and the discount, you can determine the rate of discount as follows:

$$\text{Rate of Discount} = \frac{\text{Discount}}{\text{Marked Price}}$$

Exercise 13.14

The marked price of a camcorder is $900 with a sales price of $540. What is the rate of discount?

Solution 13.14

We must do two things.

Step 1: Determine the discount:

Discount = Marked Price − Sales Price
= $900 − $540
= $360

Step 2: Determine the rate of discount:

$$\text{Rate of Discount} = \frac{\text{Discount}}{\text{Marked Price}}$$

$$= \frac{\$360}{\$900}$$

$$= 0.40$$

$$= 40\%$$

The rate of discount is 40%.

Exercise 13.15

The Fashion Outlet is having an end-of-season sale. The marked price of a coat is $370. The sale price is $259. What is the rate of discount?

Solution 13.15

Step 1: Determine the discount:

$$\text{Discount} = \text{Marked Price} - \text{Sale Price}$$
$$= \$370 - \$259$$
$$= \$111$$

Step 2: Determine the rate of discount:

$$\text{Rate of Discount} = \frac{\text{Discount}}{\text{Marked Price}}$$
$$= \frac{\$111}{\$370}$$
$$= 0.30$$
$$= 30\%$$

The rate of discount is 30%.

Exercise 13.16

A furniture store advertised furniture on sale for up to 40% off the marked price. A chair is on sale for $289. The marked price is $419. Is this an honest sale? Why or why not?

Solution 13.16

Step 1: First, determine the discount:

$$\text{Discount} = \text{Marked Price} - \text{Sale Price}$$
$$= \$419 - \$289$$
$$= \$130$$

Step 2: Then, determine rate of discount:

$$\text{Rate of Discount} = \frac{\text{Discount}}{\text{Marked Price}}$$
$$= \frac{\$130}{\$419}$$
$$= 0.31026$$
$$= 31.026\%$$

Although 31.026% < 40%, this is an honest sale, because the advertisement indicated *up to* 40% off.

13.3 SIMPLE INTEREST

Money saved on the purchase of an item is called discount. The price paid for the use of borrowed money is called **interest.** If you borrow money for a specified period of time, the interest is paid at the *end* of that time. The amount of money borrowed is called the **principal.** In this section, we examine simple interest problems.

When you save money in a bank or other financial institution, the money that you earn on your savings is also called interest. The interest for borrowed money or saved money is calculated using three factors:

- The **principal,** P, which is the amount borrowed or saved.
- The **rate,** r, which is given as a percent per period of time.
- The **time,** t, given in years (or part of a year).

INTEREST

The following formula is for **simple interest** with the interest calculated once:

$$\text{Interest} = \text{Principal} \times \text{Rate} \times \text{Time}$$

or

$$I = P \times r \times t$$

Usually, the interest on money borrowed or saved is calculated frequently, such as monthly, so that interest is paid on the accumulated interest as well as the principal. This is called **compound interest.** Because we use only simple interest in this section, we refer to it as just "interest."

Exercise 13.17

Determine the interest when $800 is borrowed (or saved) at 7% per year for one year.

Solution 13.17

$$\begin{aligned}
\text{Interest} &= \text{Principal} \times \text{Rate} \times \text{Time} \\
&= \$800 \times 7\% \times 1 \\
&= (\$800)(0.07)(1) \\
&= \$56
\end{aligned}$$

The interest is $56.

Exercise 13.18

Determine the interest when $1,400 is borrowed (or saved) at 8% per year for nine months.

Solution 13.18

In this problem, the time = 9 months = $\frac{9}{12} = \frac{3}{4}$ years. Hence,
Interest = Principal × Rate × Time

$$= \$1,400 \times 8\% \times \frac{3}{4}$$
$$= (\$1,400)(0.08)(0.75)$$
$$= \$84$$

The interest is $84.

Exercise 13.19

Determine the interest when $1,500 is borrowed (or saved) at 8.5% per year for thirty months.

Solution 13.19

The time = 30 months = $\frac{30}{12} = \frac{5}{2} = 2.5$ years. Hence,

Interest = Principal × Rate × Time

$$= \$1,500 \times 8.5\% \times \frac{5}{2}$$
$$= (\$1,500)(0.085)(2.5)$$
$$= \$318.75$$

The interest is $318.75.

Exercise 13.20

Determine the interest when $600 is borrowed (or saved) at 9.5% per year for sixty days.

Solution 13.20

The time = 60 days = $\frac{60}{360} = \frac{1}{6}$ years. Hence,

Interest = Principal × Rate × Time

$$= \$600 \times 9.5\% \times \frac{1}{6}$$
$$= (\$600)(0.095) \times \frac{1}{6}$$
$$= \$9.50$$

The interest is $9.50.

In Exercise 13.20, we used 360 days in one year for what is called **ordinary** interest. However, bankers generally use 365 days for one year.

Exercise 13.21
Determine the interest when $1,200 is borrowed (or saved) at 7% per year for forty-two months.

Solution 13.21
The time = 42 months = $\dfrac{42}{12} = \dfrac{7}{2} = 3.5$ years.

$$\begin{aligned}\text{Interest} &= \text{Principal} \times \text{Rate} \times \text{Time}\\ &= \$1{,}200 \times 7\% \times 3.5\\ &= (\$1{,}200)(0.07)(3.5)\\ &= \$294.00\end{aligned}$$

The interest is $294.00.

RATE
If you know the principal, the time, and the interest, you can determine the rate as follows:

$$\text{Rate} = \dfrac{\text{Interest}}{\text{Principal} \times \text{Time}}$$

or

$$r = \dfrac{I}{P \times t}$$

Exercise 13.22
Determine the yearly rate of interest if $2,000 is borrowed (or saved) for eight months with interest of $120.

Solution 13.22
Here, time = 8 months = $\dfrac{8}{12} = \dfrac{2}{3}$ year. Hence,

$$\begin{aligned}\text{Rate} &= \dfrac{\text{Interest}}{\text{Principal} \times \text{Time}}\\ &= \dfrac{\$120}{\$2{,}000(2 \div 3)}\\ &= \dfrac{\$120}{\$4{,}000 \div 3}\\ &= \dfrac{120}{1} \times \dfrac{3}{4{,}000}\\ &= \dfrac{360}{4{,}000}\\ &= \dfrac{9}{100}\end{aligned}$$

$= 9\%$ yr

The rate is 9% per year.

Exercise 13.23

Determine the yearly rate of interest if $2,400 is borrowed (or saved) for six months with interest of $78.

Solution 13.23

Here, time = 6 months = $\dfrac{6}{12} = \dfrac{1}{2}$ year. Hence,

$$\text{Rate} = \dfrac{\text{Interest}}{\text{Principal} \times \text{Time}}$$

$$= \dfrac{\$78}{\$2,400 \times (1 \div 2)}$$

$$= \dfrac{\$78}{\$1,200}$$

$$= 0.065$$

$$= 6.5\%$$

The rate is 6.5% per year.

Exercise 13.24

Determine the yearly rate of interest if $1,800 is borrowed (or saved) for eighteen months with interest of $297.

Solution 13.24

Here, time = 18 months = 1.5 years. Hence

$$\text{Rate} = \dfrac{\text{Interest}}{\text{Principal} \times \text{Time}}$$

$$= \dfrac{297}{\$1800 \times 1.5}$$

$$= 0.11$$

$$= 11\%$$

The rate is 11% per year.

PRINCIPAL

If you know the interest, the rate, and the time, you can determine the principal as follows:

$$\text{Principal} = \frac{\text{Interest}}{\text{Rate} \times \text{Time}}$$

or

$$P = \frac{I}{r \times t}$$

Exercise 13.25

Determine the principal if $135 interest is paid on a loan at 6% per year for nine months.

Solution 13.25

Here, time = 9 months = $\frac{9}{12} = \frac{3}{4}$ year. Hence,

$$\begin{aligned}
\text{Principal} &= \frac{\text{Interest}}{\text{Rate} \times \text{Time}} \\
&= \frac{\$135}{0.06 \times .75} \\
&= \frac{\$135}{0.045} \\
&= \$3,000
\end{aligned}$$

The principal is $3,000.

Exercise 13.26

Determine the principal if $437.50 interest is earned on a deposit at 7% per year for thirty months.

Solution 13.26

Here, time = 30 months = $\frac{30}{12}$ = 2.5 years. Hence,

$$\text{Principal} = \frac{\text{Interest}}{\text{Rate} \times \text{Time}}$$

$$= \frac{\$437.50}{0.07 \times 2.5}$$

$$= \frac{\$437.50}{0.175}$$

$$= \$2,500$$

The principal is $2,500.

Exercise 13.27

Determine the principal if $450 interest is paid on a loan (or deposit) at 12% per year for six months.

Solution 13.27

Here, time = 6 months = 0.5 years. Hence,

$$\text{Principal} = \frac{\text{Interest}}{\text{Rate} \times \text{Time}}$$

$$= \frac{\$450}{0.12 \times 0.5}$$

$$= \$7,500$$

The principal is $7,500.

AMOUNT

When you add the interest to the principal, the sum is called the **amount**. We have:

$$\text{Amount} = \text{Principal} + \text{Interest}$$

Exercise 13.28

B.J. borrowed $1,100 at 7% per year for six months. At the end of six months, how much does she owe?

Solution 13.28
We must do two things.

Step 1: First, determine the interest:
$$\text{Interest} = \text{Principal} \times \text{Rate} \times \text{Time}$$
$$= \$1{,}100 \times 0.07 \times 0.5$$
$$= \$38.50$$

Step 2: Then, determine the amount:
$$\text{Amount} = \text{Principal} + \text{Interest}$$
$$= \$1{,}100 + \$38.50$$
$$= \$1{,}138.50$$

The total amount owed at the end of the six months is $1,138.50.

Exercise 13.29
Determine the total amount of a loan of $2,200 at 9% for eight months.

Solution 13.29
We must do two things.

Step 1: First, determine the interest:
$$\text{Interest} = \text{Principal} \times \text{Rate} \times \text{Time}$$
$$= \$2{,}200 \times 0.09 \times (8 \div 12)$$
$$= \$132$$

Step 2: Then, determine the amount:
$$\text{Amount} = \text{Principal} + \text{Interest}$$
$$= \$2{,}200 + \$132$$
$$= \$2{,}332$$

The total amount owed at the end of the eight months is $2,332.

13.4 SALES OF PROPERTY

In this section, we introduce some of the mathematics and terminology associated with buying a house.

AFFORDING A HOUSE

You must first decide how much you can afford to pay for a house and what the down payment will be. As a general rule, the purchase price for a house should not exceed four times your annual salary.

Exercise 13.30

Mr. Brown and his wife have an annual salary of $37,500. They wish to purchase their first house. Based upon their salary, how much can they afford to pay for a house?

Solution 13.30

Mr. and Mrs. Brown can afford to pay four times their annual salary for a house. We have:

$$4 \times \$37{,}500 = \$150{,}000$$

Therefore, Mr. and Mrs. Brown can afford to pay $150,000 for a house, based upon their annual salary.

Another way to determine how much you can afford to pay for a house is the following:

Step 1: Determine your monthly salary.
Step 2: Subtract the amount of your monthly bills (those that you will not be able to pay within six months).
Step 3: Multiply the figure from Step 2 by 35%.

Exercise 13.31

Mr. and Mrs. Smith have a gross monthly income of $2,950. Their monthly bills amount to $675. Determine the maximum monthly payment (to the nearest dollar) that they can afford for a house.

Solution 13.31

Step 1: The gross monthly income is $2,950.
Step 2: Subtract the amount of the monthly bills ($675) from the gross monthly income: $2,950 − $675 = $2,275.
Step 3: Multiply $2,275 by 35%: 35% × $2,275 = $796.25.

Therefore, to the nearest dollar, the maximum house payment should be $796 per month.

Exercise 13.32

If your gross monthly income is $4,300 and your monthly bill payments are $1,380, determine the maximum monthly payment (to the nearest dollar) that you can afford for a house payment.

Solution 13.32

Step 1: Your gross monthly income is $4,300.
Step 2: Subtract your monthly bill payments ($1,380) from $4,300: $4,300 − $1,380 = $2,920.
Step 3: Multiply $2,920 by 35%: 35% × $2,920 = $1,022.

Therefore, the maximum monthly payment that you can afford for a house is $1,022.

DOWN PAYMENT

Once you find a house that you would like to buy (usually working with a real estate agent), you sign a **sales contract** or **purchase agreement** and give a **down payment**, called **earnest money.** Your lender (usually a bank) will require a minimum down payment depending upon the type of loan, called a **mortgage,** and the appraised value of the property. The minimum down payment may be as low as 5% but is usually about 20%.

Exercise 13.33
Determine the down payment for a house with a purchase price of $83,000 if the down payment rate is 15%.

Solution 13.33
The down payment is the purchase price times the rate. We have:
$$15\% \times \$83,000 = \$12,450$$
Therefore, the required down payment is $12,450.

Exercise 13.34
Determine the down payment for a house with a purchase price of $185,000 if the down payment rate is 20%.

Solution 13.34
The down payment is the purchase price times the rate. We have:
$$20\% \times \$185,000 = \$37,000$$
Therefore, the required down payment is $37,000.

ORIGINATION FEES AND POINTS

In addition to the down payment, the lender may also charge an **origination fee** or points or both. An origination fee is a one-time charge made by the lender to cover the costs of processing your loan. This may be a percentage of the loan amount or a flat fee. **Points** are also one-time charges that generally reflect market conditions. Each point is usually equal to 1% of the amount of your loan.

Exercise 13.35
You obtain a mortgage loan in the amount of $75,000 and your lender charges 4 points. What is the fee charged?

Solution 13.35
The fee charged is 4 points or 4% of $75,000. We have:
$$4\% \times \$75,000 = \$3,000$$
Therefore, the fee charged for the 4 points is $3,000.

Exercise 13.36
You obtain a mortgage loan in the amount of $90,000 and your lender charges an origination fee of 1%. What is the fee charged?

Solution 13.36
The fee charged is 1% of $90,000. We have:
$$1\% \times \$90,000 = \$900$$
Therefore, the origination fee is $900.

Exercise 13.37
Mr. and Mrs. Newcomer have applied for a mortgage to purchase a house selling for $117,000. The lender requires a down payment of 20%. In addition, there is an origination fee of $550 and 4.5 points. What is the amount of the mortgage and the total amount of fees to be paid?

Solution 13.37
There are several parts to this problem.

Step 1: First, we determine the amount of the mortgage by subtracting the down payment from the purchase price. The down payment is 20% of the purchase price. We have:
$$20\% \times \$117,000 = \$23,400$$
Next, we subtract the down payment from the purchase price to obtain the amount of the mortgage. We have:
$$\$117,000 - \$23,400 = \$93,600$$
Therefore, the amount of the mortgage is $93,600.

Step 2: The points are paid on the amount of the mortgage. Hence, we have:
$$4.5\% \times \$93,600 = \$4,212$$
Step 3: The total amount of fees paid is the origination fee plus the points. We have:

$4,212   points
+550   origination fee
$4,762

Therefore, the total amount of fees is $4,762.

Exercise 13.38
A house has a purchase price of $127,000. A lender has agreed to give a buyer a mortgage in the amount of 75% of the purchase price. The lender will charge 5 points and an origination fee of 1% of the mortgage. What is the amount of the down payment and the total amount of fees that the buyer must pay?

Solution 13.38

Again, there are several parts to this problem.

Step 1: We have:
Down payment = Purchase price − Mortgage
= $127,000 − (75% × $127,000)
= $127,000 − $95,250
= $31,750

Therefore, the down payment will be $31,750.

Step 2: The total amount of fees is the sum of the points and the origination fee. Hence:
a) Points = 5% × Mortgage
= 5% × $95,250
= $4,762.50

b) Origination fee = 1% × Mortgage
= 1% × $95,250
= $952.50

c) Total amount of fees:
$4,762.50   points
+ 952.50   origination fee
$5,715.00

Therefore, the total amount of fees will be $5,715.

13.5 CONSUMER CREDIT

We close out this chapter with a discussion of consumer credit. There are basically two types of consumer credit: installment loan and line of credit.

INSTALLMENT LOANS

An **installment loan** is an agreement with a merchant to pay for a purchase by making (usually equal) payments at regular time intervals (usually monthly) over a definite period of time. For this type of credit, there is a finance charge called **add-on interest.** This is a simple interest amount that is added to the cost of the purchase. The total of both the purchase price and the interest is paid for over the term of the loan.

Add-on interest = Amount of purchase financed × rate × time.

Exercise 13.39

Determine the amount of add-on interest for the purchase of a TV set that costs $795, if the total amount is financed over three years at 12% interest per year.

Solution 13.39

Add-on interest = Amount of purchase financed × rate × time
= $795 × 12% × 3
= $286.20

The add-on interest for the TV set is $286.20.

Exercise 13.40

Determine the amount of add-on interest for the purchase of a washer/dryer set that costs $1,150, if the total amount is financed over thirty months at 15% interest per year.

Solution 13.40

Add-on interest = Amount of purchase financed × rate × time
= $1,150 × 15% × 2.5 (30 months = 2.5 years)
= $431.25

The add-on interest for the washer/dryer set is $431.25.

The add-on interest must be added to the amount financed to determine the total amount to be repaid. We have:

Total amount to be repaid = Amount financed + add-on interest

Exercise 13.41

Determine the total amount to be repaid if a living room set is purchased for $3,900 and the total amount is financed at 11.5% for twenty-four months.

Solution 13.41

There are two parts to this problem.

1. Determine the add-on interest as follows:

   Add-on interest = Amount of purchase financed × rate × time
   = $3,900 × 11.5% × 2
   = $897

2. Determine the total amount to be repaid as follows:

   Total amount to be repaid = Amount financed + add-on interest
   = $3,900 + $897
   = $4,797

The total amount to be repaid for the purchase of the set is $4,797.

Exercise 13.42

Determine the total amount to be repaid if a microwave oven is purchased at $679 and the total amount is financed at 13% per year for eighteen months.

Solution 13.42
There are two parts to this problem.

Step 1: Determine the add-on interest as follows:
Add-on interest = Amount of purchase financed × rate × time
= $679 × 13% × 1.5 (18 months = 1.5 years)
= $132.41 (rounded up)

Step 2: Determine the total amount to be repaid as follows:
Total amount to be repaid = Amount financed + add-on interest
= $679 + $132.41
= $811.41

The total amount to be repaid for the purchase of the microwave oven is $811.41.

If you know the total amount of repayment and also the term of the loan, you can compute the amount of each payment as follows:

$$\text{Amount of each payment} = \frac{\text{Amount to be repaid}}{\text{Number of payments}}$$

Note: Bankers and other lenders use formulas and tables to compute monthly payments.

Exercise 13.43
Determine the monthly payment for the purchase of the living room set in Exercise 13.41.

Solution 13.43

$$\text{Monthly payment} = \frac{\text{Amount to be repaid}}{\text{Number of payments}}$$

$$= \frac{\$4,797}{24}$$

= $199.88 (Rounded up)

The monthly payments for the purchase of the living room set would be $199.88. (Actually, the payment of $199.88 would be made for the first twenty-three months, and the last payment would be $199.76.)

Exercise 13.44
Determine the monthly payments for the purchase of the microwave oven in Exercise 13.42.

Solution 13.44

$$\text{Monthly payment} = \frac{\text{Amount to be repaid}}{\text{Number of payments}}$$

$$= \frac{\$811.41}{18}$$

$$= \$45.08 \text{ (rounded up)}$$

The monthly payments will be $45.08. (Actually, the payments will be $45.08 for the first forty-one months with the last payment being $45.05.)

Exercise 13.45

A sun room is to be built for $9,600. Determine the monthly payments if the total amount is to be financed at 9.5% per year for five years.

Solution 13.45

There are three parts to this problem.

Step 1: Determine the add-on interest as follows:

Add-on interest = Amount of purchase financed × rate × time
= $9,600 × 9.5% × 5
= $4,560

Step 2: Determine the amount to be repaid as follows:

Amount to be repaid = Amount financed + add-on interest
= $9,600 + $4,560
= $14,160

Step 3: Determine the monthly payments as follows:

$$\text{Monthly payment} = \frac{\text{Amount to be repaid}}{\text{Number of payments}}$$

$$= \frac{\$14,160}{60}$$

$$= \$236$$

The monthly payments would be $236.

Exercise 13.46

Determine the monthly payments for the purchase of a riding mower that costs $1,495 if the total purchase is financed at 10.5% per year for forty-two months.

Solution 13.46

There are three parts to this problem.

Step 1: Determine the add-on interest as follows:
Add-on interest = Amount of purchase financed × rate × time
= $1,495 × 10.5% × 3.5 (42 months = 3.5 years)
= $549.41 (rounded down)

Step 2: Determine the total amount to be repaid as follows:
Total amount to be repaid = Amount financed + add-on interest
= $1,495 + $549.41 = $2,044.41

Step 3: Determine the monthly payments as follows:

$$\text{Monthly payment} = \frac{\text{Amount to be repaid}}{\text{Number of payments}}$$

$$= \frac{\$2,044.41}{42}$$

$$= \$48.68$$

The monthly payments will be $48.68. (Actually, the payments will be $48.68 for the first forty-one months with the last payment being $48.53.)

LINE OF CREDIT

The second type of consumer credit is the **line of credit.** This type of credit involves the use of credit cards issued by American Express, VISA, MasterCard, and others, as well as by department stores and oil companies. For these credit cards, you pay an interest rate called the **annual percentage rate (APR).** The APR is stated usually as a daily or a monthly rate.

Exercise 13.47

You have a credit card with a stated interest rate of 1.5% per month. Determine the APR.

Solution 13.47

Because the stated interest rate is a monthly rate and because there are twelve months in a year, we have:
APR = monthly rate × 12
= 1.5%/month × 12 months
= 18%

The APR for your credit card is 18%.

Exercise 13.48

Joyce has a credit card with a stated interest rate of 0.04789% per day. Determine the APR.

Solution 13.48

Because the stated interest rate is a daily rate and because there are 365 days in a year, we have:

APR = daily rate × 365
= 0.04789% × 365
= 17.47985%

Rounded to the nearest tenth, the APR for Joyce's card is 17.5%.

The simple interest charged for the use of a credit card is calculated according to different methods. The most commonly accepted methods are the previous balance, the adjusted balance, and the average daily balance.

PREVIOUS BALANCE METHOD

When using the **previous balance method,** the interest is calculated by using the simple interest formula with P = the previous balance, r = the annual rate, and $t = \dfrac{1}{12}$ (for one month).

Exercise 13.49

Using the previous balance method, calculate the monthly interest on a $1,250 credit card charge with a 17.4% APR.

Solution 13.49

For this exercise, we have:
$P = \$1{,}250$, $r = 0.174$, and $t = \dfrac{1}{12}$. Hence,
$I = Prt$

$I = \$1{,}250 \times 0.174 \times \dfrac{1}{12}$
$= \$18.125$

The interest charge for the month would be $18.13 (rounded up).

Exercise 13.50

Using the previous balance method, calculate the monthly interest on a $2,740 credit card charge with a 19% APR.

Solution 13.50

For this exercise, we have:
$P = \$2{,}740$, $r = 0.19$, and $t = \dfrac{1}{12}$. Hence,
$I = Prt$

$I = \$2{,}740 \times 0.19 \times \dfrac{1}{12}$
$= \$43.39$ (rounded up)

The interest charge for the month would be $43.39.

ADJUSTED BALANCE METHOD

When using the **adjusted balance method,** the interest is calculated by using the simple interest formula with P = the adjusted balance, r = the annual rate, and $t = \frac{1}{12}$ (for one month).

Using the adjusted balance method to calculate the monthly interest, credit is given for payments received before the due date.

Exercise 13.51

Using the adjusted balance method, calculate the monthly interest on a $1,675 credit card bill with a 21% APR, and assume that a payment of $65 was sent in and received by the due date.

Solution 13.51

Step 1: Determine the adjusted balance as follows:
   Previous balance: $1,675
   Less payment: − 65
   Adjusted balance: $1,610

Step 2: Use $I = Prt$ with $P = \$1,610$, $r = 0.21$, and $t = \frac{1}{12}$. Hence,

$$I = \$1,610 \times 0.21 \times \frac{1}{12}$$
$$= \$28.175$$

The interest charge for the month would be $28.18 (rounded up).

Exercise 13.52

Using the adjusted balance method, calculate the monthly interest charge on a $3,815 credit card bill with an 18% APR, assuming that a payment of $525 was sent in and received by the due date.

Solution 13.52

Step 1: Determine the adjusted balance as follows:
   Previous balance: $3,815
   Less payment: − 525
   Adjusted balance: $3,290

Step 2: Use $I = Prt$ with $P = \$3,290$, $r = 0.18$, and $t = \frac{1}{12}$. Hence,

$$I = \$3,290 \times 0.18 \times \frac{1}{12}$$
$$= \$49.35$$

The interest charge for the month would be $49.35.

AVERAGE DAILY BALANCE METHOD

When using the **average daily balance method,** the interest is calculated by using the simple interest formula with P = the average daily balance, r = the annual rate, and

$$t = \frac{\text{Number of days in the billing period (in years)}}{365}$$

To use the average daily balance method, you must first determine the average daily balance by dividing the total of the daily balances by the number of days in the billing period. For these exercises, assume that all previous balances are posted on the first of each month and all payments are due by the first of the next month.

Exercise 13.53

On the first day of May, the balance on your credit card was $1,960. You sent in a payment of $120 that was received and posted on the 12th of May. Determine your average daily balance for May.

Solution 13.53

For the first twelve days of May, your balance was $1,960. Then, for the rest of May (nineteen days), your balance was $1,960 − $120 = $1,840. We now add the balance for each day as follows:

```
12 days at $1,960:    $23,520
19 days at $1,840:   + 34,960
Total:                $58,480
```

To get the average daily balance, divide $58,480 by 31 (the number of days in May) as follows: $58,480 ÷ 31 = $1,886.45 (to the nearest cent). The average daily balance for the month of May is $1,886.45.

Exercise 13.54

On the first day of September, your credit card balance was $2,085. You sent in a payment of $150 that was received and posted on the 9th day of September. You also sent in a payment of $315 that was received and posted on the 20th day of September. What was your average daily balance for the month of September?

Solution 13.54

For the first nine days of September, your daily balance was $2,085. For the next eleven days, your daily balance was $2,085 − $150 = $1,935. For the rest of the month, your daily balance was $1,935 − $315 = $1,620. We now add the daily balances for the month as follows (there are 30 days in September):

```
9 days at $2,085:     $18,765
11 days at $1,935:     21,285
10 days at $1,620:   + 16,200
Total:                $56,250
```

To get the average daily balance for the month, divide $56,250 by 30 (the number of days in September) as follows: $56,250 ÷ 30 = $1,875. The average daily balance for the month of September is $1,875.

Exercise 13.55

If the APR for your credit card in Exercise 13.53 is 20.5%, determine the interest for the month of May.

Solution 13.55

Using the average daily balance method, we have:

$I = Prt$ with $P =$ the average daily balance ($1,886.45), $r = 0.205$ and $t = \dfrac{31}{365}$, because there are thirty-one days in May. Hence,

$I = Prt$

$$I = \$1,886.45 \times 0.205 \times \dfrac{31}{365}$$

$= \$32.85$ (rounded up)

The interest charge for the month of May would be $32.85.

Exercise 13.56

If the APR for your credit card in Exercise 13.54 is 19.8%, determine the interest for the month of September.

Solution 13.56

Using the average daily balance method, we have:

$I = Prt$, with $P =$ the average daily balance ($1,875), $r = 0.198$ and $t = \dfrac{30}{365}$, because there are thirty days in September. Hence, $I = Prt$

$$I = \$1,875 \times 0.198 \times \dfrac{30}{365}$$

$= \$30.52$ (rounded up)

The interest charge for the month of September would be $30.52.

In this chapter, we introduced some real-life applications that use percents. These included commission, discount, simple interest, sales of property, and consumer credit.

TEST YOURSELF

Whenever possible, use your calculator to help you arrive at your solutions.

1) Paula works for a straight commission and receives a commission rate of 6.5% on the total amount of sales. If her total sales for a week amounted to $6,750, what was her commission?

2) During one month, Dick had gross sales of $65,715, and his sales returns amounted to $2,035. He works for a 5.5% commission on his net sales. What was the amount of Dick's commission for the month?

3) Patti works on a straight commission of 6% of her net sales. If her commission for a given week was $432, what was the amount of her net sales?

4) Randy earns a salary of $1,400 per month plus a commission. If his total pay for a month was $4,628 and his commission rate is 12%, what was the amount of his net sales for the month?

5) Joel works at an automotive supply store. He receives a salary of $225 per week plus a commission of 8.5% of net sales. If Joel has gross sales of $4,697 for the week and returns of $863, what is his total pay for the week?

6) Joyce works for a straight commission of 7.3% of net sales. If her commission for the week was $551.15, what was the amount of her net sales?

7) Laura sells real estate for a commission of 6.5% and undeveloped land for 10.5% commission. If she sells a house for $185,000 and two tracts of land for $65,000 and $95,000, what will be her commission for the three sales?

8) A plasma TV is advertised at 20.5% off the regular price. If the TV is marked at $2,250, determine the discount and the sales price.

9) The marked price of a home computer with printer is $2,700 with a sales price of $1,620. What is the rate of discount?

10) Determine the interest when $1,500 is borrowed at 8.5% per year for eighteen months.

11) Determine the yearly rate of interest if $4,800 is borrowed for two years with interest of $624.

12) Determine the principal if $405 interest is paid on a loan at 8% per year for twenty-seven months.

13) Erin borrowed $1,475 at 8.6% for 18 months. At the end of the 18 months, what is the amount that Erin owes?

14) Determine the time for a $7,500 loan (or deposit) at 12% if the interest is $450.

15) Fred has a gross monthly income of $5,450 and average monthly expenses of $2,685. Determine the maximum monthly payment (to the nearest dollar) that Fred can afford for a house payment.

16) If your gross monthly income is $3,970 and your monthly bill payments are $1,610, determine the maximum monthly payment (to the nearest dollar) that you can afford for a house payment.

17) You have applied for a mortgage to purchase a house selling for $126,500. The lender requires a down payment of 15%. In addition, there is an origination fee of 1% of the mortgage and 3 points. What is the amount of down payment and fees that must be paid?

18) Determine the amount of add-on interest for the purchase of a refrigerator that costs $995, if the total amount is financed over three years at 11% per year.

19) Determine the total amount to be repaid for the purchase of an item that sells for $2,600, if the total amount is financed at 9.5% per year for eighteen months.

20) Determine the monthly payment for the purchase of the refrigerator in Exercise 18.

21) Determine the monthly payment for the purchase of the item in Exercise 19.

22) Using the adjusted balance method, calculate the monthly interest on a $1,925 credit card bill with a 19% APR, and assume that a payment of $255 was sent in and received by the due date.

23) Using the adjusted balance method, calculate the monthly interest charge on a $4,609 credit card bill with a 17.5% APR, assuming that a payment of $640 was sent and received by the due date.

24) On the first day of October, the balance on your credit card was $2,140. You sent in a payment of $175 that was received and posted on the 10th day of October. You also sent in a payment of $120 that was received and posted on the 22nd day of October. What was your average daily balance for the month of October?

25) Determine the interest due for the month of October if the APR for your credit card in Exercise 24 is 19.5% of the average daily balance.

TEST YOURSELF ANSWERS

1) Commission = (0.065)($6,750) = $438.75
2) Net sales = $65,715 − $2,035 = $63,688; Commission = (0.055)($63,688) = $3,502.40
3) Net sales = $432 ÷ 0.06 = $7,200
4) Total pay − salary = $4,628 − $1,400 = $3,228; Net sales = $3,228 ÷ 0.12 = $26,900
5) Gross sales − returns = $4,697 − $863 = $3,834; Commission = (0.085)($3,834) = $325.89; Total weekly pay = $225 + $325.89 = $550.89
6) Amount of sales = $\dfrac{\$551.15}{0.073} = \$7,550$
7) Commission for house = (0.065)($185,000) = $12,025; Commission for land = (0.105)($65,000 + $95,000) = $16,800; Total commission = $12,025 + $16,800 = $28,825
8) Discount = (0.205)($2,250) = $461.25; Sale price = $2,250 − $461.25 = $1,788.75
9) Discount = $2,700 − $1,620 = $1,080

   Rate of discount = $\dfrac{\$1,080}{\$2,700} = 40.40\%$
10) I = ($1,500)(0.085)(1.5) = $191.25
11) $r = \dfrac{\$624}{\$4,800 \times 2} = 6.5\%$
12) $P = \dfrac{\$405}{0.08 \times 2.25} = \$2,250$
13) Interest = (0.086)($1,475)(1.5) = $190.28

    Amount = $1,475 + $190.28 = $1,665.28
14) $t = \dfrac{\$450}{(\$7,500)(0.12)} = 0.5$ yr or 6 mo.
15) (0.35)($5,450 − $2,685) = $967.75
16) (0.35)($3,970 − $1,610) = $826
17) Down payment = (0.15)($126,500) = $18,975; Origination fee = (0.01)($126,500 − $18,975) = $1,075.25; Points = (0.03)($126,500 − $18,975) = $3,225.75; Down payment + fees = $18,975 + $1,075.25 + $3,225.75 = $23,276
18) ($995)(0.11)(3) = $328.35
19) $2,600 + [($2,600)(0.095)(1.5)] = $2,970.50
20) ($995 + $328.35) ÷ 36 ≈ $36.76
21) $2,970.50 ÷ 18 ≈ $165.03
22) $1,925 − $255 = $1,670; ($1,670)(0.19) ÷ 12 ≈ $26.45
23) Adjusted balance = $4,609 − $640 = $3,969; I = ($3,969)(0.175)(1/12) ≈ $57.88

24) 10 days @ $2,140: $21,400
    12 days @ ($2,140 − $175) = $1,965: $23,580
    9 days @ ($1,965 − $120) = $1,845: $16,605
                                                                          $61,585

    Average daily balance = $61,585 ÷ 31 ≈ $1,986.62

25) ($1,986.62)(0.195)(31/365) ≈ $32.91

Index

A

Accuracy 313
    Exercise 313
    Solution 313
Adding mixed numbers 102
    Exercise 102–103
    Solution 102–104
Adding decimal numbers *see* decimals, addition of
Addition of whole numbers 4
    Properties of 4
        Associative property 4
        Commutative property 4
        Exercise 4–9
        Solution 4–9
Addition and subtraction with 0 (zero) 35
    Example 35
Additive identity 4, 35

Add-on interest 410
    Exercise 410–13
    Solution 411–14
Adjusted balance method *see* consumer credit
Algebraic expression 317
    Constants 317
    Variables 317
Angles 224
    Classification of 224
    Side of 224
    Vertex of 224
Annual percentage rate 414
    Exercise 414
    Solution 414–15
Apparent limit 372
Applications involving decimals 138
Applications of ratios and proportions 161
APR *see* annual percentage rate
Arithmetic average *see* mean

Associate property of addition *see* addition, properties of

Associative property for multiplication *see* multiplication, properties of

Average daily balance method *see* consumer credit

B

Bar graph 376
    Exercise 376–78
    Solution 377–79

Bimodal distribution 386

Borrowing 11
    Exercise 11–16
    Solution 11–16

C

Central tendency, measure of 384

Circle graph 381
    Exercise 382–83
    Solution 382–84

Circles 220
    Area of 234
        Exercise 235–37
        Solution 235–37
    Centre of 220

Circumference of 220, 234
    Exercise 221, 235–37
    Formula 220
    Solution 221–22, 235–37
    Diameter of 220
    Radius of 220

Coefficient 318
    Numerical 318

Commission 391
    Exercise 391–96
    Solution 391–96

Commutative property of addition *see* addition, properties of

Commutative property for multiplication *see* multiplication, properties of

Composite Number 48
    Exercise 48–50
    Solution 48–50

Compound interest *see* interest

Constants *see* algebraic expression

Consumer credit 410–18
    Adjusted balance method 416
        Exercise 416
        Solution 416
    Average daily balance method 417
        Exercise 417–18
        Solution 417–18
    Installment loan 410
        Exercise 410–13
        Solution 411–14
    Line of credit 414
        Exercise 414
        Solution 414–15
    Previous balance method 415
        Exercise 415
        Solution 415

Converting percents to decimals 176
    Exercise 176–78
    Solution 176–78
Converting percents to fractions 175
    Exercise 175
    Solution 175–76
Cost problems *see* decimals, applications

D

Decimals 117–44
    Addition of 121
        Exercise 121
        Solution 121–22
    Applications 138
        Cost problems 138
            Exercise 138–44
            Solution 138–44
    Conversion
        Decimal to fraction 134
            Exercise 134
            Solution 134–35
        Fraction to decimal 135
            Exercise 135–36
            Solution 135–36
    Dividing by powers of ten 129
        Exercise 129–31
        Solution 130–31
    Division of 127
        Exercise 127–28
        Solution 128–29
    Multiplication of 124
        Exercise 125
        Solution 125
    Multiplying by power of ten 126
        Exercise 126
        Solution 126–27
    Place values 117
        Exercise 118
        Solution 118–19
    Reading 119
        Exercise 119–20
        Solution 121
    Remainders 131
        Exercise 131
        Solution 131–32
    Rounding after division 132
        Exercise 132
        Solution 132–33
    Rounding of 123
        Exercise 124
        Solution 124
    Subtraction of 122
        Exercise 122–23
        Solution 122–23
    Terminating and repeating 137
        Exercise 137
        Solution 137
Denominator of fractions 134
Descriptive statistics 363

Discount 397
    Exercise 397–98
    Rate of 397
        Exercise 398–99
        Solution 398–99
    Solution 397–98
Division of whole numbers 23
    Language of division 23
        Exercise 23–29
        Solution 23–29
Down payment *see* sales of property

E

Earnest money 408
Endpoint 223
English system of units 195
    Conversions within the English system 196
        Exercise 196–97
        Solution 196–98
Equality of two ratios 157
    Exercise 157
    Solution 157–58
Equations 329
    Left member 329
    Linear 331 *see also* solving linear equations
    Right member 329
    Solutions to 330
Equivalent fractions *see* fractions
Evaluating algebraic expressions 319
    Exercise 319–21
    Solution 319–21
Exponents 32

F

Factor 16, 45, 318
Factored form 45
    Example 45
Formulas in algebra 351
    Area of a trapezoid 356
        Exercise 356
        Solution 356
    Averages 353
        Exercise 353–54
        Solution 354
    Distance-rate-time 351
        Exercise 351
        Solution 351
    Gravity 354
        Exercise 354–55
        Solution 355
    IQ 352
        Exercise 353
        Solution 353
    Temperature 352
        Exercise 352
        Solution 352
Fraction to decimal conversion *see* decimals, conversion
Fractions
    Adding like fractions 91
        Exercise 91
        Solution 91
    Adding unlike fractions 92
        Exercise 92–96
        Solution 92–96

Dividing 85
    Exercise 85–86
    Solution 85–86
Equivalent fractions 73
    Exercise 73
    Solution 73–74
Fundamental principle 74
    Exercise 74–76
    Solution 74–76
Mixed numbers 69
    Dividing of 86
Mixed numbers as improper fractions 71
    Exercise 72
    Solution 72–73
Multiplying 78
    Exercise 78–79
    Solution 78–81
Multiplying mixed numbers 81
    Exercise 81–84
    Solution 82–84
Ordering of 96
    Exercise 96–97
    Solution 97–99
Proper and improper 67
    Exercise 67–69
    Solution 68–69
Reciprocals 84
    Exercise 84
    Solution 85
Simplifying 76
    Exercise 76–78
    Solution 77–78
Subtracting 99
    Exercise 99–101
    Solution 100–102
Frequency 364
Frequency distribution 364
    Exercise 365
    Grouped 366
        Exercise 368–70
        Solution 368–71
    Solution 365–66
Frequency polygon 374
    Exercise 374–76
    Solution 375–76

G

GCF *see* greatest common factor
Greater than 3, 254
    Exercise 3
    Solution 4
Greatest common factor 56
    Of two members 56
        Exercise 56
        Solution 57
    Of more than two members 57
        Exercise 57–59
        Solution 57–59
Gross sales 392
Grouped frequency distribution *see* frequency distribution

H

Histogram 371
    Exercise 372–74
    Solution 372–74
Hypotenuse 242

I

Identity 331
Identity property for multiplication *see* multiplication, properties of
Improper fractions as mixed numbers *see* fractions
Inferential statistics *see* statistics inference
Installment loan *see* consumer credit
Integers 253
    Absolute value of 256
        Exercise 257
        Solution 257–58
    Negative 254, 308
    Ordering of 254
        Exercise 255–56
        Solution 255–56
    Positive 254, 299–300, 302, 305–06, 309
Interest 400
    Amount 405
        Exercise 405–06
        Solution 406
    Compound 400
    Exercise 400–02
    Principal 400, 404
        Exercise 404–05
        Solution 404–05
    Rate of 400, 402
        Exercise 402–03
        Solution 402–03
    Simple 400
    Solution 400–02

L

Language of algebra 345
Language of multiplication 16
LCM *see* least common multiple
Least common multiple 59
    Of more than two numbers 60
        Exercise 60
        Solution 61–62
    Of two numbers 59
        Exercise 60
        Solution 60
Left member *see* equation
Length, units of 195
Less than 3, 254
    Exercise 3
    Solution 4
Like fractions 90
    Exercise 90
    Solution 90
Like terms 322
Line 222
Line graph 380
    Exercise 380–81
    Solution 380–81
Line of credit *see* consumer credit
Line segment 223

Linear equations see equations, linear
Liquid capacity, units of 203
    Exercise 204–07
    Solution 204–07

M

Marked price 397
Mean 385
    Exercise 385
    Solution 385
Median 385
    Exercise 385
    Solution 386
Metric systems of units 195
    Conversions within the metric systems 199
    Exercise 199
    Solution 199
Mixed operations 30
    Exercise 30
    Solution 30
Mixed operations with exponents 32
    Exercise 32–34
    Solution 33–35
Mixed operations with parentheses 31
    Exercise 31
    Solution 31
Mixed numbers see fractions, mixed number
Mixed numbers as improper fractions see fractions, mixed number
Mode 386
    Exercise 386–87
    Solution 386–87

Mortgage 408
Multiples 46
    Exercise 47
    Solution 47
Multiplication of whole numbers 16
    Properties of 17
        Associative property 18
        Commutative property 17
        Identity property 17
    Exercise 17–18
    Solution 17–18
Multiplication and division with 0 (zero) 36
    Exercise 36
    Solution 37
Multiplicative identity 17
Multiplying fractions see fractions, multiplying
Multiplying decimal numbers see decimals, multiplication of
Multiplying mixed numbers see fractions
Multiplying numbers with more than one digit 18
    Exercise 18–22
    Solution 19–22

N

Natural numbers 3
Negative integer exponents 309
    Exercise 309–10
    Solution 309–10
Net sales 392
    Amount of 394
        Exercise 394–95
        Solution 394–95

Numerals 1
Numerator of fraction 134
Numerical coefficient *see* coefficient

O

Ordering of whole numbers 3
Ordering of fractions *see* fractions, ordering of
Order of operations 29
Order of subtraction 10
    Exercise 10
    Solution 10–11
Origination fee *see* sales of property

P

Parallelogram 231 *see* also quadrilateral
    Exercise 231–32
    Solution 231–32
Percent 171
    Problems involving 179
        Exercise 179–82
        Solution 179–83
Percent and fractions 171
    Exercise 171–74
    Solution 172–74
Percent increase or decrease 183
    Amount of 183
        Exercise 183–84
        Solution 183–84
    Rate of 185
        Exercise 185–86
        Solution 185–87

Perimeter 218
    Definition of 218
    Exercise 218–20
    Solution 218–20
Pie chart *see* circle graph
Place value 1
    Decimals 117
    Exercise 1–3
    Solution 2–3
Points *see* sales of property
Polygon
    Definition of 217
    Classification of 217
Positive integer exponents 299
Power of a product 305
    Exercise 305–06
    Solution 305–06
Power of a quotient 306
    Exercise 306–08
    Solution 307–08
Power rule 301
    Exercise 301–02
    Solution 301–02
Powers of ten in the metric system 198
Previous balance method *see* consumer credit
Prime factorization 50
    Exercise 50
    Solution 50–51
Prime number 48
    Exercise 48
    Solution 48

Prime relative 59
Product 16, 45, 318
Product rule 300
    Exercise 300
    Solution 300–01
Proportions 157
    Basic property 158
        Exercise 158–59
        Solution 158–60
Protractor 224
Purchase agreement 408
Pythagorean formula 242
    Exercise 243–45
    Solution 243–45
Pythagorean theorem *see* Pythagorean formula

Q

Quadrilaterals 227
Quotient *see* division, language of
Quotient rule 302
    Exercise 302–04
    Solution 303–04

R

Rate of discount *see* discount
Rate of interest *see* interest
Rates 154
    Exercise 154
    Solution 155–56

Rational numbers 253
    Absolute value of 261
        Exercise 261
        Solution 261
    Classification of 258
        Exercise 258
        Solution 259
    Definition of 258
    Opposites of 259
        Exercise 259
        Solution 260
    Ordering of 260
Ratios
    Examples of 151
    Exercise 152
    Solution 152–53
Ratios multiplier 161
    Exercise 161
    Solution 161
Ratios with fractions 153
    Exercise 153–54
    Solution 153–54
Ratios with three quantities 162
    Exercise 162–65
    Solution 162–65
Ray 223
Reading decimal numbers *see* decimals, reading
Real limits 372
Reciprocals *see* fractions, reciprocals

Rectangle 227 *see also* quadrilateral
    Exercise 227–29
    Solution 227–29
Remainders *see* decimals, remainder
Right member *see* equation, right member
Rounding of decimals *see* decimals, rounding of
Rounding after division *see* decimals, rounding after

S
Sales contract 408
Sales of property 406
    Affording a house 406
        Exercise 407
        Solution 407
    Down payment 408
        Exercise 408
        Solution 408
    Origination fee 408
        Exercise 408–09
        Solution 408–10
    Points 408
Scientific notation 311
    Rewriting a number 311
        Exercise 311–12
        Solution 311–12
Signed numbers 253
    Addition and subtraction of 262
        Exercise 262–72
        Solution 262–73

Mixed operations, applications 284
    Exercise 284–91
    Solution 284–92
Multiplication and division 273
    Exercise 274–80
    Solution 274–82
Simplification of 282
    Exercise 283
    Solution 284
Significant digits 312
Simple interest *see* interest
Simplifying algebraic expressions 322
    Combining like terms 323
        Exercise 323–28
        Solution 323–29
    Exercise 322
    Removing symbols of grouping 324
        Exercise 325
        Solution 325
    Solution 322
Solving linear equations 332
    Addition rule 332
        Exercise 332–33
        Solution 332–33
    Combined operations 341
        Exercise 341–44
        Solution 341–45
    Division rule 338
        Exercise 338–40
        Solution 339–41

Multiplication rule 336
    Exercise 336–38
    Solution 336–38
Subtraction rule 334
    Exercise 334–35
    Solution 334–36
Square 230 *see also* quadrilateral
    Exercise 230
    Solution 230–31
Square roots 243
Statistical inference 363
Statistics 363
Straight commission 391
Subtraction of whole numbers 9
    Exercise 9
    Solution 9
Subtracting decimal numbers *see* decimals, subtracting of
Subtracting fractions *see* fractions, subtracting
Subtracting mixed numbers
    Exercise 104–107
    Solution 104–108
Sum 318 *see also* addition

T

Term 318
Test for divisibility 51
    By 2   51
    By 3   52
    By 4   52
    By 5   53
    By 6   53
    By 7   54
    By 8   54
    By 9   55
    By 10 55
    By 11 55
    Exercise 51–55
    Solution 51–56
Time, units of 208
    Exercise 208–11
    Solution 208–11
Translating word expression into algebraic expression 345
    Exercise 345–47
    Solution 346–47
Trapezoid 233
    Exercise 233–34
    Solution 233–34
Triangles
    Area of 225
    Exercise 225–26
    Perimeter of 225
    Solution 225–27

U

Units
- Of length *see* length, units of
- Of liquid capacity *see* liquid capacity, units of
- Of time *see* time, units of
- Of weight *see* weight, units of

Unlike fraction 90
- Exercise 90
- Solution 90

Unlike terms 322

V

Variable *see* algebraic expression

Volume 237
- Of cube 239
  - Exercise 240
  - Solution 240
- Of rectangular solid 238
  - Exercise 238–39
  - Solution 238–39
- Of sphere 240
  - Exercise 241
  - Solution 241–42

W

Weight, units of 200
- Exercise 201–03
- Solution 201–03

Whole number 3
- Addition of *see* addition of whole numbers
- Division of *see* division of whole numbers
- Multiplication of *see* multiplication of whole numbers
- Ordering of *see* ordering of whole numbers
- Subtracting of *see* subtraction of whole numbers

Word expression 345

Word problems 347
- Exercise 348–50
- Solution 348–51

Z

Zero exponent 308